Autodesk
AutoCAD Architecture 2024
Fundamentals

Elise Moss

AUTODESK
Authorized Developer

SDC
PUBLICATIONS

SDC Publications
P.O. Box 1334
Mission, KS 66222
913-262-2664
www.SDCpublications.com
Publisher: Stephen Schroff

ISBN-13: 978-1-63057-594-6
ISBN-10: 1-63057-594-1

Printed and bound in the United States of America.

Preface

No textbook can cover all the features in any software application. This textbook is meant for beginning users who want to gain a familiarity with the tools and interface of AutoCAD Architecture before they start exploring on their own. By the end of the text, users should feel comfortable enough to create a standard model, and even know how to customize the interface for their own use. Knowledge of basic AutoCAD and its commands is helpful but not required. I do try to explain as I go, but for the sake of brevity I concentrate on the tools specific to and within AutoCAD Architecture.

The files used in this text are accessible from the Internet from the book's page on the publisher's website: www.SDCpublications.com. They are free and available to students and teachers alike.

We value customer input. Please contact us with any comments, questions, or concerns about this text.

Elise Moss
elise_moss@mossdesigns.com

Acknowledgments from Elise Moss

This book would not have been possible without the support of some key Autodesk employees.

The effort and support of the editorial and production staff of SDC Publications is gratefully acknowledged. I especially thank Stephen Schroff for his helpful suggestions regarding the format of this text.

Finally, truly infinite thanks to Ari for his encouragement and his faith.

- Elise Moss

About the Author

Elise Moss has worked for the past thirty years as a mechanical designer in Silicon Valley, primarily creating sheet metal designs. She has written articles for Autodesk's Toplines magazine, engineering.com, AUGI's PaperSpace, DigitalCAD.com and Tenlinks.com. She is President of Moss Designs, creating custom applications and designs for corporate clients. She has taught CAD classes at Laney College, DeAnza College, SFSU, Silicon Valley College, and for Autodesk resellers. Autodesk has named her as a Faculty of Distinction for the curriculum she has developed for Autodesk products. She holds a baccalaureate degree in Mechanical Engineering from San Jose State.

She is married with two sons. Her husband, Ari, is retired from a distinguished career in software development.

Elise is a third-generation engineer. Her father, Robert Moss, was a metallurgical engineer in the aerospace industry. Her grandfather, Solomon Kupperman, was a civil engineer for the City of Chicago.

She can be contacted via email at elise_moss@mossdesigns.com.

More information about the author and her work can be found on her website at www.mossdesigns.com.

Authorized Developer

Other books by Elise Moss
Autodesk Revit Architecture 2023 Basics
AutoCAD 2023 Fundamentals

Table of Contents

Lesson 3: Floor Plans

Lesson 4: Space Planning

Lesson 5: Roofs

Lesson 6: Floors, Ceilings, and Spaces

Lesson 7: Views, Schedules and Legends

Lesson 8: Spaces & Sheets

Lesson 1:
ACA Overview

AutoCAD Architecture (ACA) enlists object-oriented process systems (OOPS). That means that ACA uses intelligent objects to create a building. This is similar to using blocks in AutoCAD. Objects in AutoCAD Architecture are blocks on steroids. They have intelligence already embedded into them. A wall is not just a collection of lines. It represents a real wall. It can be constrained, has thickness and material properties, and is automatically included in your building schedule.

AEC is an acronym for Architectural/Engineering/Construction.
BID is an acronym for Building Industrial Design.
BIM is an acronym for Building Information Modeling.
AIA is an acronym for the American Institute of Architects.
MEP is an acronym for Mechanical/Electrical/Plumbing.

The following table describes the key features of objects in AutoCAD Architecture:

Feature Type	Description
AEC Camera	Create perspective views from various camera angles. Create avi files.
AEC Profiles	Create AEC objects using polylines to build doors, windows, etc.
Anchors and Layouts	Define a spatial relationship between objects. Create a layout of anchors on a curve or a grid to set a pattern of anchored objects, such as doors or columns.
Annotation	Set up special arrows, leaders, bar scales.
Ceiling Grids	Create reflected ceiling plans with grid layouts.
Column Grids	Define rectangular and radial building grids with columns and bubbles.
Design Center	Customize your AEC block library.
Display System	Control views for each AEC object.
Doors and Windows	Create custom door and window styles or use standard objects provided with the software.
Elevations and Sections	An elevation is basically a section view of a floor plan.
Layer Manager	Create layer standards based on AIA CAD Standards. Create groups of layers. Manage layers intelligently using Layer Filters.
Masking Blocks	Store a mask using a polyline object and attach to AEC objects to hide graphics.
Model Explorer	View a model and manage the content easily. Attach names to mass elements to assist in design.
Multi-view blocks	Blocks have embedded defined views to allow you to easily change view.

Feature Type	Description
Railings	Create or apply different railing styles to a stair or along a defined path.
Roofs	Create and apply various roof styles.
Floorplate slices	Generate the perimeter geometry of a building.
Spaces and Boundaries	Spaces and boundaries can include floor thickness, room height, and wall thickness.
Stairs	Create and apply various stair types.
Tags and Schedules	Place tags on objects to generate schedules. Schedules will automatically update when tags are modified, added, or deleted.
Template Files	Use templates to save time. Create a template with standard layers, text styles, linetypes, dimension styles, etc.
Walls	Create wall styles to determine material composition. Define end caps to control opening and end conditions. Define wall interference conditions.

AutoCAD Architecture sits on top of AutoCAD. It is helpful for users to have a basic understanding of AutoCAD before moving to AutoCAD Architecture. Users should be familiar with the following:

- ❑ AutoCAD toolbars and tab on the ribbons
- ❑ Zoom and move around the screen
- ❑ Manage blocks
- ❑ *Draw* and *Modify* commands
- ❑ *Model* and *Paper space* (layout)
- ❑ Dimensioning and how to create/modify a dimension style

If you are not familiar with these topics, you can still move forward with learning AutoCAD Architecture, but you may find it helpful to have a good AutoCAD textbook as reference in case you get stuck.

The best way to use AutoCAD Architecture is to start your project with massing tools (outside-in design) or space planning tools (inside-out design) and continue through to construction documentation.

If you use the **QNEW** tool, it will automatically use the template set in the Options dialog.

Express Tools

Express Tools are installed with AutoCAD Architecture. If you do not see Express Tools on the menu bar, try typing EXPRESSTOOLS to see if that loads them.

Exercise 1-1:

Setting the Template for QNEW

Drawing Name: New
Estimated Time: 15 minutes

This exercise reinforces the following skills:

- ❑ Use of templates
- ❑ Getting the user familiar with tools and the ACA environment

1. Launch ACA.

2. 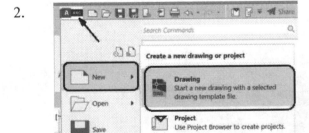 Select the large A to access the Application Menu.

   Select **New→Drawing**.

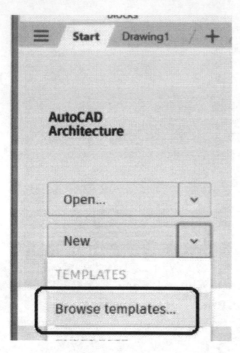

You can also select **New** from the Start tab.

Click the down arrow next to New and select **Browse templates…**

3.

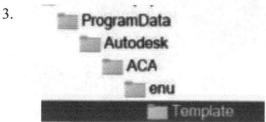

Browse to the *Template* folder under *ProgramData/Autodesk/ACA 2024/enu.*

4.

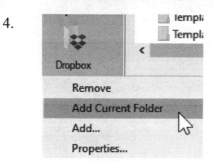

Right click on the left panel of the dialog.

Select **Add Current Folder**.

This adds the template folder to the palette.

5. 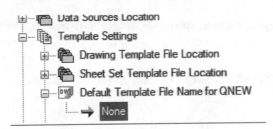 *Select the desired template to use.*

Each template controls units, layers, and plot settings.

Select *Aec Model (Imperial.Ctb).*

This template uses Imperial units (feet and inches and a Color Table to control plot settings).

Click **Open.**

6. Place your cursor on the command line.

7. 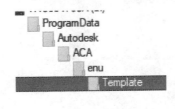 Right click the mouse.

Select **Options** from the short-cut menu.

8. Select the **Files** tab.
Locate the *Template Settings* folder.

- ⊞ ▣ Data Sources Location
- ⊟ ▣ Template Settings
  - ⊞ 📁 Drawing Template File Location
  - ⊞ 📁 Sheet Set Template File Location
  - ⊟ 📄 Default Template File Name for QNEW
    - ➡ None

9. Click on the + symbol to expand.
Locate the Default Template File Name for QNEW.

10. ⋯➡ None   Highlight the path and file name listed or None (if no path or file name is listed) and select the **Browse** button.

11. Browse to the *Template* folder or select it on the palette if you added it.

This should be listed under *Program Data/Autodesk/ ACA 2024/enu*

ProgramData
Autodesk
ACA
enu
Template

12. 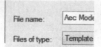 Locate the *Aec Model (Imperial Ctb).dwt [Aec Model (Metric Ctb).dwt]* file.

Click **Open**.

Note that you can also set where you can direct the user for templates. This is useful if you want to create standard templates and place them on a network server for all users to access.

You can also use this setting to help you locate where your template files are located.

13. Click **Apply** and **OK**.

14. Select the **QNEW** tool button.

15. Note that you have tabs for all the open drawings.

You can switch between any open drawing files by selecting the folder tab. You can also use the + tab to start a new drawing. The + tab acts the same way as the QNEW button and uses the same default template.

16. Select the **Model** tab of the open drawing.

17. 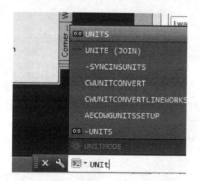 Type **units** on the *Command* line.

Notice how you get a selection of commands as you type.

Click ENTER to select UNITS.

18.  Note that the units are in Architectural format.

The units are determined by the template you set to start your new drawing.

19. Close the drawing without saving.

Templates can be used to preset layers, property set definitions, dimension styles, units, and layouts.

AEC Drawing Options

Menu	Tools→Options
Command line	Options
Context Menu→Options	Place mouse in the graphics area and right click
Shortcut	Place mouse in the command line and right click

Access the Options dialog box.

ACA's Options include five additional AEC specific tabs.
They are AEC Editor, AEC Content, AEC Project Defaults, AEC Object Settings, and AEC Dimension.

AEC Editor

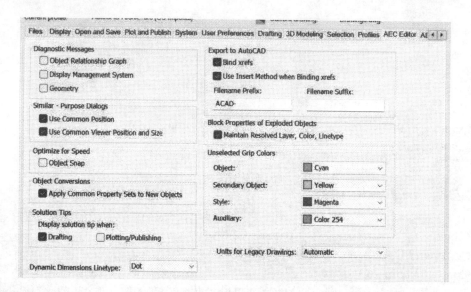

Diagnostic Messages	All diagnostic messages are turned off by default.
Similar-Purpose Dialogs	Options for the position of dialog boxes and viewers.
Use Common Position	Sets one common position on the screen for similar dialog boxes, such as door, wall, and window add or modify dialog boxes. Some dialog boxes, such as those for styles and properties, are always displayed in the center of the screen, regardless of this setting.
Use Common Viewer Position and Sizes	Sets one size and position on the screen for the similar-purpose viewers in AutoCAD Architecture. Viewer position is separately controlled for add, modify, style, and properties dialog boxes.
Optimize for Speed	Options for display representations and layer loading in the Layer Manager dialog.
Object Snap	Enable to limit certain display representations to respond only to the Node and Insert object snaps. This setting affects stair, railing, space boundary, multi-view block, masking block, slice, and clip volume result (building section) objects.

Apply Common Property Sets to New Objects	Property sets are used for the creation of schedule tables. Each AEC object has embedded property sets, such as width and height, to be used in its schedule. AEC properties are similar to attributes. Size is a common property across most AEC objects. Occasionally, you may use a tool on an existing object and the result may be that you replace the existing object with an entirely different object. For example, if you apply the tool properties of a door to an existing window, a new door object will replace the existing window. When enabled, any property sets that were assigned to the existing window will automatically be preserved and applied to the new door provided that the property set definitions make sense.
Solution Tips 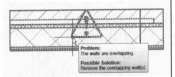	Users can set whether they wish a solution tip to appear while they are drafting or plotting. A solution tip identifies a drafting error and suggests a solution.
Dynamic Dimensions Linetype	Set the linetype to be used when creating dynamic dimensions. This makes it easier to distinguish between dynamic and applied dimensions. You may select either continuous or dot linetypes.
Export to AutoCAD 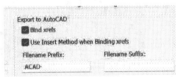	Enable Bind Xrefs. If you enable this option, the xref will be inserted as a block and not an xref. Enable Use Insert Method when binding Xrefs if you want all objects from an xref drawing referenced in the file you export to be automatically exploded into the host drawing. If you enable this option, the drawing names of the xref drawings are discarded when the exported drawing is created. In addition, their layers and styles are incorporated into the host drawing. For example, all exploded walls, regardless of their source (host or xref) are located on the same layer. Disable Use Insert Method when binding Xrefs if you want to retain the xref identities, such as layer names, when you export a file to AutoCAD or to a DXF file. For example, the blocks that define walls in the host drawing are located on A-Wall in the exploded drawing. Walls in an attached xref drawing are located on a layer whose name is created from the drawing name and the layer name, such as Drawing1|WallA.

	Many architects automatically bind and explode their xrefs when sending drawings to customers to protect their intellectual property.
	Enter a prefix or a suffix to be added to the drawing filename when the drawing is exported to an AutoCAD drawing or a DXF file.
	In order for any of these options to apply, you have to use the Export to AutoCAD command. This is available under the Files menu.
Unselected Grip Colors	Assign the colors for each type of grip.

Units for Legacy Drawings	Determines the units to be used when opening an AutoCAD drawing in AutoCAD Architecture.
	    Automatic – uses the current AutoCAD Architecture units setting
	    Imperial – uses Imperial units
	    Metric – uses Metric units

Option Settings are applied to your current drawing and saved as the default settings for new drawings. Because AutoCAD Architecture operates in a Multiple Document Interface, each drawing stores the Options Settings used when it was created and last saved. Some users get confused because they open an existing drawing and it will not behave according to the current Options Settings.

AEC Content

| 3D Modeling | Selection | Profiles | AEC Editor | AEC Content | AEC Project Defaults | AEC Object Settings | AEC Dimension |

Architectural / Documentation / Multi-Purpose Object Style Path:

C:\ProgramData\Autodesk\ACA 2020\enu\Styles\Imperial   [Browse...]

AEC DesignCenter Content Path:

C:\ProgramData\Autodesk\ACA 2020\enu\AEC Content   [Browse...]

Tool Catalog Content Root Path:

C:\ProgramData\Autodesk\ACA 2020\enu\   [Browse...]

IFC Content Path:

C:\ProgramData\Autodesk\ACA 2020\enu\Styles\   [Browse...]

Detail Component Databases: [Add/Remove...]

Keynote Databases: [Add/Remove...]

☑ Display Edit Property Data Dialog During Tag Insertion

Architectural/Documentation/Multipurpose Object Style Path	Allows you to specify a path of source files to be used to import object styles; usually used for system styles like walls, floors, or roofs.
AEC DesignCenter Content Path	Type the path and location of your content files or click Browse to search for the content files.
Tool Catalog Content Root Path	Type the path and location of your Tool Catalog files or click Browse to search for the content files.
IFC Content Path	Type the path and location of your Tool Catalog files or click Browse to search for the content files.
Detail Component Databases	Select the **Add/Remove** button to set the search paths for your detail component files.
Keynote Databases	Select the **Add/Remove** button to set the search paths for your keynote database files.
Display Edit Schedule Data Dialog During Tag Insertion	To attach schedule data to objects when you insert a schedule tag in the drawing, this should be ENABLED.

IFC Content is an object-based building data model that is non-proprietary. It uses open-source code developed by the IAI (International Alliance for Interoperability) to promote data exchange between CAD software. IFC Content might be created in Autodesk Revit, Inventor, or some other software for use in your projects.

Using AEC Content

AutoCAD Architecture uses several pre-defined and user-customizable content including:

- ❏ Architectural Display Configurations
- ❏ Architectural Profiles of Geometric Shapes
- ❏ Wall Styles and Endcap geometry
- ❏ Door styles
- ❏ Window styles
- ❏ Stair styles
- ❏ Space styles
- ❏ Schedule tables

Standard style content is stored in the AEC templates subdirectory. You can create additional content, import and export styles between drawings.

AEC Object Settings

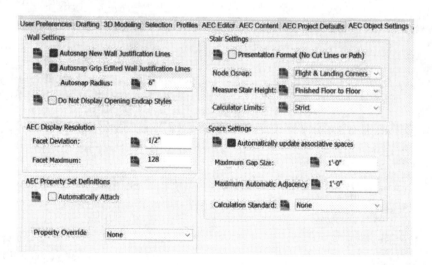

Wall Settings	
Autosnap New Wall Justification Lines	Enable to have the endpoint of a new wall that is drawn within the Autosnap Radius of the baseline of an existing wall automatically snap to that baseline. If you select this option and set your Autosnap Radius to 0, then only walls that touch clean up with each other.
Autosnap Grip Edited Wall Justification Lines	Enable to snap the endpoint of a wall that you grip edit within the Autosnap Radius of the baseline of an existing wall. If you select this option and set your Autosnap Radius to 0, then only walls that touch clean up with each other.
Autosnap Radius	Enter a value to set the snap tolerance.
Do not display opening end cap styles	Enable Do Not Display Opening Endcap Styles to supClick the display of endcaps applied to openings in walls. Enabling this option boosts drawing performance when the drawing contains many complex endcaps.
Stair Settings	
Presentation Format (No Cut Lines or Path)	If this is enabled, a jagged line and directional arrows will not display.

Node Osnap	Determines which snap is enabled when creating stairs:
	    Vertical Alignment
	    Flight and Landing Corners

Measure Stair Height	Rough floor to floor – ignore offsets Finished floor to floor – include top and bottom offsets
Calculator Limits	Strict – stair will display a defect symbol when an edit results in a violation of the Calculation rules. Relaxed – no defect symbol will be displayed when the stair violates the Calculation rules. The Calculation rules determine how many treads are required based on the riser height and the overall height of the stairs.

AEC Display Resolution 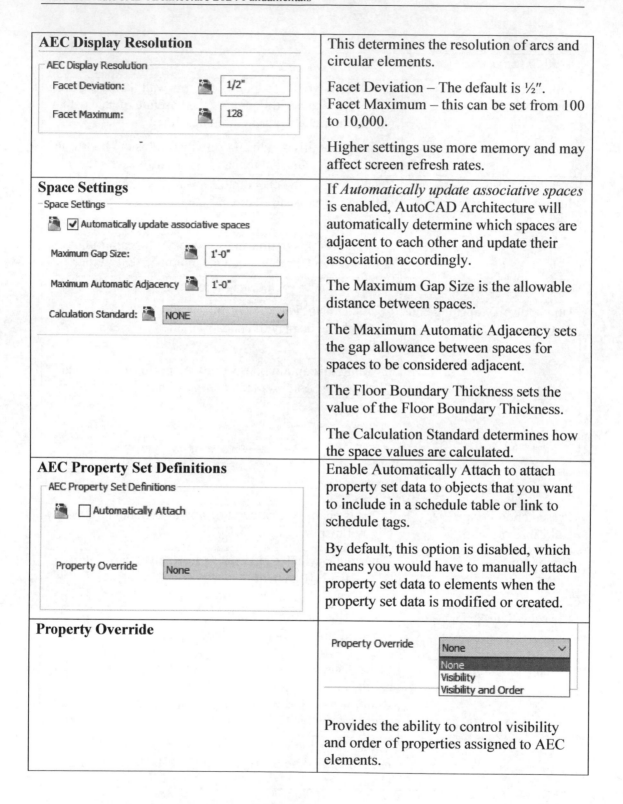	This determines the resolution of arcs and circular elements. Facet Deviation – The default is ½″. Facet Maximum – this can be set from 100 to 10,000. Higher settings use more memory and may affect screen refresh rates.
Space Settings	If *Automatically update associative spaces* is enabled, AutoCAD Architecture will automatically determine which spaces are adjacent to each other and update their association accordingly. The Maximum Gap Size is the allowable distance between spaces. The Maximum Automatic Adjacency sets the gap allowance between spaces for spaces to be considered adjacent. The Floor Boundary Thickness sets the value of the Floor Boundary Thickness. The Calculation Standard determines how the space values are calculated.
AEC Property Set Definitions	Enable Automatically Attach to attach property set data to objects that you want to include in a schedule table or link to schedule tags. By default, this option is disabled, which means you would have to manually attach property set data to elements when the property set data is modified or created.
Property Override	Provides the ability to control visibility and order of properties assigned to AEC elements.

AEC Dimension

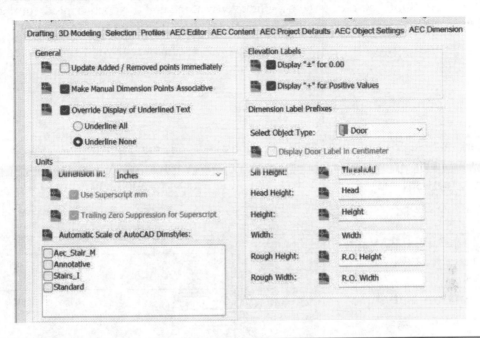

General	
Update Added/Removed points immediately	Enable to update the display every time you add or remove a point from a dimension chain.
Make Manual Dimension Points Associative	Enable if you want the dimension to update when the object is modified.
Override Display of Undefined Text	Enable to automatically underline each manually overridden dimension value.
Underline All	Enable to manually underline overridden dimension values.
Underline None	Enable to not underline any overridden dimension value.
Units	
Dimension in:	Select the desired units from the drop-down list.
Use Superscript mm	If your units are set to meters or centimeters, enable superscripted text to display the millimeters as superscript. 4.34^{25}

Trailing Zero SupClickion for Superscript	To supClick zeros at the end of superscripted numbers, select Trailing Zeros SupClickion for Superscript. You can select this option only if you have selected Use Superscript mm and your units are metric. 4.12^3
Automatic Scale of AutoCAD Dimstyles	Select the dimstyles you would like to automatically scale when units are reset.
Elevation Labels	Select the unit in which elevation labels are to be displayed. This unit can differ from the drawing unit.
Display +/- for 0.00 Values	
Display + for Positive Values	
Dimension Label Prefixes	Select Object Type: Door / Door / Window / Opening / Stair Dimension Label Prefixes are set based on the object selected from the drop-down. You can set label prefixes for Doors, Windows, Openings, and Stairs. Select the object, and then set the prefix for each designation. The designations will change based on the object selected.

AEC Project Defaults

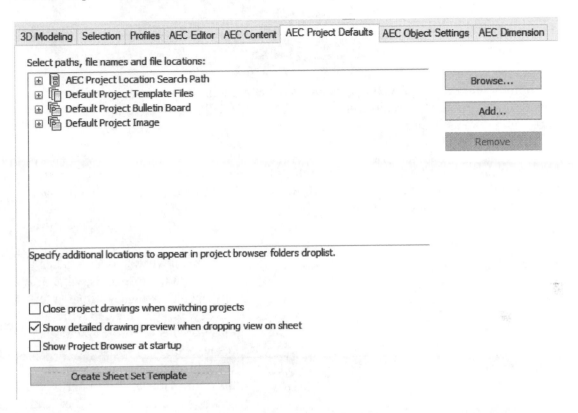

Most BID/AEC firms work in a team environment where they divide a project's tasks among several different users. The AEC Project Defaults allow the CAD Manager or team leader to select a folder on the company's server to locate files, set up template files with the project's title blocks, and even set up a webpage to post project information.

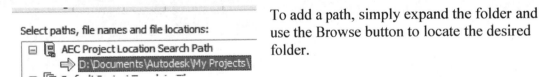

To add a path, simply expand the folder and use the Browse button to locate the desired folder.

You can add more than one path to the AEC Project Location Search Path. All the other folders only allow you to have a single location for template files, the project bulletin board, and the default project image.

I highly recommend that you store any custom content in a separate directory as far away from AutoCAD Architecture as possible. This will allow you to back up your custom work easily and prevent accidental erasure if you upgrade your software application.

You should be aware that AutoCAD Architecture currently allows the user to specify only ONE path for custom content, so drawings with custom commands will only work if they reside in the specified path.

☐ Close project drawings when switching projects	If enabled, when a user switches projects in the Project Navigator, any open drawings not related to the new project selected will be closed. This conserves memory resources.
☑ Show detailed drawing preview when dropping view on sheet	If enabled, then the user will see the entire view as it is being positioned on a sheet. If it is disabled, the user will just see the boundary of the view for the purposes of placing it on a sheet.
☐ Show Project Browser at startup	If enabled, you will automatically have access to the Project Browser whenever you start AutoCAD Architecture.

Architectural Profiles of Geometric Shapes

You can create mass elements to define the shape and configuration of your preliminary study, or mass model. After you create the mass elements you need, you can change their size as necessary to reflect the building design.

- **Mass element**: A single object that has behaviors based on its shape. For example, you can set the width, depth, and height of a box mass element, and the radius and height of a cylinder mass element.

Mass elements are parametric, which allows each of the shapes to have very specific behavior when it comes to the manipulation of each mass element's shape. For example, if the corner grip point of a box is selected and dragged, then the width and depth are modified. It is easy to change the shape to another form by right-clicking on the element and selecting a new shape from the list.

Through Boolean operations (addition, subtraction, intersection), mass elements can be combined into a mass group. The mass group provides a representation of your building during the concept phase of your project.

- **Mass group**: Takes the shape of the mass elements and is placed on a separate layer from the mass elements.
- **Mass model**: A virtual mass object, shaped from mass elements, which defines the basic structure and proportion of your building. A marker appears as a small box in your drawing to which you attach mass elements.

As you continue developing your mass model, you can combine mass elements into mass groups and create complex building shapes through addition, subtraction, or intersection of mass elements. You can still edit individual mass elements attached to a mass group to further refine the building model.

To study alternative design schemes, you can create a number of mass element references. When you change the original of the referenced mass element, all the instances of the mass element references are updated.

The mass model that you create with mass elements and mass groups is a refinement of your original idea that you carry forward into the next phase of the project, in which you change the mass study into floor plates and then into walls. The walls are used to start the design phase.

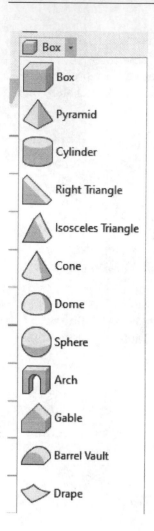

Mass Elements are accessed from the Home tab on the ribbon.

Mass Elements that can be defined from the Home tab include Arches, Gables, Cones, etc.

Exercise 1-2:
Creating a New Geometric Profile

Drawing Name: New
Estimated Time: 15 minutes

This exercise reinforces the following skills:

- Use of AEC Design Content
- Use of Mass Elements
- Use of Views
- Visual Styles
- Modify using Properties

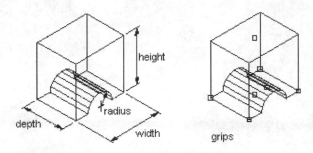

Creating an arch mass element

1. Start a New Drawing.

2. Select the **QNEW** tool from the Standard toolbar.

3. Because we assigned a template in Exercise 1-1, a drawing file opens without prompting us to select a template.

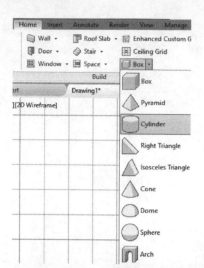

All the massing tools are available on the **Home** ribbon in the Build section.

4.

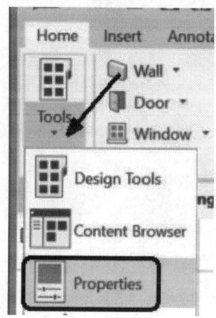

If you don't see the Properties palette:

Launch the **Properties** palette under Tools.

5. Arch   Select the **Arch** tool from the Massing drop down list.

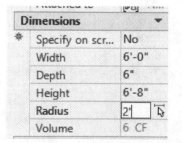

Expand the Dimensions category in the Properties palette.
Set Specify on screen to **No**.
Set the Width to **6′ [183]**.
Set the Depth to **6″ [15]**.
Set the Height to **6′ 8″ [203]**.
Set the Radius to **2′ [61]**.

Units in brackets are centimeters.

You need to enter in the ' and " symbols.

6.

    Pick a point anywhere in the drawing area.

    Click ENTER to accept a Rotation Angle of 0 or right click and select **ENTER**.

    Click **ENTER** to exit the command.

7.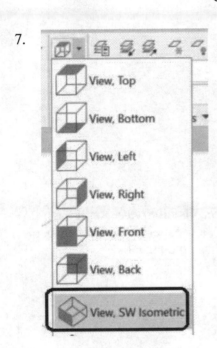

    Switch to an isometric view.
    To switch to an isometric view, simply go to **View** panel on the Home ribbon. Under the Views list, select **SW Isometric**.

    Or select the Home icon next to the Viewcube.

Our model so far.

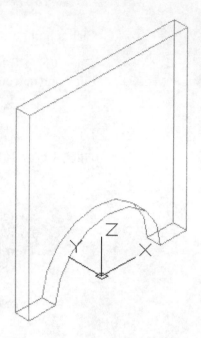

8.   Select **Visual Styles, Realistic** on the View panel on the Home ribbon to shade the model.

You can also type **SHA** on the command line, then **R** for Realistic.

9. To change the arch properties, select the arch so that it is highlighted. If the properties dialog doesn't pop up automatically, right click and select **Properties**.

10.

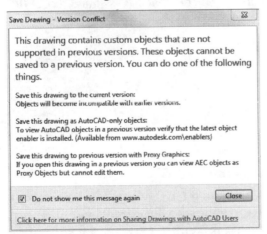

Dimensions	▼
Width	6'-0"
Depth	6"
Height	6'-8"
Radius	9"
Volume	20 CF

Change the Radius to **9"** [15].

Pick into the graphics window and the arch will update.

11. Click **ESC** to release the grips on the arc.

12. Save the drawing as *ex1-2.dwg*.

Save Drawing - Version Conflict ⊠

This drawing contains custom objects that are not supported in previous versions. These objects cannot be saved to a previous version. You can do one of the following things.

Save this drawing to the current version:
Objects will become incompatible with earlier versions.

Save this drawing as AutoCAD-only objects:
To view AutoCAD objects in a previous version verify that the latest object enabler is installed. (Available from www.autodesk.com\enablers)

Save this drawing to previous version with Proxy Graphics:
If you open this drawing in a previous version you can view AEC objects as Proxy Objects but cannot edit them.

☑ Do not show me this message again [Close]

Click here for more information on Sharing Drawings with AutoCAD Users

If you see this dialog, it means that the mass element just created (the arch) cannot be saved as a previous version file.

Click **Close**.

The Style Manager

Access the Style Manager from the Manage ribbon.

The Style Manager is a Microsoft® Windows Explorer-based utility that provides you with a central location in Autodesk AutoCAD Architecture where you can view and work with styles in drawings or from Internet and intranet sites.

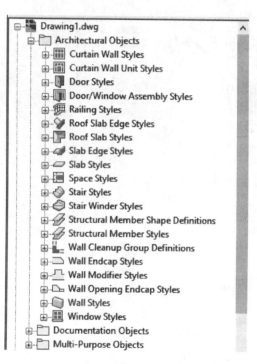

Styles are sets of parameters that you can assign to objects in Autodesk AutoCAD Architecture to determine their appearance or function. For example, a door style in Autodesk AutoCAD Architecture determines what door type, such as single or double, bi-fold or hinged, a door in a drawing represents. You can assign one style to more than one object, and you can modify the style to change all the objects that are assigned that style.

Depending on your design projects, either you or your CAD Manager might want to customize existing styles or create new styles. The Style Manager allows you to easily create, customize, and share styles with other users. With the Style Manager, you can:

- Provide a central point for accessing styles from open drawings and Internet and intranet sites

- Quickly set up new drawings and templates by copying styles from other drawings or templates

- Sort and view the styles in your drawings and templates by drawing or by style type

- Preview an object with a selected style

- Create new styles and edit existing styles
- Delete unused styles from drawings and templates
- Send styles to other Autodesk AutoCAD Architecture users by email

Objects in Autodesk AutoCAD Architecture that use styles include 2D sections and elevations, AEC polygons, curtain walls, curtain wall units, doors, endcaps, railings, roof slab edges, roof slabs, schedule tables, slab edges, slabs, spaces, stairs, structural members, wall modifiers, walls, window assemblies, and windows.

Additionally, layer key styles, schedule data formats, and cleanup group, mask block, multi-view block, profile, and property set definitions are handled by the Style Manager.

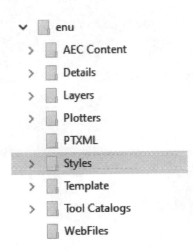

Most of the objects in Autodesk AutoCAD Architecture have a default Standard style. In addition, Autodesk AutoCAD Architecture includes a starter set of styles that you can use with your drawings. The Autodesk AutoCAD Architecture templates contain some of these styles. Any drawing that you start from one of the templates includes these styles.

You can also access the styles for doors, end caps, spaces, stairs, walls, and windows from drawings located in *ProgramData\Autodesk\ACA 2024\enu\Styles*.

Property set definitions and schedule tables are located in the *ProgramData\Autodesk\ACA 2024\enu\AEC Content* folder.

Exercise 1-3:
Copying a Style to a Template

Drawing Name: Styles.dwg, Imperial (Metric).ctb.dwt
Estimated Time: 20 minutes

This exercise reinforces the following skills:

- Style Manager
- Use of Templates
- Properties dialog

1. 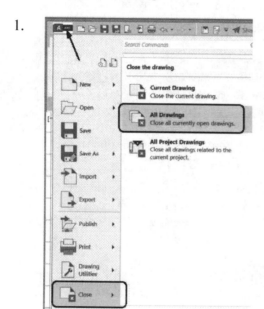 Close any open drawings.

   Go to the Application menu and select **Close → All Drawings**.

   You can also type CLOSEALL.

2. **Start** [+] Start a New Drawing using plus tab.

3. Activate the **Manage** ribbon.

   Style Manager Select the **Style Manager** tool.

4. *Style Manager* Select the **File Open** tool in the Style Manager dialog.

   File Edit View

5. File name: Styles   Locate the *styles.dwg*.
   Files of type: Drawing (*.dwg)   *This file is downloaded from the publisher's website.*
   Click **Open**.

Note: Make sure the Files of type is set to dwg or you will not see your files.

6. 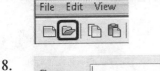 Styles.dwg is now listed in the Style Manager.

7.  Select the **File Open** tool.

8. 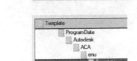 Set the Files of type to ***Drawing Template (*.dwt).***

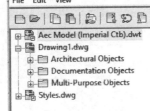 *You should automatically be directed to the Template folder.*

If not, browse to the Template folder.

9. Locate the *AEC Model Imperial (Metric) Ctb.dwt* file. Click **Open**.

This is the default template we selected for QNEW.

10. You should see files open in the Style Manager.

11. Locate the **Brick_Block** Wall Style in the *Styles.dwg.*

12. 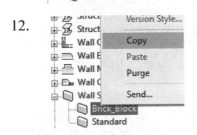 Right click and select **Copy**.

13. 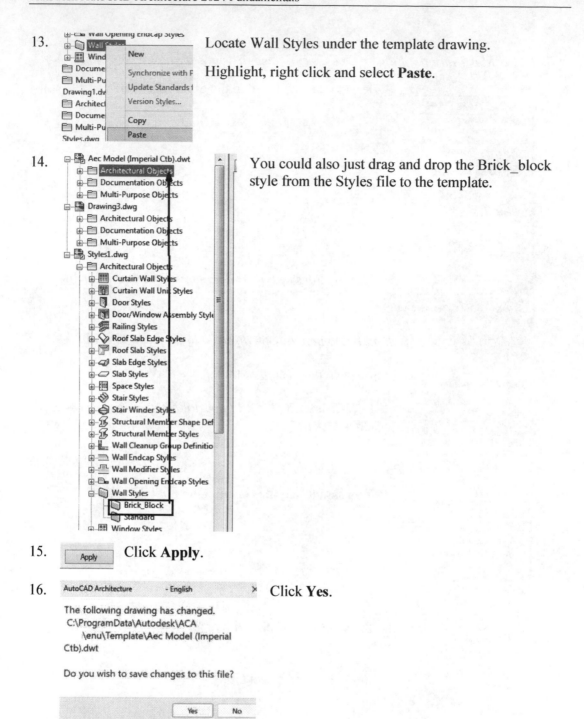 Locate Wall Styles under the template drawing.

Highlight, right click and select **Paste**.

14. You could also just drag and drop the Brick_block style from the Styles file to the template.

15. Apply   Click **Apply**.

16. Click **Yes**.

AutoCAD Architecture   - English

The following drawing has changed.
C:\ProgramData\Autodesk\ACA
      \enu\Template\Aec Model (Imperial
Ctb).dwt

Do you wish to save changes to this file?

Yes   No

17. Close the Style Manager.

18. Drawing1*   Look at the drawing folder tabs. The Styles.dwg and the template drawing are not shown. These are only open in the Style Manager.

19. **Start** [+]   Select the **plus** tab to start a new drawing.

20. Home | Insert
    Wall ·
    Door ·
    Window ·
    Activate the Home tab on the ribbon.
    Select the **Wall** tool.

21. Search...
    Style | Standard
    Bound spaces | Brick_Block
    Cleanup automati... | Standard
    Cleanup group d... | Standard
    The Properties dialog should pop up when the WallAdd tool is selected.

    In the Style drop-down list, note that Brick_Block is available.

    If you don't see the wall style in the drop-down list, verify that you set the default template correctly in Options.

22. Click Escape to end the WallAdd command.

23. Close all drawing files without saving.

The Styles Browser

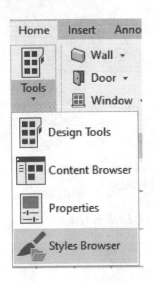

The Styles Browser can be launched from the Home ribbon under the Tools drop-down.

The Styles Browser displays the styles available for various elements

You can see the styles that are available in the Content Library as well as the active drawings.

You can use the Styles Browser to copy and paste styles between drawings.

Exercise 1-4:

Copying a Style using the Style Browser

Drawing Name: Stylesbrowser.dwg
Estimated Time: 10 minutes

This exercise reinforces the following skills:

- Style Browser
- Use of Styles
- Importing a Style into a Drawing

1.

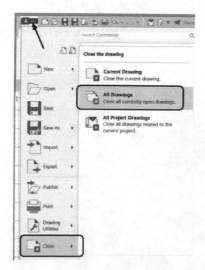

Close any open drawings.

Go to the Application menu and select **Close → All Drawings**.

You can also type CLOSEALL.

2. 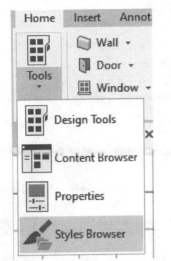 Start a New Drawing using plus tab.

3. Switch to the Home ribbon.

Click **Styles Browser** under Tools to launch the Styles Browser.

4. 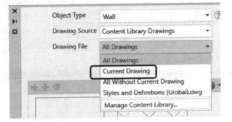 On the Drawing Source field, click the down arrow and select **Current Drawing**.

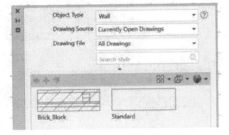 Only one wall style is listed.

If you modified the template used by QNEW, you will see two wall styles.

5. 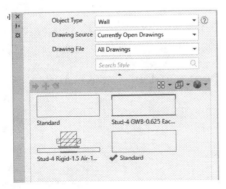 Open the *stylesbrowser.dwg*.

This file is downloaded from the publisher's website.

The Styles Browser shows all the wall styles available in this file.

6.

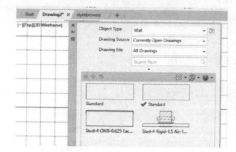

Click on the **Drawing2** tab (the new drawing you created at the start of the exercise).

Notice the Drawing Source has changed to Currently Open Drawings.

The green check indicates any style being used in any of the open drawings.

7.

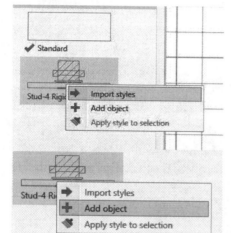

Highlight the Stud-4 Rigid-1.5 Air wall style.

Right click and select **Import styles**.

Note you can hold down the CTL key to select more than one style to be imported.

Close the *stylesbrowser.dwg* without saving.

8.

Right click on the Stud-4 Rigid-1.5 Air wall style.

Click **Add object**.

9.

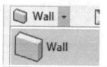

Click a point to start placing a wall.
Drag the mouse to the right.
Click to place an end point for the wall.
Click ENTER to exit the command.

Close the Style Browser.

10.

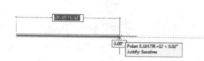

Click the **Wall** tool on the Home ribbon.

11.

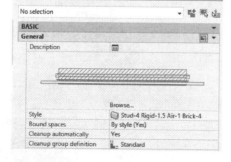

Note that the imported wall style is available in the drawing now.

Place a wall using the imported style.

12.

Close without saving.

Visual Styles are stored with the drawing.

Visual Styles

Visual styles determine the display of edges, lighting, and shading in each viewport.

You can choose a predefined visual style in the upper-left corner of each viewport. In addition, the Visual Styles Manager displays all styles available in the drawing. You can choose a different visual style or change its settings at any time. ACA comes with several pre-defined visual styles or you can create your own.

Exercise 1-5:

Creating a New Visual Style

Drawing Name: visual_style.dwg
Estimated Time: 15 minutes

This exercise reinforces the following skills:

- ❑ Workspaces
- ❑ Use of Visual Styles
- ❑ Controlling the Display of Objects

1. Open *visual_style.dwg*.

2.    Activate the **View** ribbon.

3. 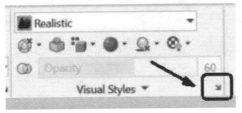   Launch the Visual Styles Manager.

     To launch, select the small arrow in the lower right corner of the Visual Styles panel.

4.

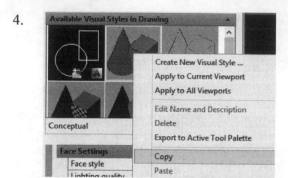

Highlight the **Conceptual** tool.

Right click and select **Copy**.

5.

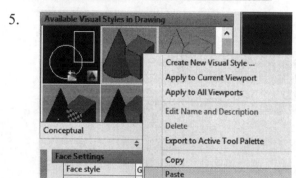

Right click and select **Paste**.

6.

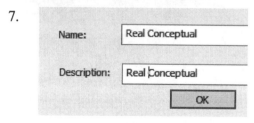

Highlight the copied tool.
Right click and select **Edit Name and Description**.

7.

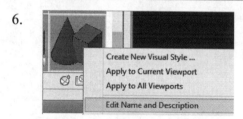

Change the name and description to **Real Conceptual**.

Click **OK**.

8.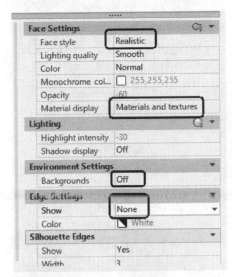

Under Face Settings:

Set the Face Style to **Realistic**.

Set Material display to **Materials and textures**.

Under Environment Settings:

Set the Backgrounds to **Off**.

Under Edge Settings:

Set the Show to **None**.

9.

Verify that the Real Conceptual Style is highlighted/ selected.

Right click and select the **Apply to Current Viewport** tool.

[−][SW Isometric][Real Conceptual]

Note that the active visual style is listed in the upper right corner of the drawing.

10. Save the drawing as *ex1-6.dwg*.

Exercise 1-6:

Creating an Autodesk Cloud Account

Drawing Name: (none, start from scratch)
Estimated Time: 5 Minutes

This exercise reinforces the following skills:

❏   Create an account on Autodesk.

> *You must have an internet connection to access the Autodesk server.
> Many government agencies do not allow the use of cloud accounts due to
> security concerns. Autodesk's cloud account is useful for students, home
> users, and small shops.*

1.    Next to the Help icon at the top of the screen:

     Select **Sign In to Autodesk account**.

2.   If you have an existing Autodesk ID, you can use it.

     If not, select the link that says **Create Account.**

     *You may have an existing Autodesk ID if you have
     registered Autodesk software in the past.
     Autodesk's cloud services are free to students and
     educators. For other users, you are provided a
     small amount of initial storage space for free.*

     *I like my students to have an on-line account
     because it will automatically back up their work.*

3.

Create account

First name

[   ] | Last name

Email

Confirm email

Password

☐ I agree to the Autodesk Terms of Use and acknowledge the Privacy Statement.

CREATE ACCOUNT

ALREADY HAVE AN ACCOUNT? SIGN IN

Fill in the form to create an Autodesk ID account.

Be sure to write down the ID you select and the password.

4.

Account created

This single account gives you access to all your Autodesk products

✓

☐ Check this box to receive electronic marketing communications from Autodesk on news, trends, events, special offers and research surveys. You can manage your preferences or unsubscribe at any time. To learn more, see the Autodesk Privacy Statement.

DONE

You should see a confirmation window stating that you now have an account.

5. ⟳ elise_moss · ☒ ⑦ ·

You will see the name you selected as your Autodesk ID in the Cloud sign-in area.

Tips Tricks

Use your Autodesk account to create renderings, back up files to the Cloud so you can access your drawings on any internet-connected device, and to share your drawings.

Exercise 1-7:

Tool Palettes

Drawing Name: tool_palettes.dwg
Estimated Time: 15 minutes

This exercise reinforces the following skills:

- ❑ Use of AEC Design Content
- ❑ Use of Wall Styles

1. Open the tool_palettes.dwg.

2. Activate the Home ribbon.

   Select **Tools→Design Tools** to launch the tool palette.

3. Right click on the Design Tools palette title bar.

   Click **New Palette**.

4. 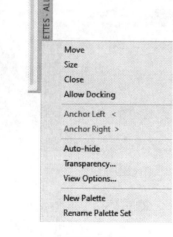 Rename the new palette **Favorite Walls**.

5.

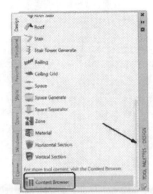

Click the **Design** tab on the tool palette.

Use the scroll bar to scroll to the bottom of the tool palette.

Locate the Content Browser icon.

Click on the **Content Browser** icon to launch.

6.

Design Tool Catalog - Imperial

Locate the **Design Tool Catalog – Imperial.**

Click on the **Design Tool Catalog – Imperial** to open.

7.

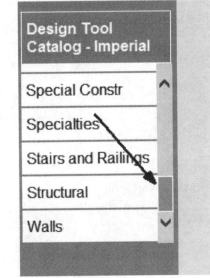

Use the scroll bar to scroll down.

Click on **Walls**.

8.

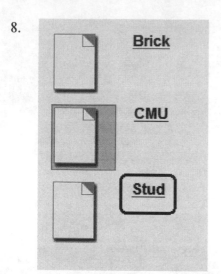

Click on the **Stud** link to look at the stud walls.

9.

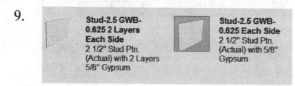

Locate the **Stud – 2.5 GWB – 0.625 2 Layers Each Side**.

10.

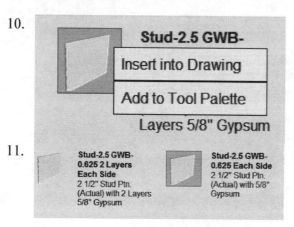

Click on the wall icon.
Select **Add to Tool Palette**.

11.

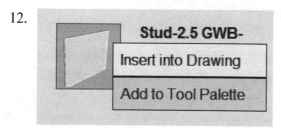

Locate the **Stud – 2.5 GWB – 0.625 Each Side**.

12.

Click on the wall icon.
Select **Add to Tool Palette**.

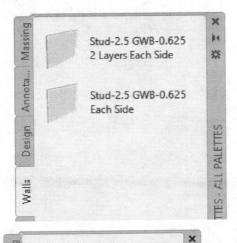

The two wall styles are now listed on the Walls palette.

13.

Right click below the second wall style.

Click **Add Separator**.

View Options...
Sort By   >
Paste
Add Text
Add Separator
Delete Palette
Rename Palette

14.

A horizontal line is added to the tool palette.

15.

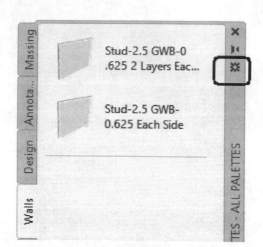

On the Favorite Walls tool palette:

Select the **Properties** button.

16.

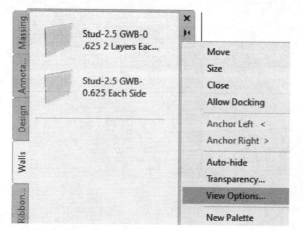

Select **View Options**.

17.

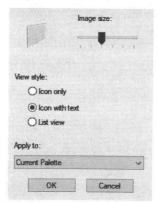

Adjust the Image Size and select the desired View style.

You can apply the preference to the Current Palette (the walls palette only) or all palettes.

Click **OK**.

18.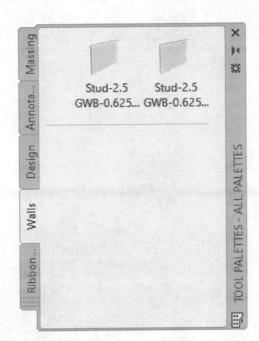

Verify that the tool palette is open with the Favorite Walls palette active.

19. Activate the **Manage** ribbon.

Select the **Style Manager** tool.

20. Locate the **Brick-Block** wall style you created in the Style Manager.

21.

Drag and drop the wall style onto your Favorite Walls palette.

Click **OK** to close the Style Manager.

22.

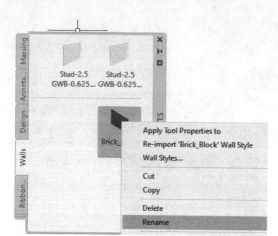

Highlight the **Brick_Block** tool.
Right click and select **Rename**.

23.

Change the name to **8″ CMU- 3.5″ Brick [200 mm CMU – 90 mm]**.

24. Select the new tool and draw a wall.
You may draw a wall simply by picking two points.
If you enable the ORTHO button on the bottom of the screen, your wall will be straight.

25. Select the wall you drew.

Right click and select **Properties**.

Note that the wall is placed on the A-Wall Layer and uses the Brick-Block Style you defined.

26. Save your file as *Styles3.dwg*.

The wall style you created will not be available in any new drawings unless you copy it over to the default template.

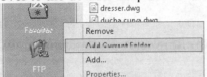 To make it easy to locate your exercise files, browse to where the files are located when you save the file. Right click on the left pane and select Add Current Folder.

A fast way to close all open files is to type CLOSEALL on the command line. You will be prompted to save any unsaved files.

Layer Manager

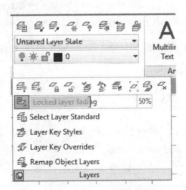

Back in the days of vellum and pencil, drafters would use separate sheets to organize their drawings, so one sheet might have the floor plan, one sheet the site plan, etc. The layers of paper would be placed on top of each other and the transparent quality of the vellum would allow the drafter to see the details on the lower sheets. Different colored pencils would be used to make it easier for the drafter to locate and identify elements of a drawing, such as dimensions, electrical outlets, water lines, etc.

When drafting moved to Computer Aided Design, the concept of sheets was transferred to the use of Layers. Drafters could assign a Layer Name, color, linetype, etc. and then place different elements on the appropriate layer.

AutoCAD Architecture has a Layer Management system to allow the user to implement AIA standards easily.

Layer Manager

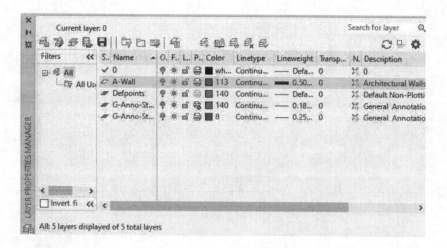

The Layer Manager helps you organize, sort, and group layers, as well as save and coordinate layering schemes. You can also use layering standards with the Layer Manager to better organize the layers in your drawings.

When you open the Layer Manager, all the layers in the current drawing are displayed in the right panel. You can work with individual layers to:

- Change layer properties by selecting the property icons
- Make a layer the current layer
- Create, rename, and delete layers

If you are working with drawings that contain large numbers of layers, you can improve the speed at which the Layer Manager loads layers when you open it by selecting the Layer Manager/Optimize for Speed option in your AEC Editor options.

The Layer Manager has a tool bar as shown.

	Layer Standards You can import Layer Standards from an existing drawing or Export Layer Standards using the current drawing. 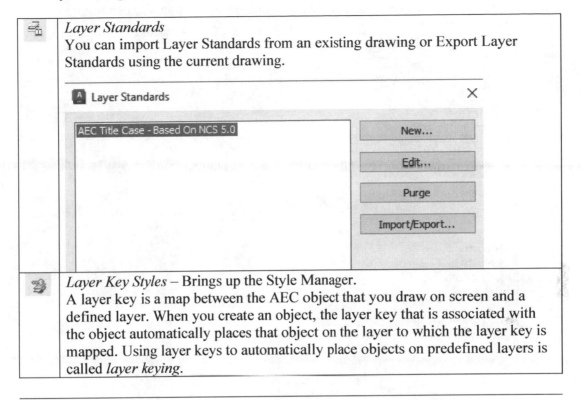
	Layer Key Styles – Brings up the Style Manager. A layer key is a map between the AEC object that you draw on screen and a defined layer. When you create an object, the layer key that is associated with the object automatically places that object on the layer to which the layer key is mapped. Using layer keys to automatically place objects on predefined layers is called *layer keying*.

The layer key styles in Autodesk AutoCAD Architecture, Release 3 and above replace the use of LY files. If you have LY files from S8 or Release 1 of AutoCAD Architecture that you want to use, then you can create a new layer key style from an LY file. On the command line enter **-AecLYImport** to import legacy LY files.

	Layer Key Overrides You can apply overrides to any layer keys within a layer key style that is based on a layer standard. The structure of the layer name of each layer that each layer key maps an object to is determined by the descriptive fields in the layer standard definition. You can override the information in each field according to the values set in the layer standard definition. You can allow overrides on all the layer keys within a layer key style, or you can select individual layer keys that you want to override. You can also choose to allow all of the descriptive fields that make up the layer name to be overridden, or you can specify which descriptive fields you want to override.
	Load Filter Groups This tool allows you to load a saved filter group.

🖫	*Save Filter Groups* This tool allows you to save a defined filter group, so that you can use it in other drawings.
🗀	New Property Filter
🗀	New Group Filter
🗔	New Standards Filter
🗐	*New Layer* Creates a new layer.
📖	*New Layer from Standard* Creates a new layer based on an existing layer standard. 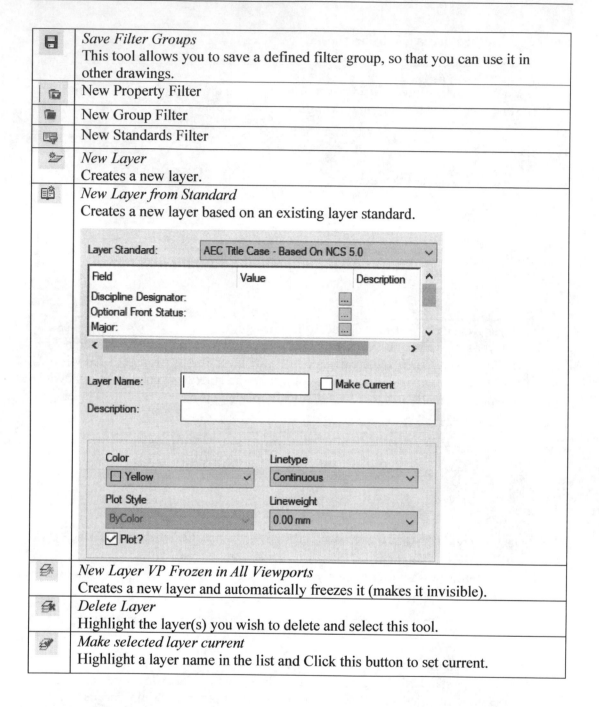
🗐	*New Layer VP Frozen in All Viewports* Creates a new layer and automatically freezes it (makes it invisible).
🗐	*Delete Layer* Highlight the layer(s) you wish to delete and select this tool.
🖉	*Make selected layer current* Highlight a layer name in the list and Click this button to set current.

The following are the default layer keys used by Autodesk AutoCAD Architecture when you create AEC objects.

Default layer keys for creating AEC objects

Layer Key	Description	Layer Key	Description
ANNDTOBJ	Detail marks	COMMUN	Communication
ANNELKEY	Elevation Marks	CONTROL	Control systems
ANNELOBJ	Elevation objects	CWLAYOUT	Curtain walls
ANNMASK	Masking objects	CWUNIT	Curtain wall units
ANNMATCH	Match Lines	DIMLINE	Dimensions
ANNOBJ	Notes, leaders, etc.	DIMMAN	Dimensions (AutoCAD points)
ANNREV	Revisions	DOOR	Doors
ANNSXKEY	Section marks	DOORNO	Door tags
ANNSXOBJ	Section marks	DRAINAGE	Drainage
ANNSYMOBJ	Annotation marks	ELEC	Electric
APPL	Appliances	ELECNO	Electrical tags
AREA	Areas	ELEV	Elevations
AREAGRP	Area groups	ELEVAT	Elevators
AREAGRPNO	Area group tags	ELEVHIDE	Elevations (2D)
AREANO	Area tags	EQUIP	Equipment
CAMERA	Cameras	EQUIPNO	Equipment tags
CASE	Casework	FINCEIL	Ceiling tags
CASENO	Casework tags	FINE	Details- Fine lines
CEILGRID	Ceiling grids	FINFLOOR	Finish tags
CEILOBJ	Ceiling objects	FIRE	Fire system equip.
CHASE	Chases	FURN	Furniture
COGO	Control Points	FURNNO	Furniture tags
COLUMN	Columns	GRIDBUB	Plan grid bubbles
GRIDLINE	Column grids	SITE	Site

Layer Key	Description	Layer Key	Description
HATCH	Detail-Hatch lines	SLAB	Slabs
HIDDEN	Hidden Lines	SPACEBDRY	Space boundaries
LAYGRID	Layout grids	SPACENO	Space tags
LIGHTCLG	Ceiling lighting	SPACEOBJ	Space objects
LIGHTW	Wall lighting	STAIR	Stairs
MASSELEM	Massing elements	STAIRH	Stair handrails
MASSGRPS	Massing groups	STRUCTBEAM	Structural beams
MASSSLCE	Massing slices	STRUCTBEAMIDEN	Structural beam tags
MED	Medium Lines	STRUCTBRACE	Structural braces
OPENING	Wall openings	STRUCTBRACEIDEN	Structural brace tags
PEOPLE	People	STRUCTCOLS	Structural columns
PFIXT	Plumbing fixtures	STRUCTCOLSIDEN	Structural column tags
PLANTS	Plants - outdoor	SWITCH	Electrical switches
PLANTSI	Plants - indoor	TITTEXT	Border and title block
POLYGON	AEC Polygons	TOILACC	Arch. specialties
POWER	Electrical power	TOILNO	Toilet tags
PRCL	Property Line	UTIL	Site utilities
PRK-SYM	Parking symbols	VEHICLES	Vehicles
ROOF	Rooflines	WALL	Walls
ROOFSLAB	Roof slabs	WALLFIRE	Fire wall patterning
ROOMNO	Room tags	WALLNO	Wall tags
SCHEDOBJ	Schedule tables	WIND	Windows
SEATNO	Seating tags	WINDASSEM	Window assemblies
SECT	Miscellaneous sections	WINDNO	Window tags
SECTHIDE	Sections (2D)		

Display Manager

The Display Manager is a utility for managing display configurations, display sets, and display representations: renaming them or deleting them, copying them between drawings, emailing them to other users, and purging unused elements from drawings.

The display system in AutoCAD Architecture is designed so that you only have to draw an architectural object once. The appearance of that object then changes automatically to meet the display requirements of different types of drawings, view directions, or levels of detail.

You can change the display settings for an element by using the Display tab on the Properties palette.

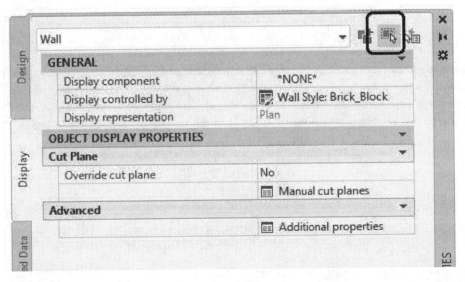

To change the display using the Display tab, click Select Components, select an object display component (like a hatch or a boundary), and then select or enter a new value for the display property you want to change (such as color, visibility, or lineweight).

Exercise 1-8:

Exploring the Display Manager

Drawing Name: display_manager.dwg
Estimated Time: 15 minutes

This exercise reinforces the following skills:

❑ Display Manager

1. Open *display_manager.dwg*.

2. 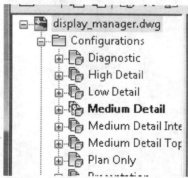 Access the **Display Manager** from the Manage ribbon.

The display system in Autodesk® AutoCAD Architecture controls how AEC objects are displayed in a designated viewport. By specifying the AEC objects you want to display in a viewport and the direction from which you want to view them, you can produce different architectural display settings, such as floor plans, reflected plans, elevations, 3D models, or schematic displays.

3. 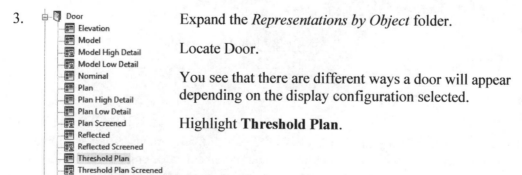 Expand the *Representations by Object* folder.

Locate Door.

You see that there are different ways a door will appear depending on the display configuration selected.

Highlight **Threshold Plan**.

4.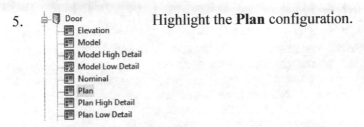

You see that the Threshold visibilities are turned off.

5.

Highlight the **Plan** configuration.

- Door
  - Elevation
  - Model
  - Model High Detail
  - Model Low Detail
  - Nominal
  - Plan
  - Plan High Detail
  - Plan Low Detail

6.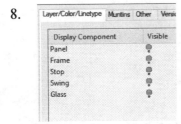

You see that some door components are turned on and others are turned off.

7.

Highlight the Model configuration.

- Display Theme
- Door
  - Elevation
  - Model
  - Model High Detail
  - Model Low Detail

8.

You see that some door components are turned on and others are turned off.

9. ⊟ ▦ ex1-8.dwg
        ⊞ ▢ Configurations
        ⊟ ▢ Sets
            ▦ Model
            ▦ Model High Detail
            ▦ Model Low Detail
            ▢ Model Presentation
            ▦ Model Structural
            ▦ **Plan**
            ▢ Plan Diagnostic
            ▦ Plan High Detail

Go to the Sets folder. The Set highlighted in bold is the active Display Representation.

Highlight **Plan**.

10.
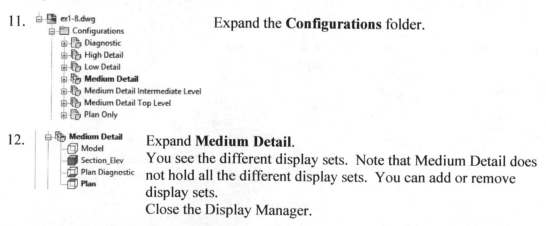

Select the Display Representation Control tab.
Locate **Door** under Objects.
Note that some display representations are checked and others are unchecked.
A check indicates that the door is visible. Which door components are visible is controlled under Representations by Object.

The three different folders – Configurations, Sets, and Representations by Object – are different ways of looking at the same elements. Each folder organizes the elements in a different way. If you change the display setting in one folder, it updates in the other two folders.

11. ⊟ ▦ ex1-8.dwg
        ⊟ ▢ Configurations
            ⊞ ▢ Diagnostic
            ⊞ ▢ High Detail
            ⊞ ▢ Low Detail
            ⊞ ▢ **Medium Detail**
            ⊞ ▢ Medium Detail Intermediate Level
            ⊞ ▢ Medium Detail Top Level
            ⊞ ▢ Plan Only

Expand the **Configurations** folder.

12. ⊟ ▢ **Medium Detail**
            ▢ Model
            ▦ Section_Elev
            ▢ Plan Diagnostic
            ▢ **Plan**

Expand **Medium Detail**.
You see the different display sets. Note that Medium Detail does not hold all the different display sets. You can add or remove display sets.
Close the Display Manager.

13. Close the drawing without saving.

Text Styles

A text style is a named collection of text settings that controls the appearance of text, such as font, line spacing, justification, and color. You create text styles to specify the format of text quickly, and to ensure that text conforms to industry or project standards.

- When you create text, it uses the settings of the current text style.

- If you change a setting in a text style, all text objects in the drawing that use the style update automatically.

- You can change the properties of individual objects to override the text style.

- All text styles in your drawing are listed in the Text Style drop-down.

- Text styles apply to notes, leaders, dimensions, tables, and block attributes.

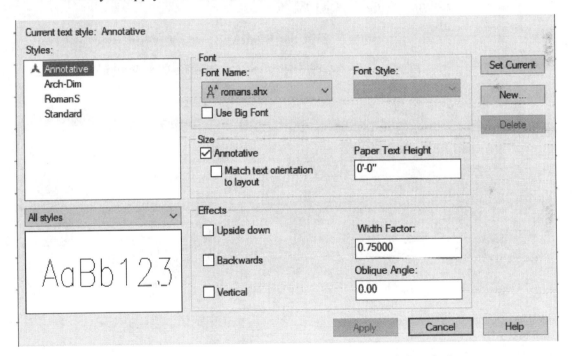

All drawings contain a STANDARD text style which can't be deleted. Once you create a standard set of text styles, you can save the styles to your template file (.dwt).

Exercise 1-9:

Creating a Text Style

Drawing Name: new using QNEW
Estimated Time: 15 minutes

This exercise reinforces the following skills:

- ❑ CUI – Custom User Interface
- ❑ Tab on the ribbon
- ❑ Panel
- ❑ Tab

1. Start a new drawing using the QNEW tool.

2. 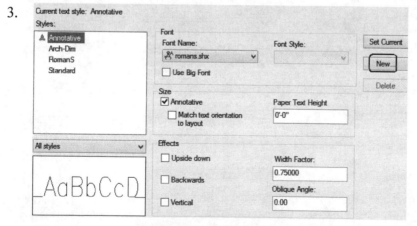 Select the **Text Style** tool on the Home ribbon.

   It is located on the Annotation panel under the drop-down.

3. Click the **New** button in the Text Styles dialog box.

4. In the New Text Style dialog box, enter **TEXT1** for the name of the new text style and Click **OK**.

5.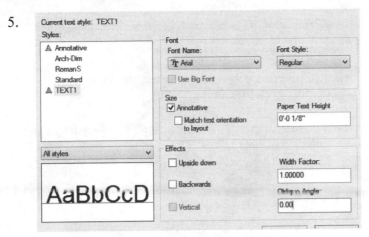

Fill in the remaining information in the Text Style dialog box as shown.

Set the Font Name to **Arial**.
Set the Font Style to **Regular**.
Set the Paper Text Height to **1/8″**.
Set the Width Factor to **1.00**.

6. Click the Apply button first and then the Close button to apply the changes to TEXT1 and then close the dialog box.

7. Save as *fonts.dwg*.

> ➢ You can use PURGE to eliminate any unused text styles from your drawing.
> ➢ You can import text styles from one drawing to another using the Design Center.
> ➢ You can store text styles in your template for easier access.
> ➢ You can use any Windows font to create AutoCAD Text.

Project Settings

The Project Navigator in AutoCAD Architecture gives us a way to Create, Access, Organize, and Coordinate all of our Project drawings in one palette.

The use of the Project Navigator begins with creating a Project in the **Project Browser**.

You can launch the Project Browser from the **Quick Access Toolbar**.

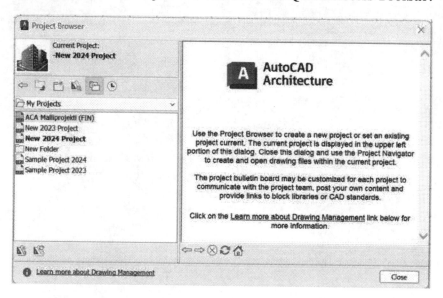

In the Project Browser, you create Projects and adjust their properties. Once a Project is created, you can right-click on it and in the pop-up menu, choose **Set Project Current**, and then click the **Close** button. This opens the **Project Navigator** palette. When you create a new project, you are also creating a folder sub-structure where all the files associated with the project will be stored. Projects rely on the use of external files all linked together to create the project. Companies that wish to use the Project Management system may opt to store all projects on a server so that team members can access the files easily. The Project Navigator also acts as a PDM server, where users can check in and out files, control file revisions, and roll-back drawings to previous versions.

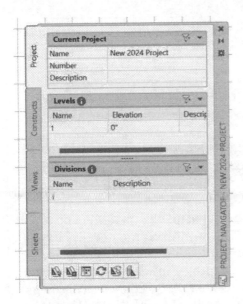

The **Project Navigator** palette has four tabs: **Project**; **Constructs**; **Views**; and **Sheets**.

The **Project** tab is where you begin.

	Here you can create the Project **Levels** and **Divisions**. Divisions are used to subdivide Levels into drawings. Levels represent the floors in a building. The construct drawings are assigned to a level of the building and the level height controls the elevation at which the drawings are inserted. Divisions represent wings or plan areas within a building. The construct drawings are assigned to a division of the building as well as the level of that division. In a small project it is possible to have only one division.

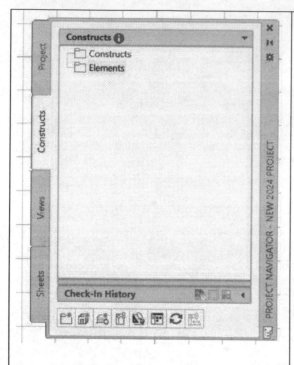

The **Constructs** tab consists of two separate areas: **Constructs**, and **Elements**.

These are your Model files. These may be further subdivided into different Categories.

Constructs are for objects that are uniquely located at a specific Level. These might be columns, walls, windows.

Elements contain objects that are not associated with a specific Level. Elements are similar to blocks and contain objects that are repeated inside the building. The elements are referenced by the constructs within a project, for example, a typical toilet room that is located in the same spot on all floors.

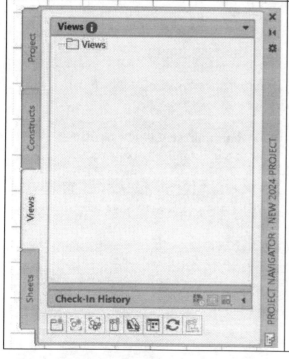

The views tab will be used to create Composite Trade Drawings which include notes, dimensions, tags and any other annotation necessary. To create one, right click on the views folder and select "New view dwg." Then select what type of view drawing you want to create. The options are general, section/ elevation or detail. Pick the type of view you need and follow the next set of prompts.

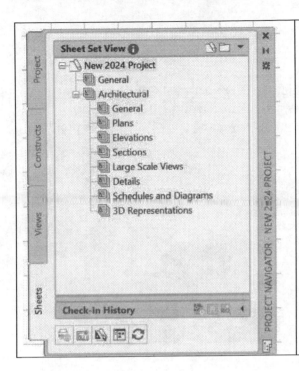

On the **Sheets** tab, you may establish **Subsets** for your drawings. For example, this project includes **General** and **Architectural** subsets. Under Architectural, there are additional Subsets. Each subset may consist of several sheets/drawings.

The Sheets Tab is where all the dimensioning and plotting is done. Each file that is created will become the layout sheets used to print. This keeps the main construct files clutter free and smaller in size. The sheets drawing files are typically split up by trade. Under each trade you will create a new sheet for each new drawing. (Example: 01-DW-INSTALL)

Tips Tricks

To quickly access the Project Navigator, use CTL+5.
You can also type PROJECTNAVIGATORTOGGLE to toggle visibility on/off.

Project Management

The Project Browser is where you perform high-level project-related tasks, such as creating a new project, selecting the currently active project, and setting project properties. The Browser allows you to create a folder where you will store your project files, so they are easily located and managed.

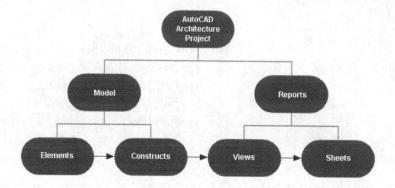

Project Tab:

This is where you will set & get all the information about the project.

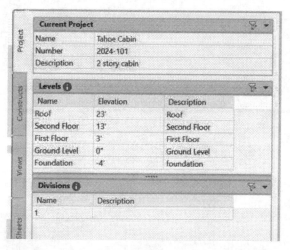

-Name
-Job Number
-Description
-Levels
-Elevations
-Divisions

***All of this should be set up before the job is started. ***

The recommended work flow in ACA looks like this:

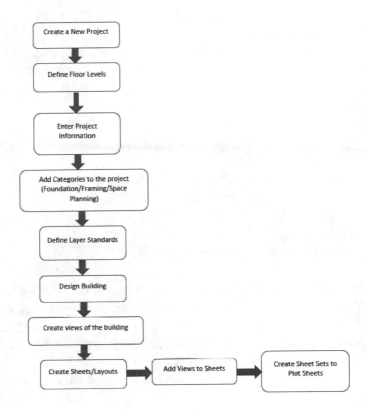

Basic Recommended Practices

- Use the Project Navigator to build your projects and manage.
- Start with project templates.
- Model in constructs. Use ACA objects (walls, doors, windows, stairs, curtainwalls, etc.)
- Dimension and annotate only in views files. Each view file should have at least one model space view.
- Sheet files should only have one layout.
- Your titleblock should be in your sheet template.
- Use fields for your sheet block annotation. The Project Navigator can populate those fields with project and sheet specific information.
- Drag model space views onto sheets.
- Use built in schedules.
- Use keynotes and tags.
- Use AEC dimension and AEC object types.
- Use grids.
- Let the layering key style build your layers. Avoid creating other layers outside of the ACA standard. Use the Project Navigator to manage your layer properties.
- Each view file should stand on its own without the need to coordinate annotation styles.
- Create your own styles but just a few per project.
- Build and use library and custom multi-view blocks for furniture, equipment, and factory components.
- Use the Styles Browser and the Styles Catalogs and StylesLibrary drawings to save time.
- Using tool palettes, especially project-based tool palettes, is a great way to manage and update content. You can create customized tool palettes to match the way you work.

After you have selected a project in the Project Browser, you use the Project Navigator to create and edit the actual building and documentation data. Here you create elements, constructs, model views, detail views, section views, and sheets, connecting them with one another.

You can define multi-story buildings using the Project Navigator. In the Navigator, you can specify the number of levels for the building. You can also use the Navigator to manage your views, sheets, and layer standards.

Note: Although project files and categories you create on the Project Navigator palette are shown as files and folders in File Explorer, you should not move, copy, delete, or rename project files from there. Such changes are not updated on the Project Navigator palette, and you could get an inconsistent view of your project data. Any changes you make to the project on the Project Navigator palette are managed and coordinated by the software. Also note that any changes you make on the Project Navigator palette (such as renaming, deleting, or changing the file's properties) cannot be undone using the AutoCAD UNDO command.

Because projects involve the use of multiple files, this can be confusing to keep track of when teaching a class or managing the files used for each lesson. To make it easier, I am going to create a zip file at the end of each lesson/chapter in the text. This file will serve as a snapshot of the textbook's project at different stages and help keep students (and instructors) on track as they gain proficiency in the software.

Exercise 1-10:
Using the Project Navigator and Project Browser

Drawing Name: new
Estimated Time: 20 minutes

This exercise reinforces the following skills:

- ❑ Project Browser
- ❑ Project Navigator
- ❑ Create levels
- ❑ Managing multi-story projects
- ❑ Archiving projects
- ❑ Closing projects

1.   Start a new drawing.

2.   Right click in the command window and select **Options**.

3.   Select the **AEC Project Defaults** tab.

4.   Set the AEC Project Location Search Path to the location where you are saving your files.

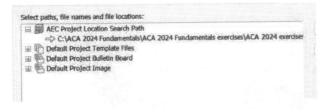

Close the dialog.

5.

Open the **Project Browser**.

6.

Select **New Project**.

7.

Type **2024-101** for the Project Number.

Type **Tahoe Cabin** as the Project Name.
Type **2-story cabin** for the Project Description.

Disable Create from template project.

Click **OK**.

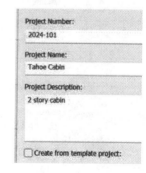

8.

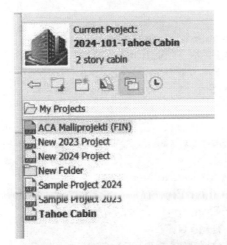

Notice that the Current Project is now the project you just defined.

Notice that the project name is in bold under My Projects.

Close the dialog.

9.

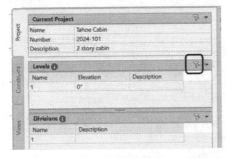

Select the **Project Navigator** tool.

10.

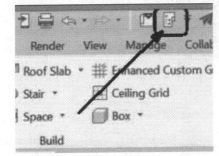

On the Project tab:

Click on **Create/Edit Levels**.

11.

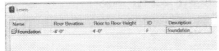

Rename the first level **Foundation**.
Change the Floor Elevation to **-4'-0"**.
Set the Floor to Floor Height to **4'-0"**.

12.

Click on **Add Level**.

13.

Type **First Floor** for the Name.
Set the Floor Elevation to **3'-0"**.
Set the Floor to Floor Height to **10'-0"**.
Set the ID to **1**.
Set the Description to **First Floor.**

Click **Add Level.**

14.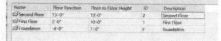

Type **Second Floor** for the Name.
Set the Floor Elevation to **13'-0"**.
Set the Floor to Floor Height to **10'-0"**.
Set the ID to **2.**
Set the Description to **Second Floor.**

Click **Add Level..**

15.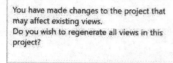

Type **Roof** for the Name.
Set the Floor Elevation to **23'-0"**.
Set the Floor to Floor Height to **10'-0"**.
Set the ID to **R.**
Set the Description to **Roof.**

Click **OK**

16.

You have made changes to the project that
may affect existing views.
Do you wish to regenerate all views in this
project?

| Yes | No |

Click **Yes**.

17.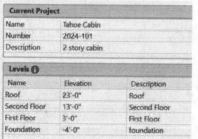

The levels are all listed.

18.

Open the **Project Browser**.

19.

Highlight the project name: **Tahoe Cabin**.

Right click and select **Show in Windows Explorer**.

20.

You can see where the project has been saved.
Note that several subfolders were automatically created where the external files referenced by the project will be saved.

21.

Highlight the project name: **Tahoe Cabin**.

Right click and select **Archive**.

22.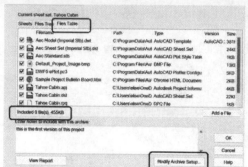

Click on the Files Table tab.

You see that nine files have already been generated.

Click on **Modify Archive Setup**.

23.

Under Archive file folder, browse to where you want to back up your project files.

Click **OK**.

24.

Enter some notes regarding your first archive.

Click **OK**.

25.

Verify that the zip file is being saved in the correct location.
Click **Save**.

26.

Highlight the project name: **Tahoe Cabin**.

Right click and select **Close Current Project**.

27.

The dialog shows that no project is active. This is useful if you elect to work outside of a project and don't want your work saved to a project.

Exercise 1-11:

Using a Template to Create a Project

Drawing Name: new
Estimated Time: 30 minutes

This exercise reinforces the following skills:

- ❏ Project Browser
- ❏ Project Navigator
- ❏ Create levels
- ❏ Managing multi-story projects
- ❏ Archiving projects
- ❏ Closing projects

1.  Start a new drawing.

2. 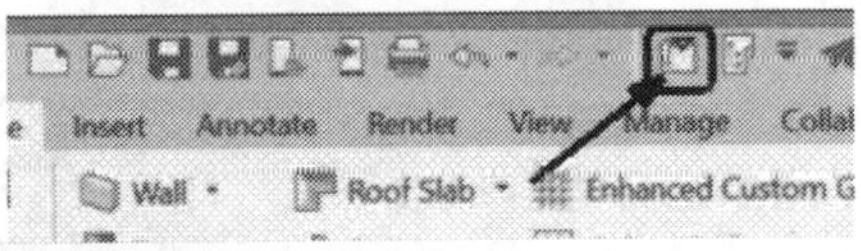 Open the **Project Browser**.

3. 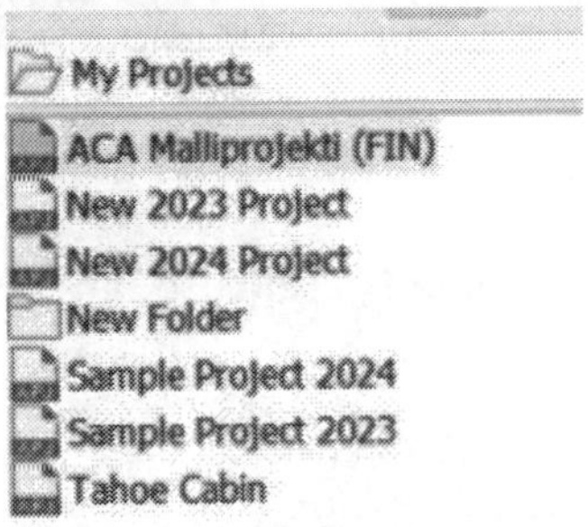 Select **New Project**.

4.

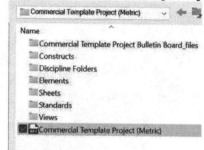

Type **2024-102** for the Project Number.

Type **Brown Medical Clinic** as the Project Name.
Type **medical clinic** for the Project Description.

Enable **Create from template project**.

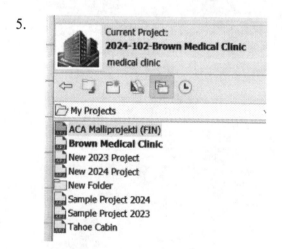

Browse to the folder and select the *Commercial Template Project (Metric).apj* file.

Click **OK**.

There will be a brief pause while ACA imports all the elements to be used in the project.

5.

The Brown Medical Clinic shows as the active/current project.

Close the Project Browser.

6.
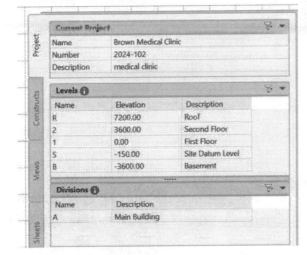
Select the **Project Navigator** tool.

7.

On the Project tab:

Notice that it is a two-story building, with a site plan and a basement.

Current Project	
Name	Brown Medical Clinic
Number	2024-102
Description	medical clinic

Levels

Name	Elevation	Description
R	7200.00	Roof
2	3600.00	Second Floor
1	0.00	First Floor
S	-150.00	Site Datum Level
B	-3600.00	Basement

Divisions

Name	Description
A	Main Building

8.
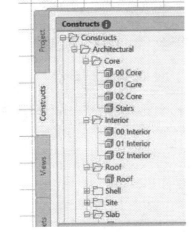

Open the Constructs tab.

Notice that categories and subcategories have already been predefined based on the selected template.

The template acts as an outline for the building project.

9.

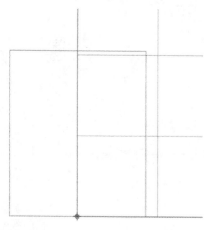

Under the Elements category, Locate the Typical Toilet Room.

Right click and select **Open**.

10.

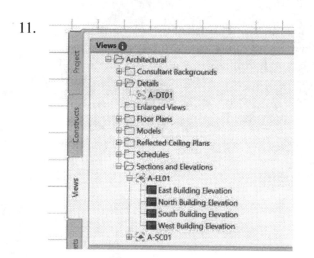

A drawing is opened that consists of a single rectangle.

The drawing is called Typical Toilet Room.

This file basically adds as a placeholder where you can add toilet stalls, sinks, etc. which can then be placed in the building model as a block.

Close the file without saving. Click on the Views tab.

11.

Expand the Details folder.

Expand the Sections and Elevations folder.

The files listed below those folders are all drawings.

12.

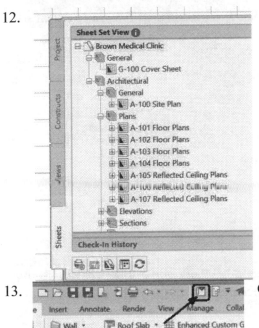

Open the Sheets tab.

These are also drawings that are automatically generated using the designated title block in the project.

13.

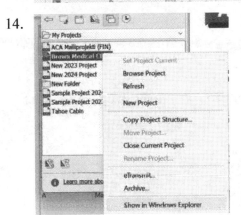

Open the **Project Browser**.

14.

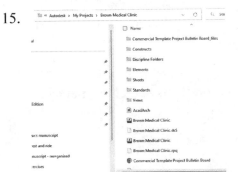

Highlight the project name: **Brown Medical Clinic**.

Right click and select **Show in Windows Explorer**.

15.

You can see where the project has been saved. Note that several subfolders were automatically created where the external files referenced by the project will be saved.

16.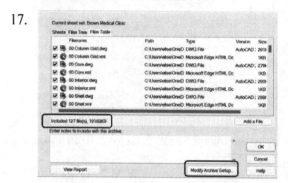

Highlight the project name: **Brown Medical Clinic**.

Right click and select **Archive**.

17.

Click on the Files Table tab.

You see that over 100 files have already been generated.

Click on **Modify Archive Setup**.

18.

Under Archive file folder, browse to where you want to back up your project files.

Click **OK**.

19.

Enter some notes regarding your first archive.

Click **OK**.

20.

Verify that the zip file is being saved in the correct location.

Click **Save**.

21.    Highlight the project name: **Brown Medical Clinic**.

Right click and select **Close Current Project**.

Close the dialog box.

Layer Keys and Layer Keying

When you create an object, the layer key that is associated with the object automatically places that object on the layer to which it is mapped. Using layer keys to automatically place objects on predefined layers is called *layer keying*. Usually, you key objects to layers that conform to a layer standard. Each layer standard names layers according to a set of rules that you can modify.

Layer Key Styles

You can create different sets of layer keys, called layer key styles, that you can use to place objects on defined layers in your drawings. Each layer key style contains a set of layer keys. Both the AIA and BS1192 layer key styles contain layer keys and layer properties, including layer name, description, color, linetype, lineweight, plot style, and plot settings for all of the AEC objects. You can import the AIA and BS1192 layer standards and layer key styles that are provided with AutoCAD Architecture 2024 toolset from the *AecLayerStd.dwg* in the *C:\ProgramData\Autodesk\ACA <version>\enu\Layers* folder. You can modify these styles, if needed, or create new layer key styles. You can also create layer key styles from LY files.

Exercise 1-12:

Modifying the ACA Layer Keys

Drawing Name: new
Estimated Time: 30 minutes

This exercise reinforces the following skills:

- ❑ Project Browser
- ❑ Project Navigator
- ❑ Create levels
- ❑ Managing multi-story projects

1.   Start a new drawing.

2.   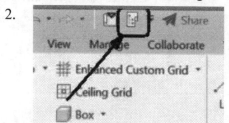   Launch the **Project Navigator**.

Notice that the Project Browser opens, not the Project Navigator.
That is because we need to open or set a project active before we can set our layer standards.

3.   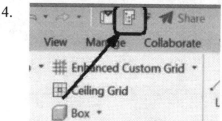   Highlight the **Brown Medical Clinic**.
Right click and select **Set Project Current**.

4.   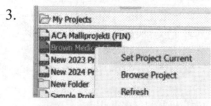   Launch the **Project Navigator**.

5.      Click on **Project Standards**.

6.

Highlight the Layer Key Styles.

Click **Attach**.

7.

Locate the *Layer Standards.dws* file in the downloaded exercise files.

Click **Open**.

8.

Click to enable the Layer Key Styles to use the Layer Standards.dws layer keys.

Click **OK**.

9.

If you like, you can add a comment about the layer standards you are using.

Click **OK**.

10.

Open the **Layer Manager**.

11.

Click on **Layer Keys**.

12.

Highlight the Layer Key Style being used.

13.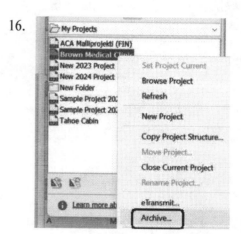

Click on the Keys tab.

You can see the layer properties for each AEC object.

14.

Review the Layer Key Styles used in your project.

Click **OK**.

15. Close the Layer Manager.

The changes that you made to the Brown Medical Clinic project will be included the next time we archive the project.

16.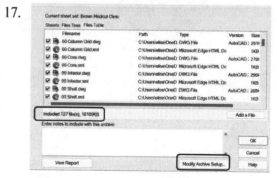

Highlight the project name: **Brown Medical Clinic**.

Right click and select **Archive**.

17.

Click on the Files Table tab.

You see that over 100 files have already been generated.

Click on **Modify Archive Setup**.

18.

Under Archive file folder, browse to where you want to back up your project files.

Click **OK**.

19.
Enter some notes regarding your archive.

Click **OK**.

20.
Verify that the zip file is being saved in the correct location.
Change the name to *Brown Medical Clinic_2.zip*.
Click **Save**.

Notes:

QUIZ 1

True or False

1. Doors, windows, and walls are inserted into the drawing as objects.

2. The Architectural Toolbars are loaded from the ACA Menu Group.

3. To set the template used by the QNEW tool, use Options.

4. If you start a New Drawing using the 'QNew' tool shown above, the Start-Up dialog will appear unless you assign a template.

5. Mass elements are used to create Mass Models.

6. The Layer Manager is used to organize, sort, and group layers.

7. Layer Standards are set using the Layer Manager.

8. You can assign materials to components in wall styles.

Multiple Choice
Select the best answer.

9. If you start a drawing using one of the standard templates, AutoCAD Architecture will automatically create _____ layouts.

    A. A work layout plus Model Space
    B. Four layouts, plus Model Space
    C. Ten layouts, plus Model Space
    D. Eleven layouts, plus Model Space

10. Options are set through:

    A. Options
    B. Desktop
    C. User Settings
    D. Template

11. The Display Manager controls:

    A. How AEC objects are displayed in the graphics window
    B. The number of viewports
    C. The number of layout tabs
    D. Layers

12. Mass Elements are created using the _____ Tool Palette.

    A. Massing
    B. Concept
    C. General Drafting
    D. Documentation

13. Wall Styles are created by:

    A. Highlighting a Wall on the Wall tab of the Tool Palette, right click and select Wall Styles
    B. Type **WallStyle** on the command line
    C. Go to **Format→Style Manager**
    D. All of the above

14. Visual Styles are used to:

    A. Determine which materials are used for rendering
    B. Controls which layers are visible
    C. Determines the display of edges, lighting, and shading in each viewport
    D. Determines the display of AEC elements in the Work layout tab

15. You can change the way AEC objects are displayed by using:

    A. Edit Object Display
    B. Edit Display Properties
    C. Edit Entity Properties
    D. Edit AEC Properties

16. The Styles Browser can be used for any of the following tasks: (select more than one)

    A. Search for object styles
    B. Import styles into your current drawing
    C. Place an element using a selected style from the browser
    D. Create new styles

17. The Project Browser may be used to:

    A. Create new projects
    B. Configure project settings and standards
    C. Select the Current Project
    D. All of the above

ANSWERS:

1) T; 2) T; 3) T; 4) F; 5) T; 6) T; 7) T; 8) T; 9) A; 10) A; 11) A; 12) A; 13) D; 14) C; 15) A; 16) A, B, and C (to create new styles you have to use the Styles Manager) 17) D

Lesson 2:
Foundations & Site Plans

Most architectural projects start with a site plan. The site plan indicates the property lines, the house location, a North symbol, any streets surrounding the property, topographical features, location of sewer, gas, and/or electrical lines (assuming they are below ground – above ground connections are not shown), and topographical features.

When laying out the floor plan for the house, many architects take into consideration the path of sun (to optimize natural light), street access (to locate the driveway), and any noise factors. AutoCAD Architecture includes the ability to perform sun studies on your models.

AutoCAD Architecture allows the user to simulate the natural path of the sun based on the longitude and latitude coordinates of the site, so you can test how natural light will affect various house orientations.

A plot plan must include the following features:

- ❑ Length and bearing of each property line
- ❑ Location, outline, and size of buildings on the site
- ❑ Contour of the land
- ❑ Elevation of property corners and contour lines
- ❑ North symbol
- ❑ Trees, shrubs, streams, and other topological items
- ❑ Streets, sidewalks, driveways, and patios
- ❑ Location of utilities
- ❑ Easements and drainages (if any)
- ❑ Well, septic, sewage line, and underground cables
- ❑ Fences and retaining walls
- ❑ Lot number and/or address of the site
- ❑ Scale of the drawing

The plot plan is drawn using information provided by the county/city and/or a licensed surveyor.

When drafting with pencil and vellum, the drafter will draw to a scale, such as ⅛″ = 1′, but with AutoCAD Architecture, you draw full-size and then set up your layout to the proper scale. This ensures that all items you draw will fit together properly. This also is important if you collaborate with outside firms. For example, if you want to send your files to a firm that does HVAC, you want their equipment to fit properly in your design. If everyone drafts 1:1 or full-size, then you can ensure there are no interferences or conflicts.

Linetypes

Architectural drafting requires a variety of custom linetypes in order to display specific features. The standard linetypes provided with AutoCAD Architecture are insufficient from an architectural point of view. You may find custom linetypes on Autodesk's website, www.cadalog.com, or any number of websites on the Internet. However, the ability to create linetypes as needed is an excellent skill for an architectural drafter.

It's a good idea to store any custom linetypes in a separate file. The standard file for storing linetypes is acad.lin. If you store any custom linetypes in acad.lin file, you will lose them the next time you upgrade your software.

Exercise 2-1:
Creating Custom Line Types

Drawing Name: New
Estimated Time: 15 minutes

This exercise reinforces the following skills:

- Creation of linetypes
- Customization

A video of this lesson is available on my MossDesigns channel on Youtube at https://www.youtube.com/watch?v=gSCyFJm6pfY

1.

Start a new drawing using **QNEW**.

Launch the **Project Browser**.

2.

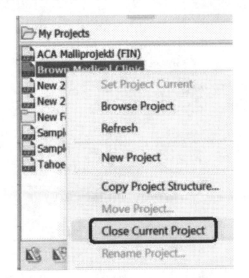

Highlight the current/active project.

Right click and select **Close Current Project**.

3.

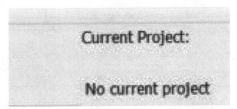

You should see No current project active in the dialog.

Close the Project Browser.

4.
_____   _____   _____   _____

Draw a property line.
This property line was created as follows:
    Set ORTHO ON.
    Draw a horizontal line 100 units long.

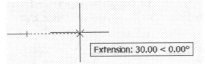

Use SNAP FROM to start the next line @30,0 distance from the previous line.
You can also simply use object tracking to locate the next line.
The short line is 60 units long.
Use SNAP FROM to start the next line @30,0 distance from the previous line.
The second short line is 60 units long.
Use SNAP FROM to start the next line @30,0 distance from the previous line.
Draw a second horizontal line 100 units long.

5.

Activate the Express Tools ribbon.

Select the **Tools** drop-down.

Go to **Make Linetype**.

6.   Browse to the folder where you are storing your work.

Create a file name called *custom-linetypes.lin* and click **Save**.

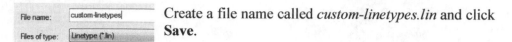

Place all your custom linetypes in a single drawing file and then use the AutoCAD Design Center to help you locate and load the desired linetype.

---

7.   When prompted for the linetype name, type property-line.
     When prompted for the linetype description, type property-line.
     Specify starting point for line definition; select the far left point of the line type.
     Specify ending point for line definition; select the far right point of the line type.
     When prompted to select the objects, start at the long line on the left and pick the line segments in order.

8.   Type **linetype** to launch the Linetype Manager.

9.

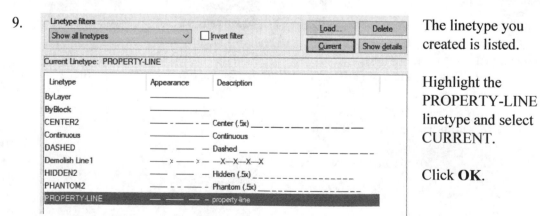

The linetype you created is listed.

Highlight the PROPERTY-LINE linetype and select CURRENT.

Click **OK**.

10. Draw some lines to see if they look OK. Make sure you make them long enough to see the dashes. If you don't see the dashes, adjust the linetype scale using Properties. Set the scale to 0.000125 or smaller.

11. 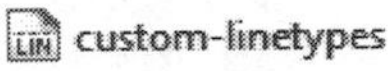  Locate the *custom-linetypes.lin* file you created.

    Open it using NotePad.

12.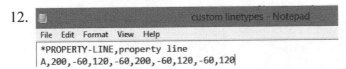
    You see a description of the property-line.

    ```
    *PROPERTY-LINE,property line
    A,200,-60,120,-60,200,-60,120,-60,120
    ```

13. Save as *ex2-1.dwg*.

Exercise 2-2:

Creating New Layers

Drawing Name: New
Estimated Time: 20 minutes

This exercise reinforces the following skills:

  ❏   Design Center
  ❏   Layer Manager
  ❏   Creating New Layers
  ❏   Loading Linetypes

1. 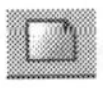   Start a new drawing using QNEW.

     Open ex2-1.dwg.

     | Start | ex2-1 | × | Drawing2* | × |

     You should see three drawing tabs:
     Start, ex2-1, and the new drawing

2.
     Activate the **View** ribbon and verify that the File Tabs is
     shaded.

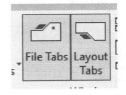

3.   Launch the Design Center.
      The Design Center is on the Insert ribbon in
      the Content section. It is also located in the
      drop-down under the Content Browser. You
      can also launch using Ctl+2.

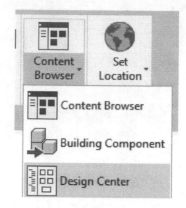

4.   Select the **Open Drawings** tab.

      Browse to the e*x2-1.dwg* file.

      Open the Linetypes folder.

      Scroll to the **Property-line** linetype
      definition.

5.   Highlight the PROPERTY-LINE
      linetype.

      Drag and drop into your active
      window or simply drag it onto the
      open drawing listed.

      *This copies the property-line
      linetype into the drawing.*

6.   Close the Design Center.

7.   Activate the **Layer Properties Manager** on the
      Home ribbon.

8. 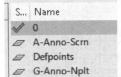 Note that several layers have already been created. The template automatically sets your Layer Standards to AIA.

9. Select the **New Layer from Standard** tool.

10. Under Layer Standard, select **Non Standard**.

Name the layer **A-Site-Property-Lines**.

Set the Color to **Green**.

Select the **Property-Line** linetype.

Set the Lineweight to **Default**.

Enable **Plot**.
Click **Apply** and **OK**.

11. Create another layer by clicking the **New Layer from Standard** tool.

Name the layer **A-Contour-Line**.

Set the color to **42**.

Select the linetype pane for the new linetype.

Click **Other**.

12. 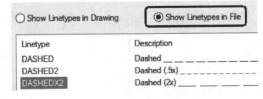 Enable **Show Linetypes in File**.

Highlight **DASHEDX2** and click **OK** to assign it to the selected layer name.

13.

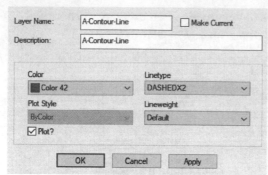

Note that the linetype was assigned properly.

Set the Lineweight to **Default**.

Enable **Plot**.

Click **OK**.

14.

S.. Name
0
A-Contour-Line
A-Site-Property-Lines
Defpoints

Set the A-Site-Property-Line layer current.

Highlight the layer name then select the green check mark to set current.

You can also double left click to set the layer current.

15. Right click on the header bar and select **Maximize all columns** to see all the properties for each layer.

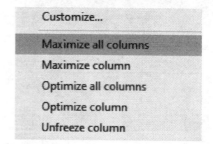

16. Use the X in the upper corner to close the Layer Manager.

17. Save as *ex2-2.dwg.*

Exercise 2-3:

Creating a Site Plan

Drawing Name: site_plan.dwg

Estimated Time: 30 minutes

This exercise reinforces the following skills:

- Use of ribbons
- Documentation Tools
- Elevation Marks
- Surveyors' Angles

1.   Open *site_plan.dwg*.

2. 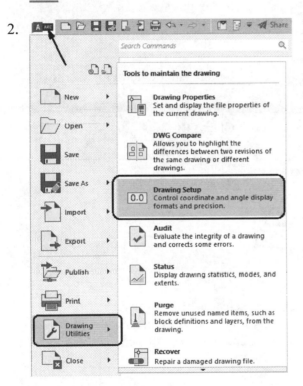 Go to **Drawing Utilities→Drawing Setup**.

This is located on the Application Menu panel.

3.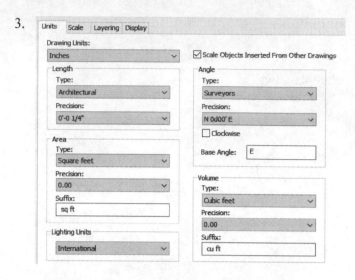

Select the **Units** tab.
Set the Drawing Units to **Inches**.
Set up the Length Type to **Architectural**.
Set the Precision to ¼″.
Set the Angle Type to **Surveyors**.
Set the Precision to **N 0d00′ E**.

4.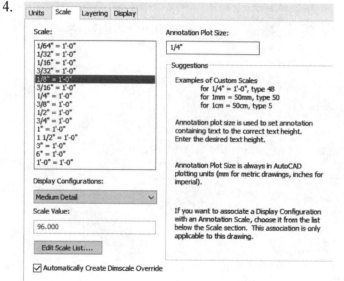

Select the **Scale** tab.

Set the Scale to **1/8" = 1'-0"**.

Set the Annotation Plot Size to ¼".

Click **OK**.

5.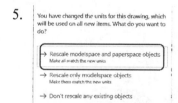

When this dialog appears, select the first option. The drawing doesn't have any pre-existing elements, but this option ensures that everything is scaled correctly.

6.  On the Status bar: verify the scale is set to **1/8" = 1'-0"**.

You can set your units as desired and then enable **Save As Default** .
Then, all future drawings will use your unit settings.

---

7. Draw the property line shown on the **A-Site-Property-Lines** layer.

    Select the Line tool.

8.

    Specify first point: 1',5'
    Specify next point: @186' 11" < s80de
    Specify next point: @75'<n30de
    Specify next point: @125'11"<n
    Specify next point: @250'<180

    This command sequence uses Surveyor's Units.

    You can use the TAB key to advance to the surveyor unit box.

```
Specify first point: 12,60
Specify next point or [Undo]: @2243<s80de
Specify next point or [Undo]: @900<n30de
Specify next point or [Close/Undo]: @1511<n
Specify next point or [Close/Undo]: @3000<180
Specify next point or [Close/Undo]:
```

Metric data entry using centimeters:

If you double click the mouse wheel, the display will zoom to fit so you can see your figure.

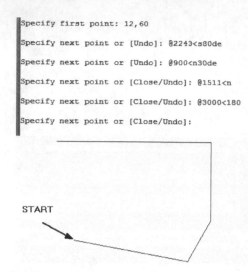

START

Turn ORTHO off before creating the arc.

9.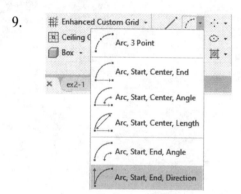

Draw an arc using **Start, End, Direction** to close the figure.

10.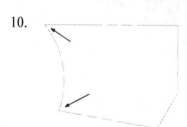

Select the bottom point as the Start point.
Right click and select the End option.
Select the top point at the End point.
Click inside the figure to set the direction of the arc.

11. Set the **A-Contour-Line layer** current.

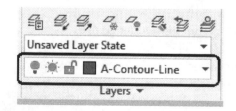

12. PTYPE   Type **PTYPE** to access the Point Style dialog.

13. Select the Point Style indicated.

Click **OK**.

Point Size: 5.0000 %

● Set Size Relative to Screen
○ Set Size in Absolute Units

14. *Use the **DIVIDE** command to place points as shown.*

You will not see the points until you set the point style.

15. Type **DIVIDE** to add points to the top horizontal line.

DIVIDE

16. DIVIDE Select object to divide: Select the top horizontal line.

17. DIVIDE Enter the number of segments or [Block]: 5 Type **5** to be divided into five equal segments. Click **ENTER**.

18. Repeat DIVIDE / Recent Input Right click and select **Repeat Divide**.
Add points to the bottom horizontal line.

19. DIVIDE Select object to divide: Select the BOTTOM horizontal line.

20. DIVIDE Enter the number of segments or [Block]: 5 Type **4** to be divided into four equal segments. Click **ENTER**.

21. Repeat DIVIDE / Recent Input Right click and select **Repeat Divide**.
Add points to the bottom horizontal line.

22. DIVIDE Select object to divide: Select the right angled line.

23. DIVIDE Enter the number of segments or [Block]: 5 Type **2** to be divided into two equal segments. Click **ENTER**.

24.

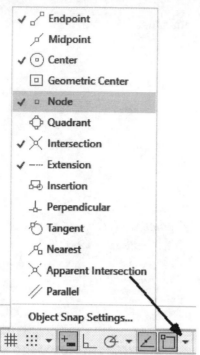

Enable the **Node** OSNAP.

25. Draw contour lines using **Pline** and **NODE** Osnaps as shown.

Be sure to place the contour lines on the contour line layer.

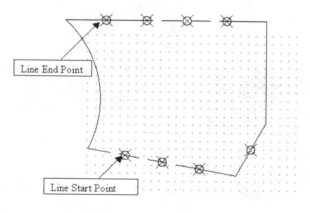

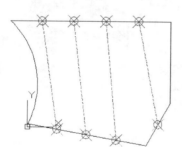

26. Type **qselect** on the *Command* line.

27.

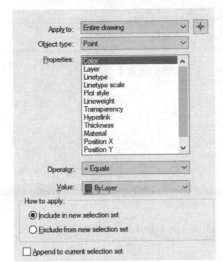

Set the Object Type to **Point**.
Set the Color = Equals **ByLayer**.

Click **OK**.

28.

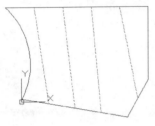

All the points are selected.
Right click and select **Basic Modify Tools→ Delete**.

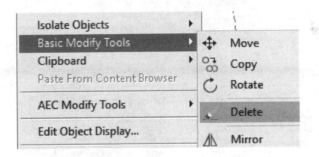

29. Type UCSICON, OFF to turn off the UCS icon.

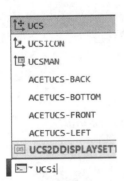

When you start typing in the command line, you will see a list of commands.

Use the up and down arrow keys on your keyboard to select the desired command.

30.

Activate the **Design Tools** palette located under the Tools drop-down on the Home ribbon.

31. Right click on the gray bar on the palette to activate the short-cut menu.

Left click on **All Palettes** to enable the other palettes.

32. Activate the Annotation tab on the palette.

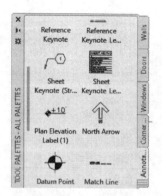

Locate the **Plan Elevation Label**, but it's not the type we need for a site plan.

Instead we'll have to add an elevation label from the Design Center to the Tool Palette.

33. Type **Ctl+2** to launch the Design Center.
Or:
Activate the **Insert** ribbon.
Select the **Design Center** tool under Content.
Or:
Type **DC**.

34. 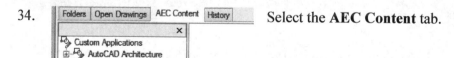 Select the **AEC Content** tab.

35. Browse to
Imperial/Documentation/Elevation Labels/ 2D Section.

[Metric/Documentation/Elevation Labels/2D Section].

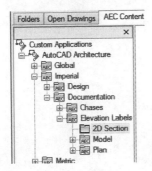

36.

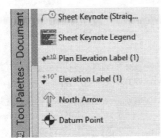

Locate the Elevation Label (1) file.

Highlight the elevation label.

Drag and drop onto the Tool Palette.

Close the Design Center.

37.

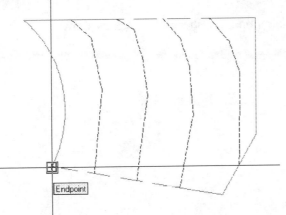

Endpoint

Select the Elevation Label on the Tool Palette.

Select the Endpoint shown.

38. Set the Elevation to **0″ [0.00]**.
Set the Prefix to **EL**.
Click **OK**.

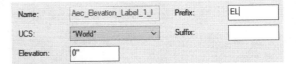

Name:	Aec_Elevation_Label_1_I	Prefix:	EL
UCS:	"World"	Suffix:	
Elevation:	0″		

39.

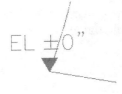

Zoom in to see the elevation label.

You can use the **Zoom Object** tool to zoom in quickly.

This is located on the View tab.

The Elevation Label is automatically placed on the G-Anno-Dims layer.

To verify this, just hover over the label and a small dialog will appear displaying information about the label.

40.   EL ±0.00"   Place an elevation label on the upper left vertex as shown.

41.   Place labels as shown.

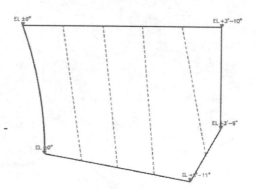

42.   To change the size of the labels, select and set the Scale to **36.00** on the Properties dialog.

Scale	
X	36.00000
Y	36.00000
Z	36.00000

43.

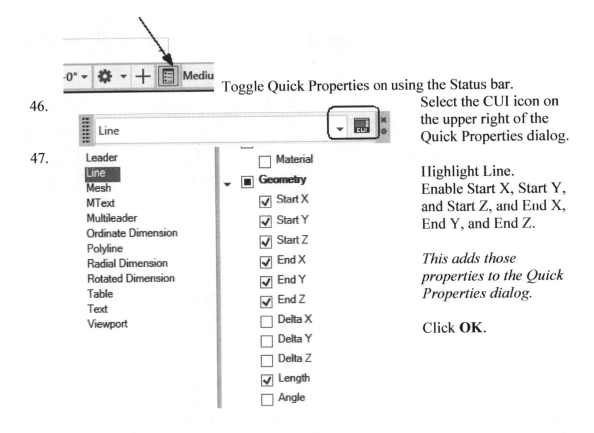

Edit Object Display...
Elevation Label Modify...
Edit Multi-View Block Definition...

To modify the value of the elevation label, select the label.

Right click and select **Elevation Label Modify**.

44.

Geometry	
Start X	1'-0"
Start Y	5' 0"
Start Z	0"
End X	185'-0 15/16"
End Y	-27'-5 1/2"
End Z	2'-11"
Delta X	184'-0 15/16"
Delta Y	-32'-5 1/2"

To change the toposurface to a 3D surface, select each line and modify the Z values to match the elevations.

45.   *A quick way to modify properties is to use Quick Properties.*

-0" ▾ ⚙ ▾ ╋ ▤ Mediu

Toggle Quick Properties on using the Status bar.

46.

Line ▾ ▣ CUI ✕ ⚙

Select the CUI icon on the upper right of the Quick Properties dialog.

47.

Leader
Line
Mesh
MText
Multileader
Ordinate Dimension
Polyline
Radial Dimension
Rotated Dimension
Table
Text
Viewport

☐ Material
▾ ▣ **Geometry**
☑ Start X
☑ Start Y
☑ Start Z
☑ End X
☑ End Y
☑ End Z
☐ Delta X
☐ Delta Y
☐ Delta Z
☑ Length
☐ Angle

Highlight Line.
Enable Start X, Start Y, and Start Z, and End X, End Y, and End Z.

This adds those properties to the Quick Properties dialog.

Click **OK**.

48.

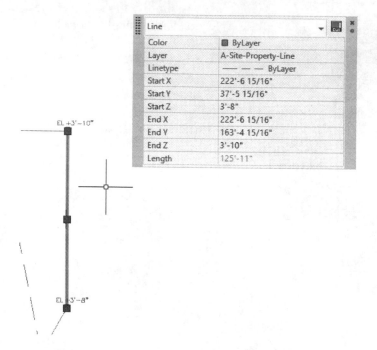

Then select each line
and change the start and
end values of Z to match
the elevation markers.

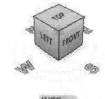

49. Use the ViewCube to inspect your toposurface and then return to a top view.

Adjust points as needed to create an enclosed area. Use ID to check the point values of each end point.

You can also type PLAN to return to a 2D view plan view.

50. Open the *fonts.dwg*. This is the file you created in a previous exercise.

51. Type **Ctl+2** to launch the Design Center.

52. Select the Open Drawings tab.

Browse to the Textstyles folder under the fonts.dwg.

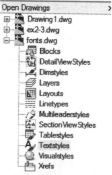

53. Drag and drop the **Text1** text style into the ex2-3 drawing.

Close the Design Center.

54.  Set **G-Anno-Dims** as the current Layer.

55. 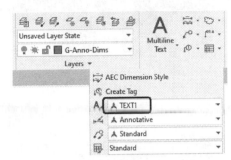 Set the Current Text Style to **TEXT1.**

56. Select the **Single Line** text tool from the Home ribbon.

57.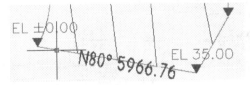

Right click and select **Justify.**
Set the Justification to **Center.**
Select the midpoint of the line as the insertion point.
Sct the rotation angle to **–10** degrees.
Use **%%d** to create the degree symbol.

58.  To create the S60... note, use the **TEXT** command.

Use a rotation angle of 60 degrees.

When you are creating the text, it will preview horizontally. However, once you left click or hit ESCAPE to place it, it will automatically rotate to the correct position.

59. Create the **Due South 125′ 11″ [Due South 3840.48]** with a rotation angle of
     90 degrees.

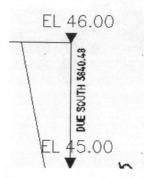

EL 46.00

DUE SOUTH 3840.48

DUE SOUTH 125′ 11″

EL 45.00

Add the text shown on the top horizontal
property line.

N 90° 250′ E

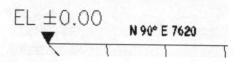

EL ±0.00

N 90° E 7620

60. Add the Chord note shown.
     Rotation angle is 90 degrees.

CHORD = 4023.36

CHORD=1584

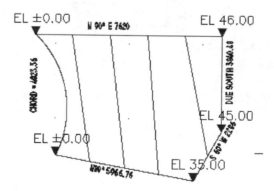

EL ±0.00   N 90° E 7620   EL 46.00

CHORD = 4023.36

DUE SOUTH 3840.48

EL 45.00

EL ±0.00

N90° E 2296

EL 35.00

N90° 5966.76

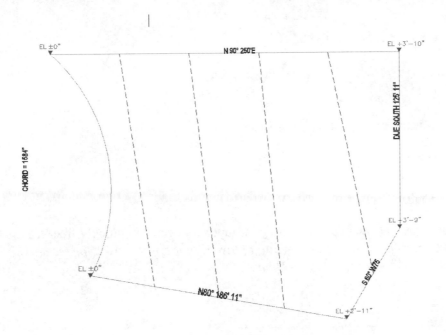

61.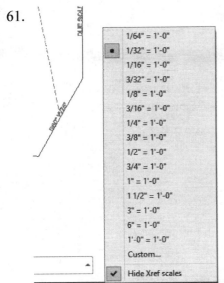

Adjust the scale to **1/32″ = 1′-0″** to see the property line labels.

62. Save the file as *ex2-3.dwg*.

Exercise 2-4:
Creating a Layer User Group

Drawing Name: layer_group.dwg
Estimated Time: 15 minutes

This exercise reinforces the following skills:

❑   Use of toolbars
❑   Layer Manager
❑   Layer User Group

A Layer User Group allows you to group your layers, so you can quickly freeze/thaw them, turn them ON/OFF, etc. *We can create a group of layers that are just used for the site plan and then freeze them when we don't need to see them.*

The difference between FREEZING a Layer and turning it OFF is that entities on FROZEN Layers are not included in REGENS.   Speed up your processing time by FREEZING layers. You can draw on a layer that is turned OFF. You cannot draw on a layer that is FROZEN.

1.   📂   Open *layer_group.dwg*.

2.   🖳   Select the **Layer Properties Manager** tool from the Home ribbon.

3.   [Filters panel menu showing:
    Visibility
    Lock
    Viewport
    Isolate Group
    Reset Xref Layer Properties
    New Properties Filter...
    New Standards Filter...
    New Group Filter
    Convert to Group Filter]
    Highlight the word **All** in the Filters pane.

    Right click and select the **New Group Filter**.

4. 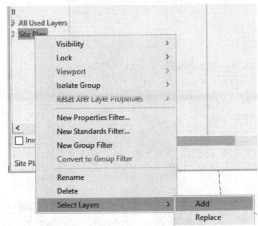 Name the group **Site Plan**.

5. Highlight the Site Plan group.
Right click and select **Select Layers →
Add**.

6. Type **ALL** on the command line.
This selects all the items in the drawing.
Click **ENTER** to finish the selection.
The layers are now listed in the Layer Manager under the Site Plan group.

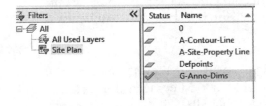

Close the Layer Manager.

7. We can use this group to quickly freeze, turn off, etc. the layers in the Site Plan group.

Save the drawing as *ex2-4.dwg* and close.

Layout Grids and Column Bubbles

Most architectural designers prefer to create a layout grid when placing structural members. The grid helps them manage the spacing of the framing and keeps the design organized. ACA comes with two tools to help with the placement of a grid: LAYOUTGRID and AECANNOGRIDBUBBLEADD.

Grids are used in floor plans to help architects and designers locate AEC elements, such as columns, walls, windows, and doors. The grid is one of the oldest architectural design tools – dating back to the Greeks. It is a useful tool for controlling the position of building elements. Grids may fill the entire design work area or just a small area.

In laying out wood framed buildings, a 16" or 24" grid is useful because that system relies on a stud spacing of 16" or 24".

In large projects, the layout design of different building sub-systems and services are usually assigned to different teams. By setting up an initial set of rules on how the grid is to be laid out, designers can proceed independently and use the grid to ensure that the elements will interact properly.

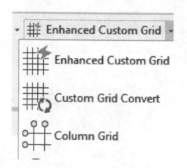

ACA comes with several grid tools.

The Enhanced Custom Grid allows you to place an entire grid using a dialog box to designate the bay spacing.

The Custom Grid Convert tool allows you to convert lines and arcs to grid lines. After the grid lines are placed, you have to manually edit the labels for each bubble.

The Column Grid tool allows you to place a rectangular grid with equal spacing between the grid lines.

Use the Enhanced Custom Grid tool if you want to use different spacing between grid lines or if you want to place radial grid lines.

Exercise 2-5:

Creating a 2D Layout Grid

Drawing Name: new
Estimated Time: 15 minutes

This exercise reinforces the following skills:

- ❑ Creating a Layout Grid
- ❑ Adding Column Bubbles

1. Left click on the + tab to add a new drawing.
   This tab uses whichever template is assigned in your options for New Drawings. You should not be in an active project for this exercise.

2.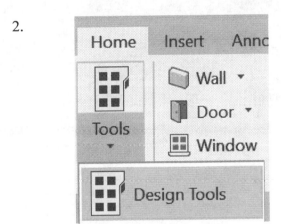
   Open the **Design Tools** palette.

3.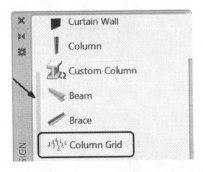
   On the Design tab:

   Scroll down until you see the Column Grid tool.

   Click to select **Column Grid**.

   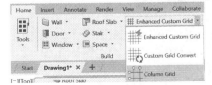
   If you do not have the Column Grid tool on the tool palette, you can locate it on the Home ribbon on the drop-down under Enhanced Custom Grid.

   You can also type LAYOUTGRID at the command prompt.

4.

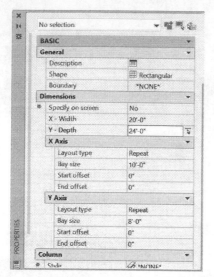

Set the X-Width to **20'-0"**.
Set the Y-Depth to **24'-0"**.

Set the X Axis Bay Size to **10'-0"**.

Set the Y Axis Bay Size to **8'-0"**.

5.

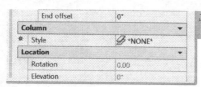

Set the Column Style to **None**.

Type **0,0** to place the grid at the Origin.
Click **ENTER** to accept a rotation angle of 0.
Left click to escape the command.

A 2D grid is placed.

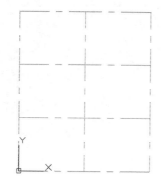

6.

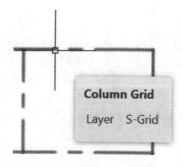

If you hover your mouse over the grid, you will see it has been placed on a new layer named **S-Grid**.

To change the color or linetype used by Layer S-Grid, use the Layer Manager.

7.

Switch the tool palette to the Document palette.

Right click on the palette bar and select **Document**.

8.

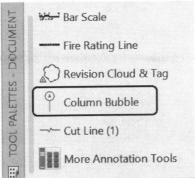

On the Annotation tab:

Locate the **Column Bubble** tool and select it.

You can also type AECANNOGRIDBUBBLEADD at the command prompt.

If you added the tool and modified it on the customized Structural tool palette, you can use the custom tool you created.

9.

Left click at the lower left corner of the grid.

Change the Label to **A**.
Enable **Apply at both ends of gridline**.
Click **OK**.

10.

The grid bubble is placed.

Left click on the next grid line to place the next bubble.

11.

Notice that the Label auto-increments to B.

Click **OK**.

12.

The grid bubble is placed.

Left click on the next grid line to place the next bubble.

13.

Click **OK**.

14.

Left click on the top line.

The Label auto-increments to D.

Click **OK**.

15.

Left click on the left line of the grid.

Change the Label to **1**.

Click **OK**.

16.

Click on the middle vertical line.

The label auto-increments to 2.

Click **OK**.

17.

Click on the right vertical line.

The label auto-increments to 3.

Click **OK**.

Click **ENTER** to exit the command.

18.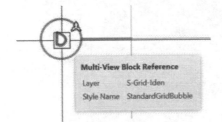

If you hover over one of the column bubbles, you can see the bubbles are placed on S-Grid-Iden layer.

Remember that Layer Key Styles control the layer properties of AEC elements.

19.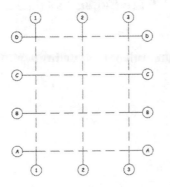

Save as *ex2-5.dwg*.

Exercise 2-6:
Creating a Custom Grid Bubble

Drawing Name: AEC Annotation Symbols (Imperial).dwg
Estimated Time: 25 minutes

This exercise reinforces the following skills:

- Tool palette
- Block Editor
- Attributes

1.

Browse to the location of the *AEC Annotation Symbols (Imperial).dwg* file.

ProgramData is a hidden folder by default. You may need to use File Explorer to make the ProgramData file visible. If you are in a classroom environment, your IT group may make this folder uneditable. If this is the case, make a copy of the file and locate it in your work folder. You will then need to set the Properties to point to the correct folder location.

2.

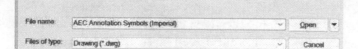

Open the file.

Switch to **Model** space.

3.

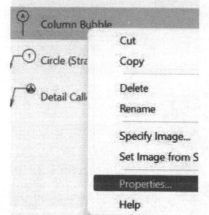

Locate the **Column Bubble** tool on the Annotation tool palette.

Right click and select **Properties**.

4.

Notice the drawing name that controls the block used.

Click **OK**.

5.

Switch to the Annotate ribbon.

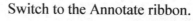

In the Text Style box, select the drop down and click on **Manage Text Styles**.

6.

Click **New**.

7.

Type **Comic Sans**.

Click **OK**.

If you prefer a different font, choose a different font.

8.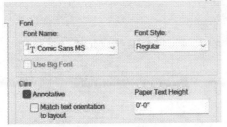

Set the Font Name to **Comic Sans MS.**

Click **Close**.

9.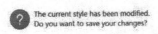

Click **Yes** to save your changes.

10.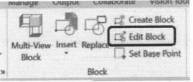

Switch to the Insert ribbon.

Click **Edit Block**.

11.

Locate the **BubbleDef** block.

Click **OK**.

12.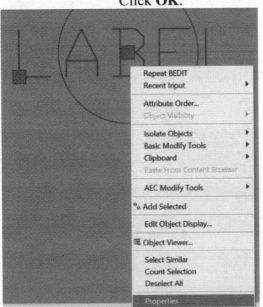

Select the attribute.
This is the text.

Right click and select **Properties**.

13.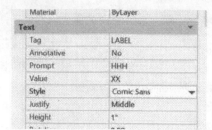

Change the Style to **Comic Sans**.

14.

Change Annotative to **No**.

Click **ESC** to release the selection.

15.

The attribute updates.

Click **Save Block** on the ribbon.

16.

Click **Close Block Editor** on the ribbon.

17.

Save the drawing file in your exercise folder with a modified name.

18.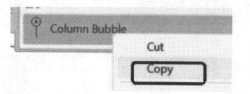

Highlight the Column Bubble on the Tool Palette.

Right click and select **Copy**.

19.

Right click and select **Paste** to add a copied column bubble.

20.

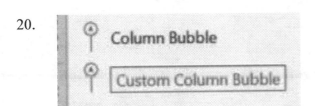

Rename the copied tool to **Custom Column Bubble**.

21.

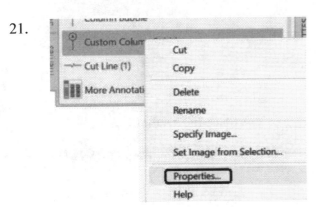

Right click on the **Custom Column Bubble** and select **Properties.**

22.

Change the symbol location to where you stored the modified bubble.

Note the symbol used is the StandardGridBubble.

Click **OK**.

23.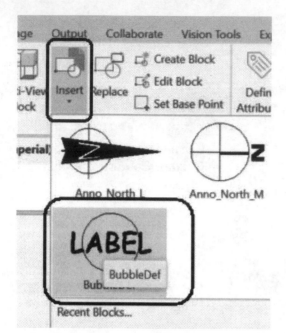

Switch to the Insert ribbon.

Select **Insert → BubbleDef** to insert the modified block.

24.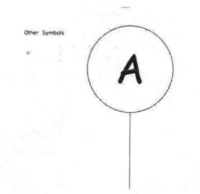

Place the symbol.

Change the attribute to A.

Add a line below the bubble.

25.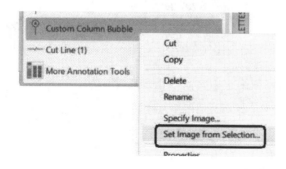

Right click on the **Custom Column Bubble** and select **Set Image from Selection.**

26.

Select the bubble and line.

Press ENTER to complete the selection.

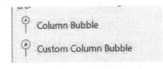

The tool image updates.

27.   Save *as AEC Annotation Symbols (Imperial)-modified.dwg*.

Exercise 2-7:
Insert Grid Lines

Drawing Name: 01 Column Grid.dwg
Estimated Time: 20 minutes

This exercise reinforces the following skills:

- Insert Drawing
- Project Navigator
- Constructs
- Column Grid
- Column Bubble
- Layer Manager

1.   Start a new drawing.
2.   Open the **Project Browser**.

3.   Verify that the Current Project is set to **Brown Medical Clinic**.

4.
5.   Open the **Project Navigator**.

6.

Open the Projects tab.

Click the **Edit** icon next to Levels.

7.

Name	Floor Elevation	Floor to Floor Height	ID	Des
Roof	3600.00	0.00	R	Roo
Level 1	0.00	3600.00	1	First
Site	-150.00	150.00	S	Site

Delete the 02 Second floor level.
Delete the 00 Level.
Rename the levels as shown.
Adjust the Site level elevation to -150.
Adjust the Level 1 elevation to 0.00.
Adjust the Roof elevation to 3600.
Click **OK**.
Open the Constructs tab.

8.

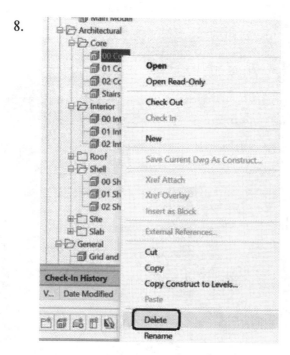

Highlight **00 Core** construct.
Right click and select **Delete**.

9.

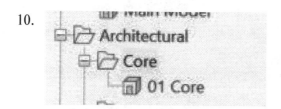

Are you sure you want to delete the Construct
'Constructs\Architectural\Core\00 Core'?

Click **Yes**.

10.

Main Model
Architectural
Core
01 Core

Delete the 02 and the Stairs constructs under Core.

11.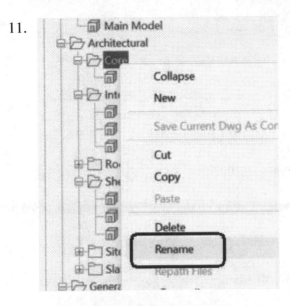

Highlight **Core**.
Right click and **Rename**.

12.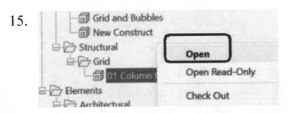

Rename **Floor Plans**.

13.

Click **Repath project now**.

14.

Are you sure you want to delete the
Construct 'Constructs\Structural\Grid\00
Column Grid'?

Yes   No

Select the **Constructs** tab.

Expand the Structural folder.
Delete the 00 Column Grid under Grid.
Click **Yes**.
Delete the 02 Column Grid under Grid.
Click **Yes**.

15.

Grid and Bubbles
New Construct
Structural
Grid
01 Column   Open
Elements   Open Read-Only
Architectural Check Out

Locate the 01 Column Grid under Grid.
Right click and select **Open**.

16.

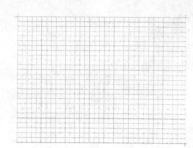

A drawing file opens.

The file doesn't have anything in it.
You may recall this project was created using a
template. The template generated more than a
hundred drawing files to act as placeholders.

Notice the name of the drawing.

17.

Set the View Scale to **1:1**.

18.

Go to the Insert ribbon.

Select **Insert Blocks from Libraries**.

19.

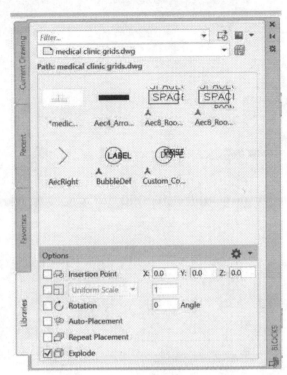

Select the Libraries tab.

Browse to the Lesson 2 downloaded files.

Locate the *medical clinic grids.dwg*. Click **Open**.

Disable Insertion Point.

20.

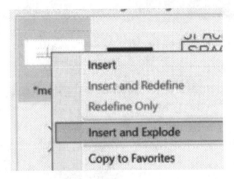

Right click on the drawing preview.

Select **Insert and Explode**.

Close the Blocks dialog.

21.

Open the Design Tools.
Activate the Annotation palette.
Locate the **Custom Column Bubble**.

22.

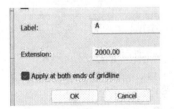

Select the bottom horizontal grid line.

Change the label to **A**.
Enable **Apply at both ends of gridline**.

Click **OK**.

23. Label the horizontal Gridlines A-D.
    Label the vertical gridlines 1-6.

    You will need to zoom in to see the bubbles.

24.  Select one of the column bubbles.
    Right click and **Select Similar.**

    All the column bubbles will be selected.

    Right click and select **Properties.**

25. 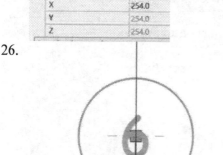 Change the Scale to **254**.

    Click **ESC** to release the selection.

26. Use the grips to reposition the bubbles so they are
    positioned outside of the grid.

27. Switch to the Home ribbon.

    Open the **Layer Manager**.

28. Change the properties for S-Grid to use the
    color 250 and DASHED linetype.
    Change the properties for S-Grid-Iden to use
    the color Cyan and Continuous linetype.

29. Close the Layer Manager.
    Save as *01 Column Grid.dwg and close.*

Exercise 2-8:
Creating Constructs

Drawing Name: foundation floorplan – Tahoe Cabin.dwg
Estimated Time: 10 minutes

This exercise reinforces the following skills:

- ❑ Project Browser
- ❑ Project Navigator
- ❑ Constructs
- ❑ Adding an existing drawing to a project

1. | floundation floorplan - Tahoe Cabin ✕ | Open *foundation floorplan – Tahoe Cabin.dwg.*

2.   Open the Project Browser.

3.   Set the **Tahoe Cabin** project current.

Sample Project 2023
Tahoe Cab

Set Project Current   Close the Project Browser.

Browse Project

Refresh

4.   Select the **Project Navigator**.

5.   Select the Constructs tab.

Constructs

Expand
New   Category
Save Current Dwg As Construct..   Construct..
Cut
Copy
Paste
Delete
Rename

Highlight **Constructs**.
Select **New→Category**.

6.      Type Architecture for the new category name.

7.   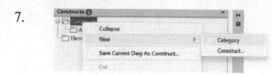   Highlight **Constructs**.
Select **New**→**Category**.

8. Name the new category **Sitework**.

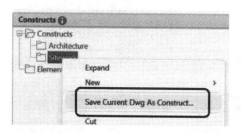

9. Highlight the Sitework category.

Right click and select **Save Current Dwg as Construct**.

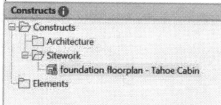

Notice that the name, description, and category are already filled in.

Name	foundation floorplan - Tahoe Cabin
Description	foundation floorplan - Tahoe Cabin
Category	Constructs\Sitework
File Name	foundation floorplan - Tahoe Cabin

10. Assign the drawing to the **Foundation Division**.

Click **OK**.

Assignments

Level	Description	Division
Roof	Roof	☐
Second Floor	Second Floor	☐
First Floor	First Floor	☐
Foundation	foundation	☑

11. *The drawing is now added to the project as a construct.*

Exercise 2-9:
Convert Polylines to Column Grids

Drawing Name: foundation floorplan – Tahoe Cabin.dwg
Estimated Time: 10 minutes

This exercise reinforces the following skills:

- ❑ Convert Polylines to Colum Grids
- ❑ Add Column Bubbles

1.   Open the **Project Navigator**.

2. 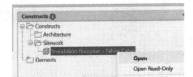 Select the Constructs tab.
    Locate the *foundation floorplan – Tahoe Cabin.dwg* under Sitework.
    Right click and select **Open**.

3. 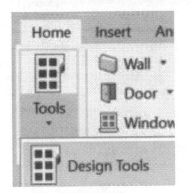 Launch the Design Tools.

4. 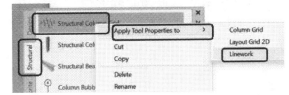 Open the Structural palette.

    Locate the **Structural Column Grid** tool.

    Right click and select **Apply Tool Properties to Linework**.

5.

Select the dashed polylines.

Click ENTER.

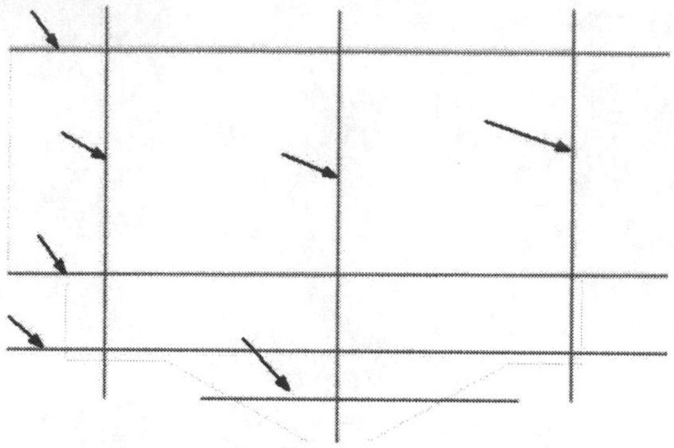

6.

Right click and select **No labels**.

Enter
Cancel
Recent Input ▶
[No labels]
🖐 Pan
±🔍 Zoom

7.

Right click and select **Yes**.

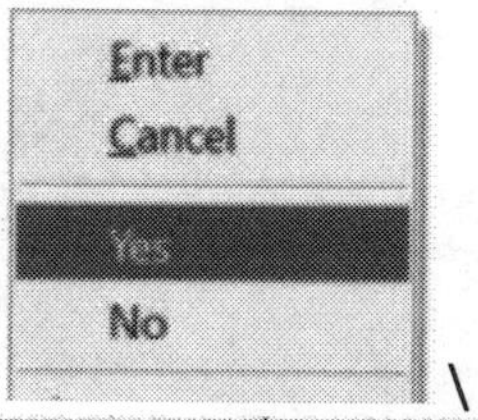

Enter
Cancel
Yes
No

8.

The Column Grid is placed.

If you hover over the grid with the mouse, you will see the element and layer.

Column Grid

Layer S-Grid

Click **ESC** to release the column grid.

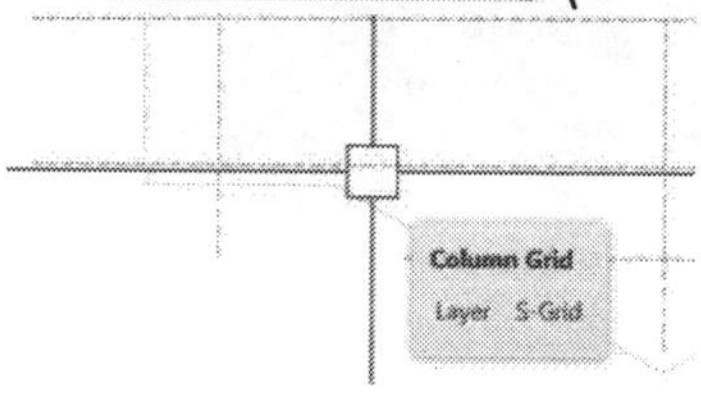

9.

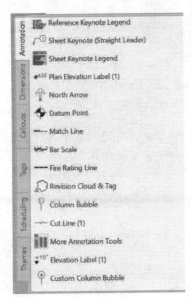

Switch to the Document Tool Palette.

Open the Annotation palette.

Locate **the Custom Column Bubble**.

10.

Label the horizontal grid lines **A-D**.

Enable **Apply at both ends of gridline**.

11.

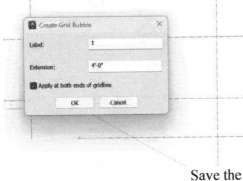

Label the vertical grid lines **1-3**.

Enable **Apply at both ends of gridline**.

The bubbles will appear small because of the scale. If you adjust the view scale, the bubbles will look larger.

12.

Save the file.

Slabs

Slabs are used to model floors and other flat surfaces. You start by defining a slab style which includes the components and materials which are used to construct the floor system.

You can create a floor slab by drawing a polyline and applying the style to the polyline or by selecting the slab tool and then drawing the boundary to be used by the slab.

Keynotes

Keynotes are used to standardize how materials are called out for architectural drawings. Most keynotes use a database. The database is created from the Construction Specifications Institute (CSI) Masterformat Numbers and Titles standard. ACA comes with several databases which can be used for keynoting or you can use your own custom database. The database ACA uses is in Microsoft Access format (*.mdb). If you wish to modify ACA's database, you will need Microsoft Access to edit the files.

Exercise 2-10:
Creating Slab Styles

Drawing Name: foundation floorplan – Tahoe Cabin.dwg
Estimated Time: 30 minutes

This exercise reinforces the following skills:

- ❏ Creating Styles
- ❏ Display Properties
- ❏ Keynotes

1.   Open the **Project Navigator**.

2. 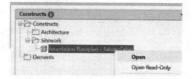 Select the Constructs tab.
    Locate the *foundation floorplan – Tahoe Cabin.dwg* under Sitework.
    Right click and select **Open**.

3.

Activate the Manage ribbon.

Select **Style Manager**.

4.

Browse to the Slab Styles category.

Highlight Slab Styles.

Right click and select **New**.

foundation floorplan - Tahoe Cabin.dwg
Architectural Objects
Curtain Wall Styles
Curtain Wall Unit Styles
Door Styles
Door/Window Assembly Styles
Railing Styles
Roof Slab Edge Styles
Roof Slab Styles
Slab Edge Styles
Slab Styles
Sta   New
Space

5.

On the General tab,

change the Name to **Gravel Fill**.

General   Components

Name:

Gravel Fill

Description:

6.

Select Keynote...

Click **Select Keynote**.

7.

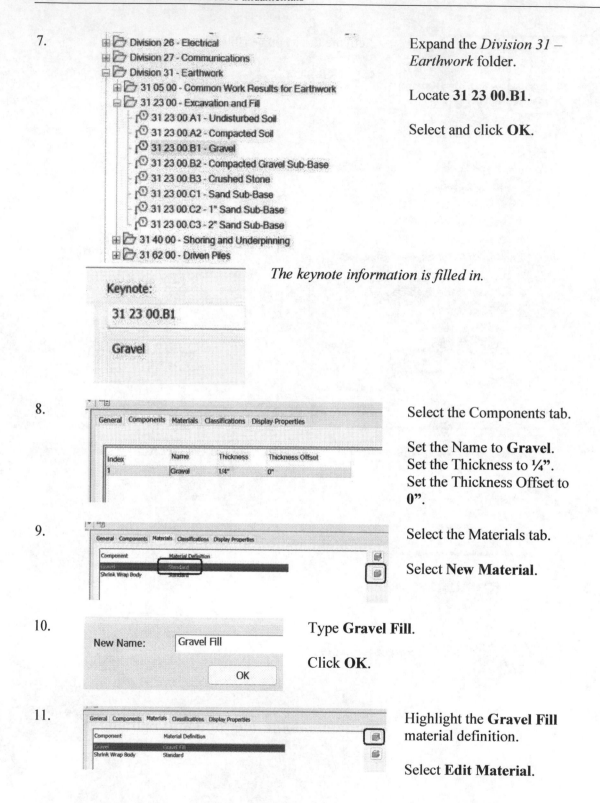

Expand the *Division 31 – Earthwork* folder.

Locate **31 23 00.B1**.

Select and click **OK**.

The keynote information is filled in.

8.

Select the Components tab.

Set the Name to **Gravel**.
Set the Thickness to ¼".
Set the Thickness Offset to **0"**.

9.

Select the Materials tab.

Select **New Material**.

10.

Type **Gravel Fill**.

Click **OK**.

11.

Highlight the **Gravel Fill** material definition.

Select **Edit Material**.

12. 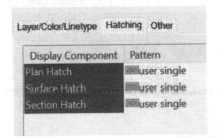 Click in the **Style Override** box or click the **Material Definition Override** icon.

13. 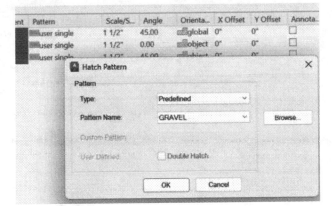 Select the Hatching tab.

    Highlight all the hatches.

14. Click on the Pattern column.

    Set the Hatch Pattern type to **Predefined**.
    Set the Pattern Name to **GRAVEL**.

    Click **OK**.

    Close all the dialogs.

15. Save the file.

Exercise 2-11:
Creating a Gravel Fill

Drawing Name: foundation floorplan – Tahoe Cabin.dwg
Estimated Time: 5 minutes

This exercise reinforces the following skills:

❑   Creating a Slab

1.   Open the **Project Navigator**.

2.   Select the Constructs tab.
Locate the *foundation floorplan – Tahoe Cabin.dwg* under Sitework.
Right click and select **Open**.

3.   Go to the Home ribbon.

Select the **Slab** tool.

4.    Change the Style to **Gravel Fill**.

5.   Trace along the elements on the gravel outline layer.

6.

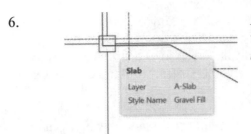

If you hover over the slab with the mouse, you will see it was placed on Layer A-Slab and uses Style Name Gravel Fill.

Save the file.

Structural Members

Architectural documentation for residential construction will always include plans (top views) of each floor of a building, showing features and structural information of the floor platform itself. Walls are located on the plan but not shown in structural detail.

Wall sections and details are used to show:

- the elements within walls (exterior siding, sheathing, block, brick or wood studs, insulation, air cavities, interior sheathing, trim)
- how walls relate to floors, ceilings, roofs, and eaves
- openings within the walls (doors/windows with their associated sills and headers)
- how walls relate to openings in floors (stairs).

Stick-framed (stud) walls usually have their framing patterns determined by the carpenters on site. Once window and door openings are located on the plan and stud spacing is specified by the designer (or the local building code), the specific arrangement of vertical members is usually left to the fabricators and not drafted, except where specific structural details require explanation.

The structural members in framed floors that have to hold themselves and/or other walls and floors up are usually drafted as framing plans. Designers must specify the size and spacing of joists or trusses, beams and columns. Plans show the orientation and relation of members, locate openings through the floor and show support information for openings and other specific conditions.

Autodesk AutoCAD Architecture includes a Structural Member Catalog that allows you to easily access industry-standard structural shapes. To create most standard column, brace, and beam styles, you can access the Structural Member Catalog, select a structural member shape, and create a style that contains the shape that you selected. The shape, similar to an AEC profile, is a 2D cross-section of a structural member. When you create a structural member with a style that you created from the Structural Member Catalog, you define the path to extrude the shape along. You can create your own structural shapes that you can add to existing structural members or use to create new structural members. The design rules in a structural member style allow you to add these custom shapes to a structural member, as well as create custom structural members from more than one shape.

All the columns, braces, and beams that you create are sub-types of a single Structural Member object type. The styles that you create for columns, braces, and beams have the same Structural Member Styles style type as well. When you change the display or style of a structural member, use the Structural Member object in the Display Manager and the Structural Member Styles style type in the Style Manager.

If you are operating with the AIA layering system as your current layer standard, when you create members or convert AutoCAD entities to structural members using the menu picks or toolbars, AutoCAD Architecture assigns the new members to layers: A-Cols, A-Cols-Brce or A-Beam, respectively. If Generic AutoCAD Architecture is your standard, the layers used are A_Columns, A_Beams and A_Braces. If your layer standard is Current Layer, new entities come in on the current layer, as in plain vanilla AutoCAD.

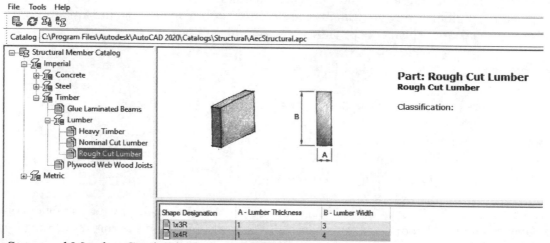

The Structural Member Catalog includes specifications for standard structural shapes. You can choose shapes from the Structural Member Catalog and generate styles for structural members that you create in your drawings.

A style that contains the catalog shape that you selected is created. You can view the style in the Style Manager, create a new structural member from the style, or apply the style to an existing member.

When you add a structural member to your drawing, the shape inside the style that you created defines the shape of the member. You define the length, justification, roll or rise, and start and end offsets of the structural member when you draw it.

You cannot use the following special characters in your style names:
- less-than and greater-than symbols (< >)
- forward slashes and backslashes (/ \)
- quotation marks (")
- colons (:)
- semicolons (;)
- question marks (?)
- commas (,)
- asterisks (*)

- vertical bars (|)
- equal signs (=)
- backquotes (`)

The left pane of the Structural Member Catalog contains a hierarchical tree view. Several industry standard catalogs are organized in the tree, first by imperial or metric units, and then by material.

	Open a catalog file - The default is the catalog that comes with ADT, but you can create your own custom catalog. The default catalog is located in the following directory path: \\Program Files\Autodesk AutoCAD Architecture R3\Catalogs\catalogs.
	Refresh Data
	Locate Catalog item based on an existing member – allows you to select a member in a drawing and then locates it in your catalog.
	Generate Member Style – allows you to create a style to be used.

Exercise 2-12:
Creating Member Styles

Drawing Name: member_styles.dwg
Estimated Time: 15 minutes

This exercise reinforces the following skills:

- ❑ Creating Member Styles
- ❑ Use of Structural Members tools

1.   Open *member_styles.dwg*.

2.   Activate the Manage ribbon.

           Go to **Style & Display → Structural Member Catalog**.

3.

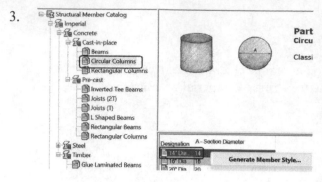

Browse to the **Imperial/Concrete/Cast in-place/Circular Columns** folder.

Highlight the **14"** Diameter size.

Right click and select **Generate Member Style.**

4.

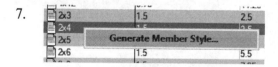

Change the Name to **14_Diamaeter Concrete Column.**

Click **OK**.

5.

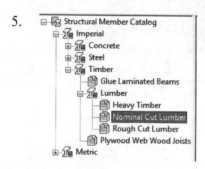

Browse to the **Imperial/Timber/Lumber/Nominal Cut Lumber** folder.

6. In the lower right pane:

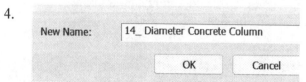

Locate the **2x4** shape designation.

7.

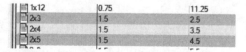

Right click and select **Generate Member Style**.

8.

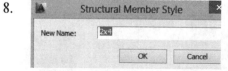

In the Structural Member Style dialog box, the name for your style automatically fills in.

Click **OK**.

9.

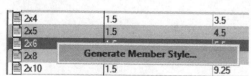

Locate the **2x6** shape designation. Right click and select **Generate Member Style**.

10.  In the Structural Member Style dialog box, the name for your style is displayed.

Click **OK**.

11.
12.  Locate the **4x4** shape designation.
Right click and select **Generate Member Style**.

13.

New Name: 4x4

In the Structural Member Style dialog box, the name for your style automatically fills in.

Click **OK**.

14. Close the dialog.
Save as *ex2-12.dwg*.

Exercise 2-13:
Creating Member Shapes

Drawing Name: member_shapes.dwg
Estimated Time: 10 minutes

This exercise reinforces the following skills:

❑   Structural Member Wizard

1. 📂   Open *member_shapes.dwg*.

2.    Activate the **Manage** ribbon.

Go to **Style & Display→Structural Member Wizard**.

3. 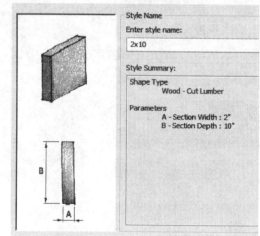 Select the **Cut Lumber** category under Wood.

Click **Next**.

4. Set the Section Width to **2″**.

Set the Section Depth to **10″**.

Click **Next**.

5. Enter the style name as **2x10**.

Click **Finish**.

6. Save as *ex2-13.dwg*.

Exercise 2-14:
Copying Structural Member Styles

Drawing Name: foundation floorplan – Tahoe Cabin.dwg,
structural member styles.dwg

Estimated Time: 10 minutes

This exercise reinforces the following skills:

- ❑ Using the Styles Manager
- ❑ Copying Structural Member Styles

1.  Open the **Project Navigator**.

2. 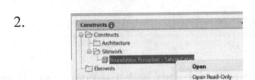 Select the Constructs tab.
Locate the *foundation floorplan – Tahoe Cabin.dwg* under Sitework.
Right click and select **Open**.

3. Go to the Manage ribbon.

Click on **Style Manager**.

4. Expand the **Structural Member Styles**.

There is one style in the current drawing: Standard.

5. 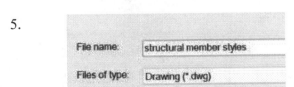 Open the *structural member styles.dwg*

6.

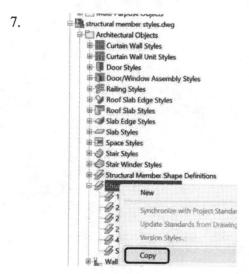

Browse to the Structural Member Styles category.

You should see all the styles that you created.

7.

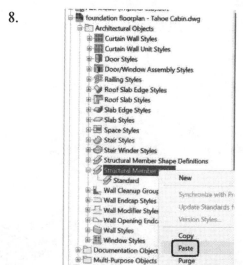

Highlight the Structural Member Styles category.

Right click and select **Copy**.

8.

Browse up to the foundation floorplan drawing.

Highlight the Structural Member Styles category.

Right click and select **Paste**.

9.

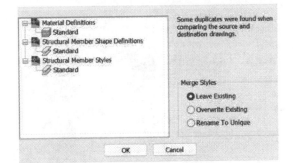

Enable **Leave Existing**.

Click **OK**.

10.

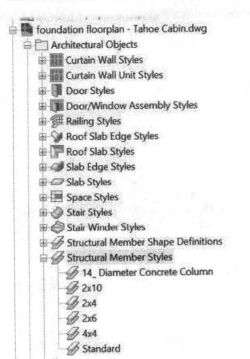

The structural members are now in the project's construct drawing.

Close the Styles Manager.

Exercise 2-15:
Create Concrete Footings

Drawing Name: foundation floorplan – Tahoe Cabin.dwg
Estimated Time: 5 minutes

This exercise reinforces the following skills:

- ❑ Project Navigator
- ❑ External References
- ❑ Create Column

1. Open the Project Navigator.

2. 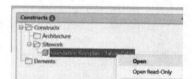 Select the Constructs tab.

   Locate the *foundation floorplan – Tahoe Cabin.dwg* under Sitework.

   Right click and select **Open**.

3. 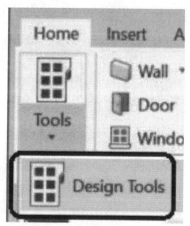 Go to the Home ribbon.

   Go to **Tools→Design Tools**.

4.

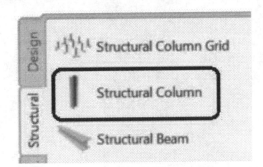

Select the Structural tab.

Select **Structural Column**.

5.

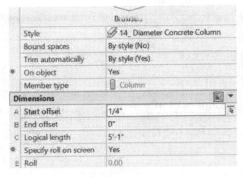

Set the Style to **14_Diameter Concrete Column**.

Set the Start Offset to ¼".

Set the Logical Length to **5' 1"**.

Place at Grid **B-1.**

6.

Save the file.

Exercise 2-16:
Create Element

Drawing Name: foundation floorplan – Tahoe Cabin.dwg
Estimated Time: 10 minutes

This exercise reinforces the following skills:

- Project Navigator
- Add a Level
- Constructs
- Elements

1.

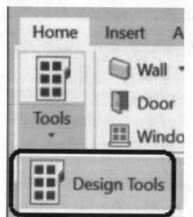

Go to the Home ribbon.

Go to **Tools→Design Tools**.

2.

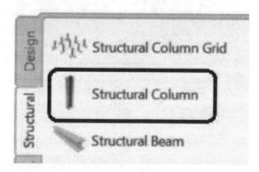

Select the Structural tab.

Select **Structural Column**.

3.

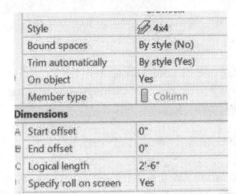

Style	🔩 4x4
Bound spaces	By style (No)
Trim automatically	By style (Yes)
On object	Yes
Member type	🪵 Column
Dimensions	
A Start offset	0"
B End offset	0"
C Logical length	2'-6"
Specify roll on screen	Yes

Set the Style to **4x4**.

Set the Start Offset to **0'-0"**.

Set the Logical Length to **2' 6"**.

Set the start point at **0,0,4'-0"**.

Place at Grid **B-1**.

Escape out of the command.

4.

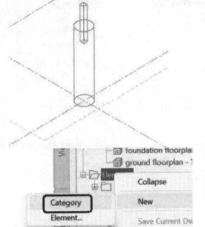

Switch to an isometric view.

The 4x4 post should appear as shown.

5.

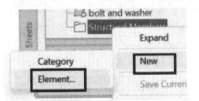

Open the Project Navigator.
Highlight **Elements**.
Right click and select **New→Category**.

6.

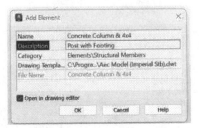

Name the Category **Structural Members**.

7.

Highlight Structural Members.
Right click and select **New→Element**.

8.

For Name, type **Concrete Column & 4x4**.
For Description, type **Post with Footing**.
Enable **Open in Drawing editor**.

Click **OK**.

9.

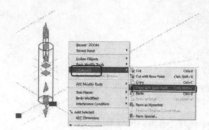

Return to the foundation floorplan – Tahoe Cabin.dwg.

Select the concrete column and 4x4.
Right click and select **Clipboard→Copy with Basepoint.**

Select the bottom center at Grid B-1.

10.

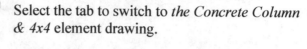

Select the tab to switch to *the Concrete Column & 4x4* element drawing.

Right click and **select Clipboard→Paste to Original Coordinates.**

11.

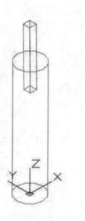

Set the basepoint to 0,0,0.

Save the drawing and close.

Exercise 2-17:
Create Main Model

Drawing Name:	ground floorplan – Tahoe Cabin.dwg
Estimated Time:	20 minutes

This exercise reinforces the following skills:

- ❑ Project Navigator
- ❑ Add a Level
- ❑ Constructs
- ❑ Elements
- ❑ External References

1. 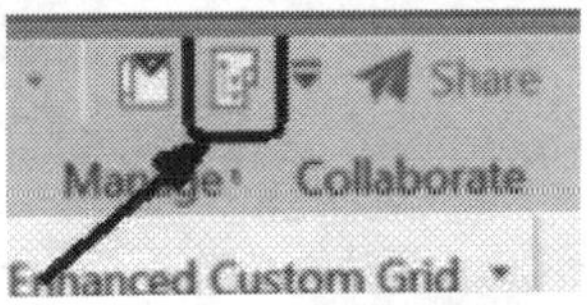 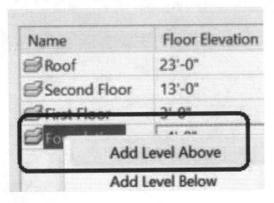 Open the Project Navigator.

2. Open the *ground floorplan – Tahoe Cabin.dwg*.

3. 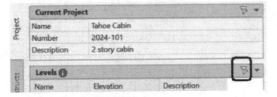 Select the **Edit Levels** tool.

4. 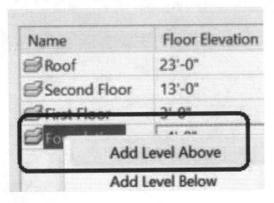 Highlight the Foundation Level.

   Right click and select **Add Level Above.**

5.

Name	Floor Elevation	Floor to Floor Height	ID	Description
Roof	23'-0"	10'-0"	R	Roof
Second Floor	13'-0"	10'-0"	2	Second Floor
First Floor	3'-0"	10'-0"	1	First Floor
Ground Level	0"	3'-0"	5	Ground Level
Foundation	-4'-0"	4'-0"	F	foundation

Add a Ground Level at **0"**.

Verify that the other levels have retained the same floor elevations. Correct if needed. Click **OK.**

6.

You have made changes to the project that may affect existing views.
Do you wish to regenerate all views in this project?

[Yes] [No]

Click **Yes.**

7. Open *ground floorplan – Tahoe Cabin.dwg.*

8.

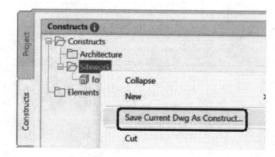

On the Project Navigator:

Open the Constructs tab.

Right click on Sitework.
Select **Save Current Dwg As Construct**.

9.

Constructs
 Architecture
 Sitework
  foundation floorplan - Tahoe Cabin
  ground floorplan - Tahoe Cabin

You see the new drawing has been added as a construct.

10.

File name: 3D Model-Master
Files of type: AutoCAD 2018 Drawing (*.dwg)

Start a new drawing.
Save as **3D Model – Master.dwg**.

11. Type **xref** to open the External Reference Manager.

12.

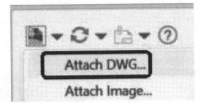

Select **Attach DWG…**

13.

Select the *ground floorplan – Tahoe Cabin.dwg*.

Place at **0,0,0**.

Click **OK**.

14.

Select **Attach DWG…**

15.

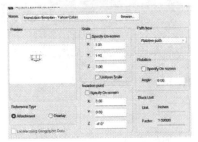

Select the *foundation floorplan – Tahoe Cabin.dwg*.

Place at **0,0,-4'-0"**.

Click **OK**.

You can see the concrete column and the wooden post.

16.

In the Project Navigator:

Highlight **Constructs**.

Right click and select **New→Category**.

17.

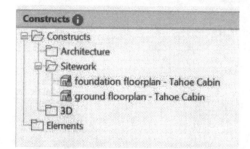

Name the new construct – **3D**.

18.

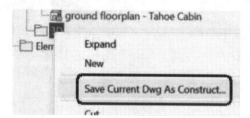

Verify that **3D Model – Master.dwg**. is the current/active window.

Right click on the 3D construct and select **Save Current Dwg As Construct**.

19.

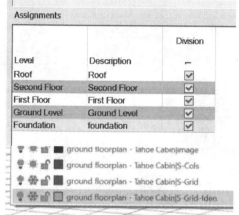

Enable all the levels.

Click **OK**.

20.

Freeze the **ground floorplan – Tahoe Cabin|S-Grid and S-Grid-Iden** layers.

21.

Save and close all the drawings.

Structural member geometry is created by a shape contained in the style of the structural member that is extruded along the axis of the member. Each shape can have different levels of detail in its geometry, allowing different displays of the structural member that the shape is defining. Each representation displays the structural shape in a different level of detail.

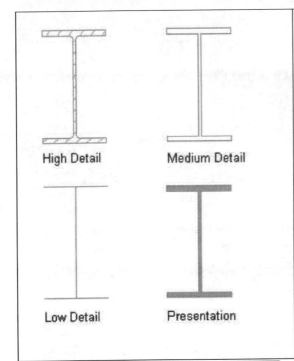

For example, you can use the Plan, Plan High Detail, and Plan Low Detail structural member representations to display a plan view of a structural member.

Most framing plans will use the Low Detail representation.

The Plan Low Detail display representation displays the member with simple lines and arcs. You can use this display representation to create a top view framing plan where you want a single line representation of a column. The Plan Low Detail display representation includes two display components, Beam Sketch and Brace Sketch. You can change the display properties of these components, such as layer, color, and linetype, to differentiate the display of beams and braces.

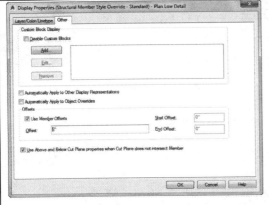

On the Other tab of the style override dialog box for this display representation, you can also add offsets between members to display them as you typically would in a top view framing plan. Note that the Use Member Offsets check box applies only to the start and end offsets. The Offset value below the check box applies only to structural braces, and allows you to offset them horizontally so that they are visible even when positioned directly underneath a beam.

Exercise 2-18:

Defining a Low Detail Representation of a Framing Plan

Drawing Name:	low_detail.dwg
Estimated Time:	10 minutes

This exercise reinforces the following skills:

- ❑ Detail Representations
- ❑ Global Cut Plane
- ❑ Layouts
- ❑ Use of Structural Members

1. 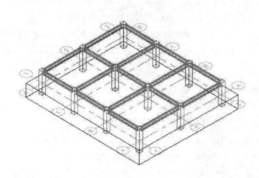 Open *low_detail.dwg*.
   Close the current project.

   The Model layout tab shows a basic frame.

   There are three components in the model: a concrete slab, some posts and some beams.

2.  Open the **Framing Plan** Layout tab.

3. 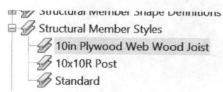 Go to the **Manage** ribbon.

   Click **Style Manager**.

4. Locate the **10in Plywood Web Wood Joist** under Structural Member Styles.

5.

Select the **Display Properties** tab.

Highlight **Plan Low Detail**.

Click **Style Overrides**.

6.

Click the **Other** tab.

Disable Use Member Offsets.

Set the Start Offset to **6"**.
Set the End Offset to **-6"**.

Click **OK**.

Close the dialog.

7.

On the status bar at the bottom of the display window:

Change the Display Representation to **Low Detail**.

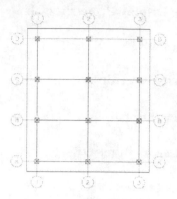

The display updates.

We still see the concrete slab. We only want to see the structural members and the column grid.

8.

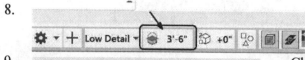

Select the **Global Cut Plane** tool on the Status Bar.

9.

Specify cut plane heights:

Display Above Range: 1000'-0"

Cut Height: 3'-6"

Display Below Range: 1'-0"

Calculate Cut Plane from Current Project: Calculate

OK   Cancel

Change the Display Below Range to **1'-0"**.
Set the Cut Height to **3' 6"**.

Click **OK**.

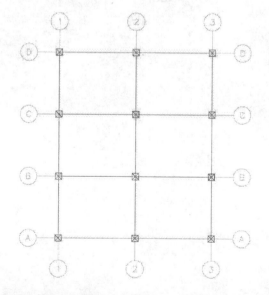

The concrete slab is no longer visible.

10. [−][SW Isometric][Realistic]

Switch back to the **Model** layout tab.

Use the controls in the upper left corner of the display window to change the display to SW Isometric and Realistic.

11.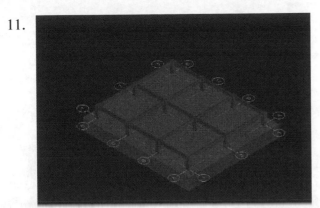

Save as ex2-17.dwg.

Structural Member Tags

C10x30	**Beam Tag by Style** Creates a beam tag based on the member style.	
	B 1	**Beam Tag by Type** Creates a beam tag based on the member type.
B1 C10x30	**Beam Tag by Type and Style** Creates a beam tag based on the member type and style.	
	C10x30	**Brace Tag by Style** Creates a brace tag based on the member style.
BR1	**Brace Tag by Type** Creates a brace tag based on the member type.	
	BR1 C10x30	**Brace Tag by Type and Style** Creates a brace tag based on the member type and style.
C10x30	**Column Tag by Style** Creates a beam tag based on the member style.	
	C1	**Column Tag by Type** Creates a column tag based on the member type.
C1 C10x30	**Column Tag by Type and Style** Creates a column tag based on the member type and style.	

ACA comes with a selection of different tags that can be used with structural members.

Simply load or add the desired tag on to your tool palette and place as desired.

To place a tag, you must be in Model space.

Exercise 2-19:
Adding Tags to Structural Members

Drawing Name: structural member tags.dwg
Estimated Time: 45 minutes

This exercise reinforces the following skills:

- ❑ Detail Representations
- ❑ Use of Structural Members
- ❑ Tags
- ❑ Tool Palettes

1. Open *structural member tags.dwg*.
   Close any open projects.

2. Open the **Framing Plan** layout tab.

3. Switch to the **Annotate** ribbon.

   Select **Annotation Tools** from the ribbon.

4. 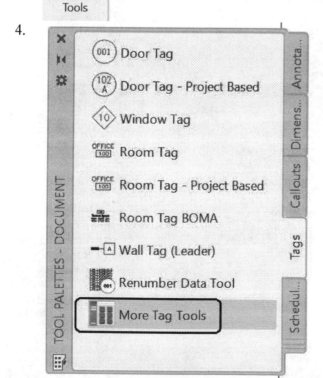 Click the **Tags** tab on the tool palette.

   Scroll down and double left click on the **More Tag Tools**.

5.

Schedule Tags

Door & Window Tags

Object Tags

Room & Finish Tags

Structural Tags

Wall Tags

Click on **Structural Tags**.

6.

C10x30   **Beam Tag by Style**
Creates a beam tag
based on the
member style.

Locate the **Beam Tag by Style** tool.

7.

Beam Tag by

C10x30   Insert into Drawing   ea

Add to Tool Palette

Hover over the tool and then select **Add to Tool Palette**.

8.

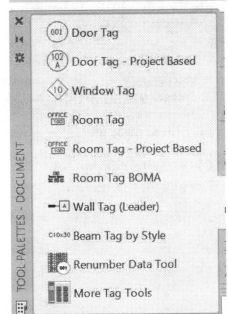

The tool will be added to the active tab on the tool palette.

Drag and drop the tool to the desired location on the tool palette.

I positioned the tool below the Wall Tag (Leader) tool.

9.

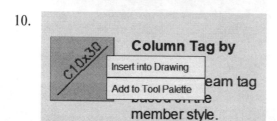

Locate the **Column Tag by Style** tool.

10.

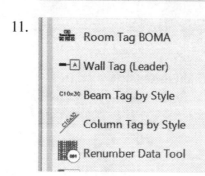

Hover over the tool and then select **Add to Tool Palette**.

11.

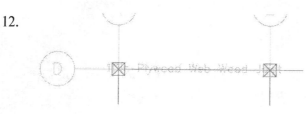

Drag and drop the new tool below the Beam Tag by Style tool.

Close the Tool Catalog.

12.

To place a tag you need to be in Model space.
Double click inside the viewport to activate Model space.
Select the **Beam Tag by Style** tool.
Select the beam located at D1-D2.
Click **ENTER** to place the tag centered on the beam.
Click **OK** when the Property Set Data dialog appears.
The tag is placed.

13.

Style Manager

Select the Manage ribbon.

Select **Style Manager**.

14.

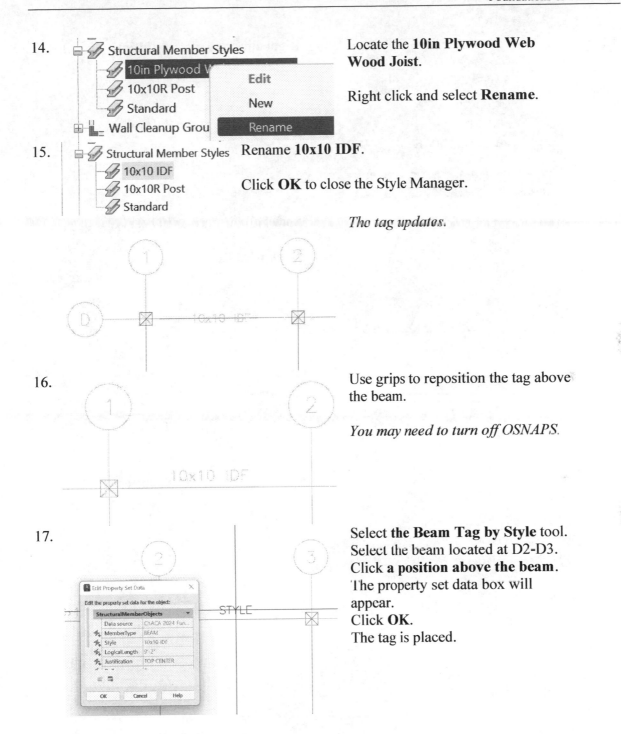

Locate the **10in Plywood Web Wood Joist**.

Right click and select **Rename**.

15. Rename **10x10 IDF**.

Click **OK** to close the Style Manager.

The tag updates.

16. Use grips to reposition the tag above the beam.

You may need to turn off OSNAPS.

17. Select **the Beam Tag by Style** tool.
Select the beam located at D2-D3.
Click **a position above the beam**.
The property set data box will appear.
Click **OK**.
The tag is placed.

18.

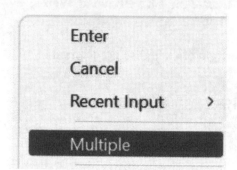

Right click and select **Multiple**.

19.

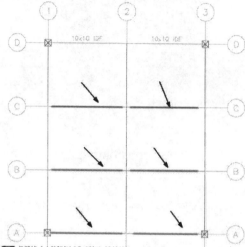

Select all the horizontal beams.

Click **ENTER**.

20.

The property set data dialog appears.

Click **OK**.

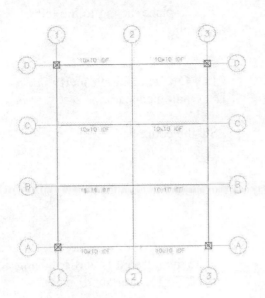

The beam tags are placed.

You are still in the tag command.

21.

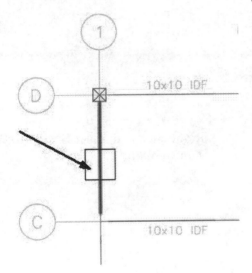

Select the vertical beam located at C1-D1.

22.

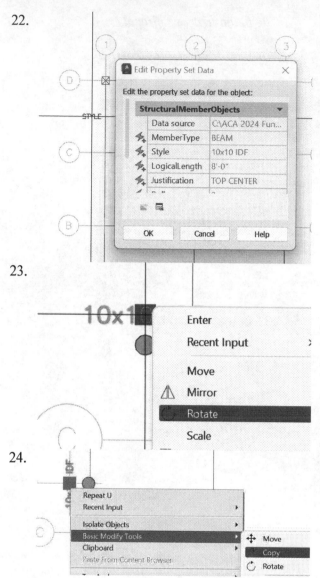

Click to place the tag to the left of the beam.

Click **OK** when the Property Set Data dialog appears.

ESC the command.

23.

Rotate the tag so it is vertical using the ROTATE command.

Then ESC to release the selection.

24.

Select the vertical tag.
Right click and select **Basic Modify Tools→Copy**.

25.

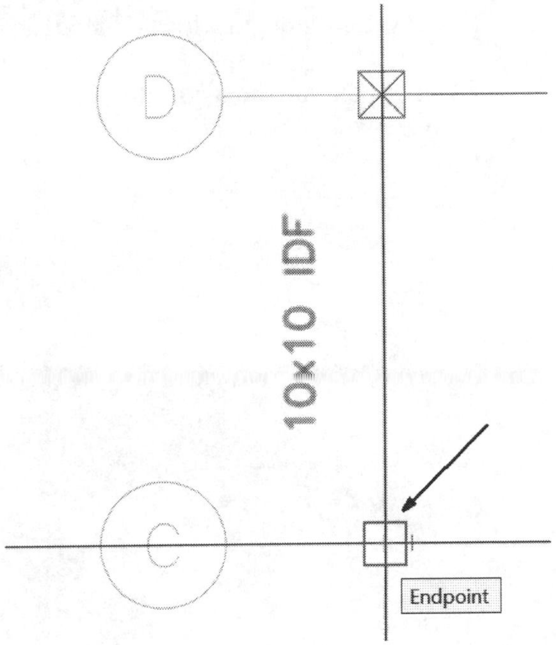

Select **C1** as the base point.

10x10 IDF

Endpoint

26.

Select the grid intersections to place the remaining vertical tags.

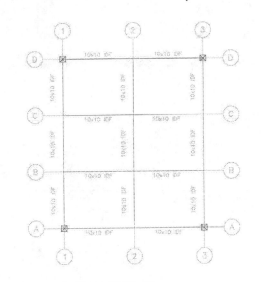

27.

Column Tag by Style

Select the **Column Tag by Style** tool.

28.

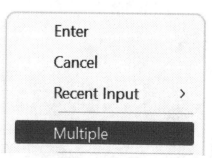

Select the column post located at D1.

Click to position the tag to the side of the post located at D1.

29.

Click **OK**.

30.

The tag is placed.

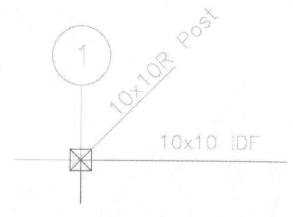

Right click and select **Multiple**.

31.

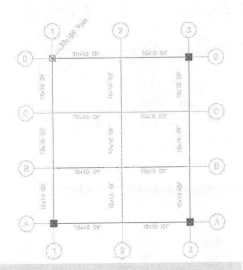

Select the remaining posts.

Click **ENTER**.

32.

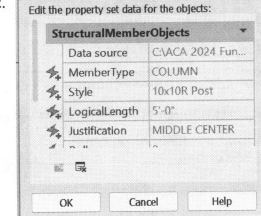

Edit the property set data for the objects:

StructuralMemberObjects	▼
Data source	C:\ACA 2024 Fun...
MemberType	COLUMN
Style	10x10R Post
LogicalLength	5'-0"
Justification	MIDDLE CENTER

OK   Cancel   Help

Click **OK**.

33.

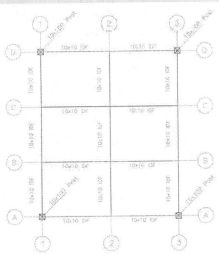

Click **ENTER** to exit the command.

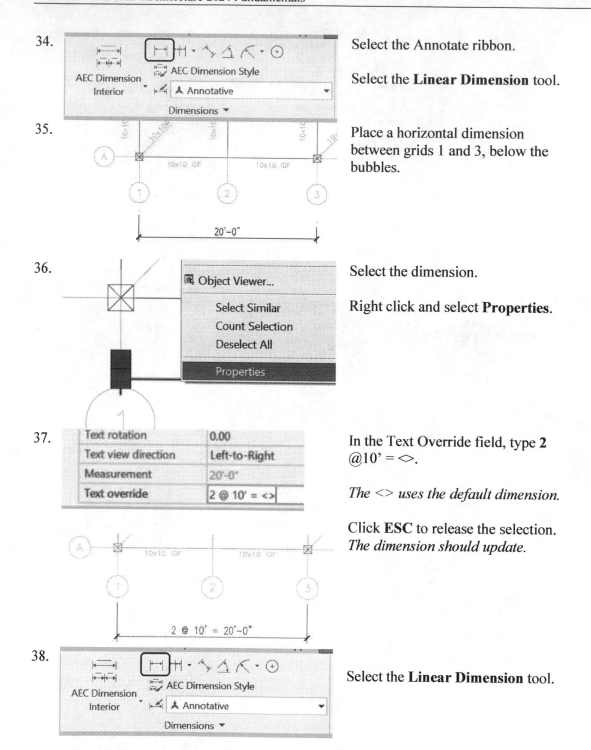

34. Select the Annotate ribbon.

    Select the **Linear Dimension** tool.

35. Place a horizontal dimension between grids 1 and 3, below the bubbles.

36. Select the dimension.

    Right click and select **Properties**.

37. In the Text Override field, type **2 @10' = <>**.

    The <> uses the default dimension.

    Click **ESC** to release the selection. *The dimension should update.*

38. Select the **Linear Dimension** tool.

39.

Place a vertical dimension between grids A and D, to the right of the bubbles.

40.

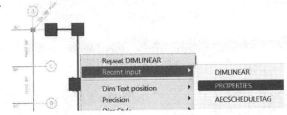

Select the dimension.

Right click and select **Recent Input→Properties**.

41.

Text view direction	Left-to-Right
Measurement	24'-0"
Text override	3 @ 8' = <>

In the Text Override field, type **3 @8' = ◇**.

Click **ESC** to release the selection.

42.

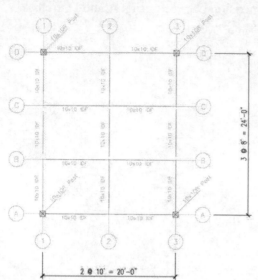

Return to Paper Space.

43. 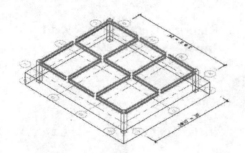 Switch to the **Model** layout tab.

44.

The linear dimensions are visible, but not the structural member tags.

You can control the appearance of the dimension using layer filter settings.

Save as *ex2-18.dwg*.

Exercise 2-20:
Adding Project Properties

Drawing Name: none
Estimated Time: 5 minutes

This exercise reinforces the following skills:

    ❑   Project Browser

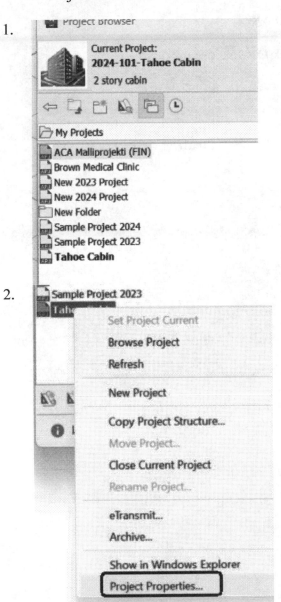

1.   Make the **Tahoe Cabin** current.

2.   Highlight the project name.

                                                          Right click and select **Project Properties**.

3.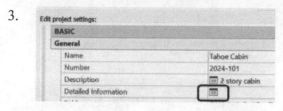

Click on **Detailed Information**.

4.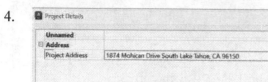

Add a Category called Address.
Add a Project Address.

Click **OK**.

Click **OK** to close the dialog.

Exercise 2-21:
Creating a View in a Project

Drawing Name: First Floor Foundation – Tahoe Cabin.dwg
Estimated Time: 20 minutes

This exercise reinforces the following skills:

- Detail Representations
- Tags
- Properties
- Constructs
- Display Order

1.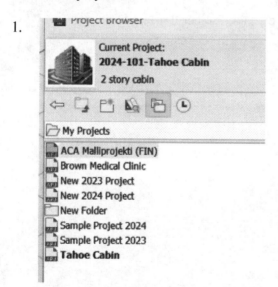

Launch the Project Browser.

Make the **Tahoe Cabin** current.

Open *First Floor Foundation – Tahoe Cabin.dwg*.

2.

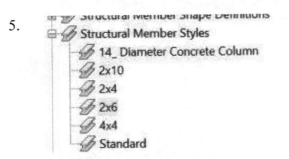

Launch the Project Navigator.

Highlight the Architecture Construct.
Right click and select **Save Current Dwg As Construct**.

3.

Name	First Floor Foundation - Tahoe Cabin
Description	First Floor Foundation - Tahoe Cabin
Category	Constructs\Architecture
File Name	First Floor Foundation - Tahoe Cabin

Assignments

Level	Description	Division
Roof	Roof	
Second Floor	Second Floor	
First Floor	First Floor	✓
Ground Level	Ground Level	
Foundation	foundation	

Enable the **First Floor**.

Click **OK**.

4.

Go to the **Manage** ribbon.

Click **Style Manager**.

5.

Locate the **2x6** under Structural Member Styles.

Structural Member Shape Definitions
Structural Member Styles
14_ Diameter Concrete Column
2x10
2x4
2x6
4x4
Standard

6.

Select the **Display Properties** tab.

Highlight **Plan Low Detail**.

Click **Style Overrides**.

7.

Offsets		
☐ Use Member Offsets	Start Offset:	6"
Offset 2"	End Offset:	-6"

Click the **Other** tab.

Disable **Use Member Offsets**.

Set the Start Offset to **6"**.
Set the End Offset to **-6"**.

Click **OK**.

Close the dialog.

8.

Diagnostic

High Detail

Low Detail

Medium Detail

Medium Detail Intermediate L

Medium Detail Top Level

Plan Only

Presentation

Reflected

Reflected Screened

Screened

Standard

Detail ▼   3'-6"   +0"

Switch to the Model tab.

On the status bar at the bottom of the display window:

Change the Display Representation to **Low Detail**.

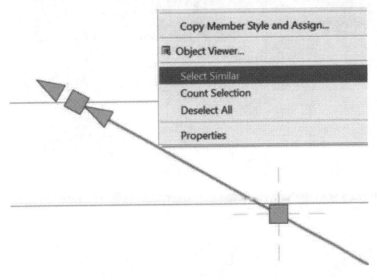

The display updates.

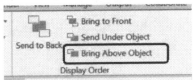

If you have difficulty seeing the beams...

Select one of the beams.

Right click and **Select Similar.**

This will select most of the beams.

9.

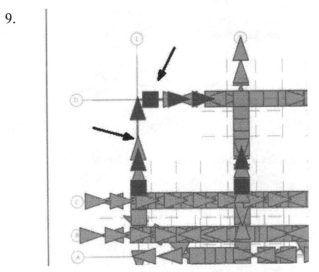

Window around the beams in the upper left corner of the grid.

10.

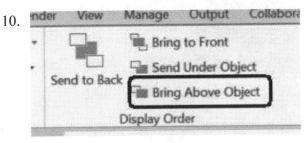

On the ribbon, **select Bring Above Object.**

11.

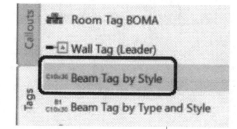

The selected elements' visibility will be turned off momentarily.

Select the Column Grid.

Hit ENTER.

12.

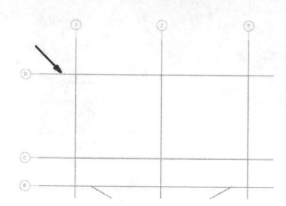

Go to the Annotate ribbon.

Launch the **Annotation Tools**.

13.

Open the Document Tool Palette.

Select the Tags palette.

Select **the Beam Tag by Style** tool.

14.

Select the beam located at D1-D2.
Click **a position above the beam**.
The property set data box will appear.
Click **OK**.
The tag is placed.

15.

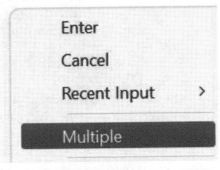

Right click and select **Multiple**.

16

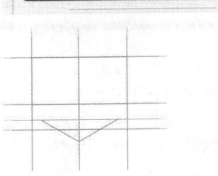

Window around all the beams.

Click **ENTER**.

17.

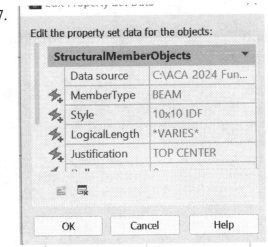

The property set data dialog appears.

Click **OK**.

The beam tags are placed.

You are still in the tag command.
ESC out of the command.

18.

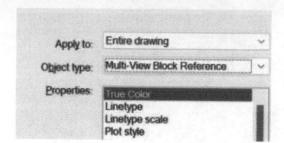

If you can't see the tags, use QSELECT to locate the tags selecting the multi-view block reference.

Type **QSELECT**.
Select the **Multi-View Block Reference** from the object type list.

Click **OK**.

19.

Zoom into one tag.

Release the selection set and select the one tag.

20.

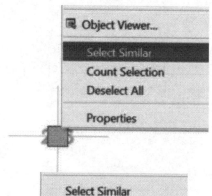

Right click and **Select Similar**.

This will select all the beam tags.

21.

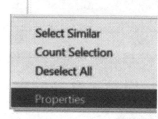

Right click and select **Properties**.

22.

Change the scale to **254**.

23.

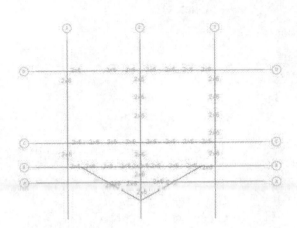

You should see all the beams and tags.

Save the file.

Project Elements

Elements are blocks that can be used anywhere in a project. By adding elements to the Project Navigator, it is faster and easier to locate favorite building elements.

Exercise 2-22:
Adding an Element to a Project

Drawing Name: bolt and washer.dwg
Estimated Time: 5 minutes

This exercise reinforces the following skills:

- ❑ Project Navigator
- ❑ Elements

1.   📂   Open *bolt and washer.dwg*.
     Open the Project Navigator.

2.

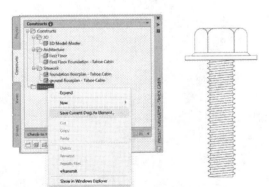

Open the Constructs tab.

Right click on **Elements**.
Select **Save Current Dwg As Element.**

3.

Name	bolt and washer
Description	bolt and washer
Category	Elements
File Name	bolt and washer

Click **OK**.

4.

ground floorpla

Elements

bolt and washer

The drawing is now added as an element.

Close the drawing.
Close the Project Navigator.

Exercise 2-23:
Structural Detail

Drawing Name: column_detail.dwg
Estimated Time: 20 minutes

This exercise reinforces the following skills:

- Detailing
- Tool Palettes
- Reference Keynote
- Cut Line
- View Title
- Edit Attributes

1. Open *column_detail.dwg*.
   Close any current project.

2.

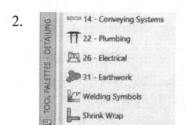

Launch the **Detailing** Tool Palette.

Select the **Basic** tab.

3. Draw a rectangle that is 2' long and 6" high.

4.  Select **31- Earthwork**.

5. Select the rectangle to identify the boundary.
Select inside the rectangle.

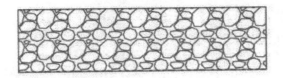

The earthwork hatch is placed.

6.  Select **03- Concrete**.

7. Set the Type to **Circular Columns**.
Set the View to **Elevation**.

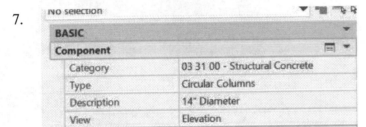

8.  Set the Justification to **Center**.

9. Select the midpoint of the top of the rectangle.

Draw the column so it is **5' 1"** high.

Use the **QUICK MEASURE** tool to verify the dimensions of the column.

10. 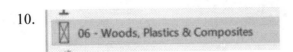 Select **06-Woods, Plastics & Composites**.

11.

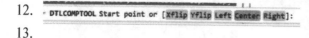

Category	06 11 00 - Wood Framing
Type	Nominal Cut Lumber
Description	2x4
View	Elevation

Select **4x4**.
Set the View to **Elevation**.

12.

DTLCOMPTOOL Start point or [Xflip Yflip Left Center Right]:

Set the Justification to **Center**.

13.

The post should be 2'-6" high and placed so it is 1'-1" embedded into the concrete column.

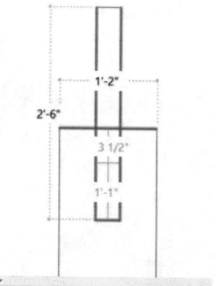

Use the **QUICK MEASURE** tool to verify the dimensions and position of the 4x4 post.

14.

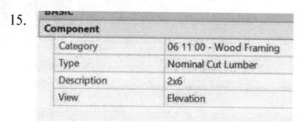

06 - Woods, Plastics & Composites

Select **06-Woods, Plastics & Composites**.

15.

Component	
Category	06 11 00 - Wood Framing
Type	Nominal Cut Lumber
Description	2x6
View	Elevation

Select **2x6**.
Set the View to **Elevation**.

16.

DTLCOMPTOOL Start point or [Xflip Yflip Left Center Right]:

Set the Justification to **Right**.

17.

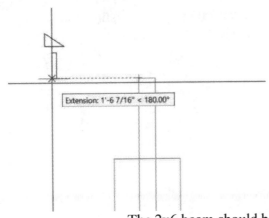

Select a starting point using the upper left corner of the 4x4 post.

18.

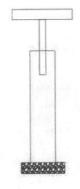

The 2x6 beam should be placed so it is on top of the 4x4 post.

You can center it on the post once it is placed.

19.

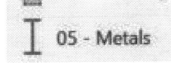

Select **05-Metals**.

20.

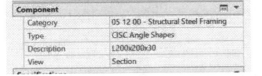

Set the Type to **CISC Angle Shapes**.
Set the Description to **L200x200x30**.
Set the View to **Section**.

21.

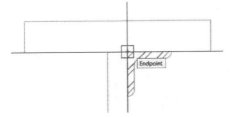

A preview will appear to help you place the angle.

22.

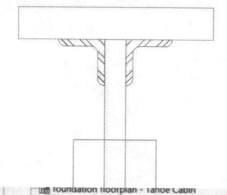

Use the MIRROR tool to mirror the angle to the other side.

23.

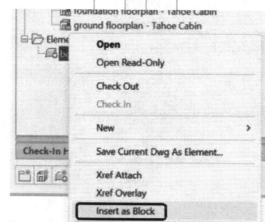

Open the Project Navigator.
Open the Constructs tab.

Right click on **bolt and washer** under **Elements**.
Select **Insert as Block**.

24.

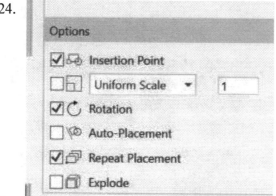

Enable **Insertion Point**.
Disable **Scale**.
Enable **Rotation**.
Enable **Repeat Placement**.

25.

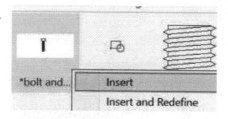

Highlight the drawing.
Right click and select **Insert**.

26.

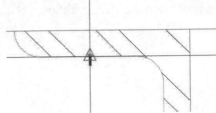

Select the midpoint of the left angle.

27.

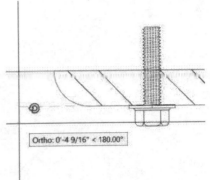

Ortho: 0'-4 9/16" < 180.00°

Rotate the block so the bolt is placed correctly.

28.

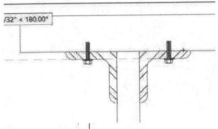

/32" < 180.00°

Place a second block.

29.

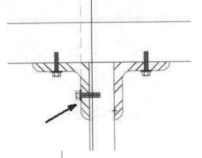

Place a third block.

30.

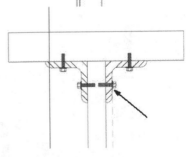

Place a fourth block.

Click ENTER to finish placing the bolts.

Close the Blocks palette.

31. 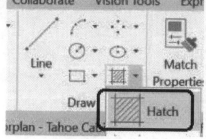 Select the **Hatch** tool from the Home ribbon.

32.  Select the **AR-CONC** hatch pattern.

33.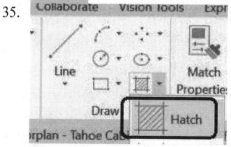

Pick inside the concrete column to place the hatch.

34. 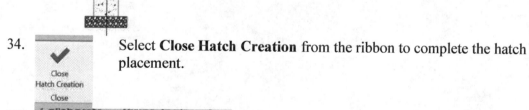 Select **Close Hatch Creation** from the ribbon to complete the hatch placement.

35. Select the **Hatch** tool from the Home ribbon.

36.  Select the **WOOD_2** hatch pattern.

37. 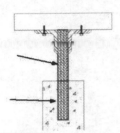 Pick inside the 4x4 post to place the hatch.

38. Select **Close Hatch Creation** from the ribbon to complete the hatch placement.

39.  Select the **Hatch** tool from the Home ribbon.

40. 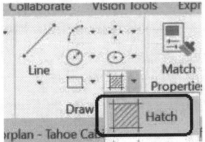 Select the **ZIGZAG30** hatch pattern.

41. 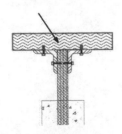 Pick inside the 2x6 beam to place the hatch.

42. Select **Close Hatch Creation** from the ribbon to complete the hatch placement.

43. Go to the Home ribbon.

Launch the **DesignTools** palette.

44.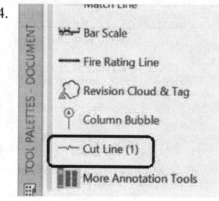

Switch to the Document Palette.
Open the Annotation tab.

Select the **Cut Line** tool.

45.

To place the cut line, specify a starting point, and an end point. Then, specify the direction of the break/cut.

Select a point above the gravel fill.
Then a point below the gravel fill.
Then a point away from the gravel fill.

Ortho: 0'-3 11/32" < 180.00°

46.

The break line is placed.

You can use the grips to adjust the size of the zig-zag.

Place cut lines as shown.

47.   Reference Keynote (Straight)

Switch to the Annotate ribbon.
Select the Reference Keynote (Straight).

48.

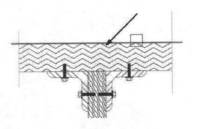

Select the 2x6 Beam.

49.

06 11 00.F1 – 2X6

Select the top side of the beam.
Select a point for the landing.
Click ENTER to accept the default text width.

50.

06 11 00.L2 – 4X4 POST

03 31 00.A1 – 14" DIAMETER CAST-IN-PLACE CIRCULAR COLUMN

05 12 23.A1 – L200X200X30

Add a keynote for the 4x4 post.
Add a keynote for the concrete column and the gravel fill.
Add a keynote for the angle.

Hint: If you set the FIELDDISPLAY to 0, the text will not be shaded.

51.

05 05 23.A8 - 1-3/8" A307 Bolt
05 05 23.A9 - 1-1/2" A307 Bolt
05 05 23.B1 - 1/2" A325 Bolt
05 05 23.B2 - 5/8" A325 Bolt

To add a keynote for the bolt, press ENTER and then select the 05 05 23.B1 keynote to be placed.

52.

The view should look similar to this.

53.

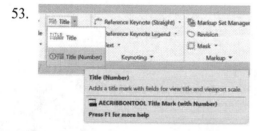

Title (Number)
Adds a title mark with fields for view title and viewport scale.

AECRIBBONTOOL Title Mark (with Number)
Press F1 for more help

Select the **Title (Number)** tool from the Annotate ribbon.

54.

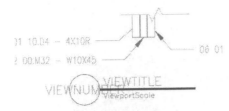

Place below the detail view.

The view title is placed.

55.

Select the View Number.
Right click and select **Edit Attributes**.

56.

Tag	Prompt	Value
NUMBER	Number	VIEWNUMBER

Value: 5

Change the View Number to **5**.

Click **OK**.

57.

Select the View Title.
Right click and select **Edit Attributes**.

58.

Attribute Text Options Properties

Tag	Prompt	Value
SCALE	Scale	1" = 1'-0"
TITLE	Title	VIEWTITLE

Value: CONCRETE COLUMN w/2x6 NAILERS

Change the Scale to **1/8" = 1'-0"**.

Change the Title to **W10X45 STEEL COLUMN w/4X10 NAILERS**.

Click **OK**.

59.

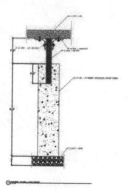

Add linear dimensions.
Save as *ex2-22.dwg*.

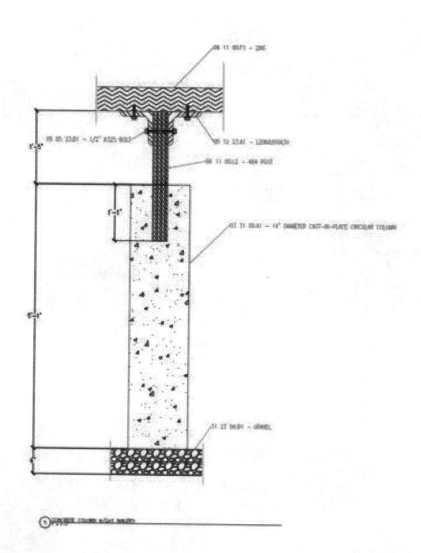

QUIZ 2

True or False

1. You must load a custom linetype before you can assign it to a layer.
2. The DIVIDE command divides a line or arc segment into smaller elements.
3. The Design Tools palette is used to access blocks and commands.
4. If you design a wall tool on the tool palette for a specific style wall, then whenever you add a wall using that design tool it will place a wall using that wall style.
5. Layer groups control which layers are displayed in the layer drop-down list.

Multiple Choice
Select the best answer.

6. If points are not visible in a view:

   A. Turn on the layer the points are placed on
   B. Bring up the Point Style dialog and set the Size Relative to Screen
   C. Change the Point Style used
   D. All of the above

7. The command to bring up the point style dialog is:

   A. PSTYLE
   B. PMODE
   C. PP
   D. STYLE

8. An easy way to change the units used in a drawing is to:

   A. Change the units on the status bar.
   B. Type UNITS
   C. Use the Drawing Setup dialog.
   D. B or C

9. To turn off the display of the USCICON:

   A. Type USCICON, OFF
   B. Type UCS, OFF
   C. Type UCSICON, 0
   D. Type UCS, 0

10. To add a Construct to a project, use the:

   A. Styles Manager
   B. Design Tools
   C. Project Browser
   D. Project Navigator

11. Slabs can be used to model: (select all that apply)
   A. Roofs
   B. Floors
   C. Ceilings
   D. Walls

12. A structural member is an object that can represent: (select all that apply)

   A. Beams
   B. Braces
   C. Columns
   D. Railings

13. To create styles for most *standard* beams, braces, and columns, you can use:

   A. Styles Manager
   B. Design Tools
   C. Structural Members Catalog
   D. Project Navigator

ANSWERS:
   1) T; 2) F; 3) T; 4) T; 5) T; 6) D; 7) A; 8) D; 9) A; 10) D; 11) A, B, and C;
   12) A, B, and C; 13) D

Lesson 3:
Floor Plans

AutoCAD Architecture comes with 3D content that you use to create your building model and to annotate your views. In ACA 2024, you may have difficulty locating and loading the various content, so this exercise is to help you set up ACA so you can move forward with your design.

The Content Browser lets you store, share, and exchange AutoCAD Architecture content, tools, and tool palettes. The Content Browser runs independently of the software, allowing you to exchange tools and tool palettes with other Autodesk applications.

The Content Browser is a library of tool catalogs containing tools, tool palettes, and tool packages. You can publish catalogs so that multiple users have access to standard tools for projects.

ACA comes with several tool catalogs. When you install ACA, you enable which catalogs you want installed with the software. By default, Imperial, Metric, and Global are enabled. The content is located in the path: C:\ProgramData\Autodesk\ACA 2024\enu\Tool Catalogs.

The floor plan is central to any architectural drawing.

A floor plan is a scaled diagram of a room or building viewed from above. The floor plan may depict an entire building, one floor of a building, or a single room. It may also include measurements, furniture, appliances, or anything else necessary to the purpose of the plan.

Floor plans are useful to help design furniture layout, wiring systems, and much more. They're also a valuable tool for real estate agents and leasing companies in helping sell or rent out a space.

If you are using Projects to organize your model, it is recommended that exterior walls are placed on Shell Constructs and interior walls are placed on Interior Constructs.

Exercise 3-1:

Creating a Floor Plan

Drawing Name:	01 Core.dwg
Estimated Time:	45 minutes

This exercise reinforces the following skills:

- ❑ Create Walls
- ❑ Wall Properties
- ❑ Wall Styles
- ❑ Style Manager
- ❑ Insert a PDF
- ❑ Insert Doors
- ❑ Insert Windows
- ❑ Materials
- ❑ Content Browser

1.  Launch the US Metric version of ACA.

   This ensures that all the content will be metric.

2. Open the Project Browser.

3. 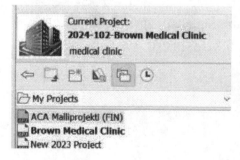 Set the **Brown Medical Clinic** Current.

   The Brown Medical Clinic project was created using a metric project template. Placeholders were automatically generated.

4.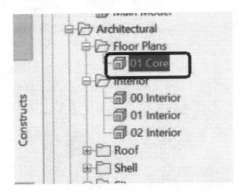

Launch the Project Navigator.

Switch to the Constructs tab.

Highlight **01 Core**.

Right click and select **Open**.

5.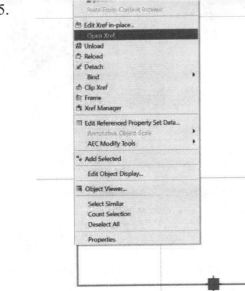

There is a rectangle in the drawing acting as a placeholder.

Select the rectangle.

Right click and select **Open Xref**.

6.

The external reference is an element called Typical Toilet Room.

Close the drawing.

7.

Switch to the Insert ribbon.

Select **PDF Import**.

8.

Locate the *medical clinic – first floor.pdf*.

Click **Open**.

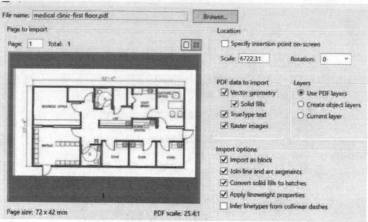

9.   Disable **Specify insertion point**.
      This will insert the PDF at 0,0,0
      Set the Scale to **6722.31**.
      This scales the PDF to the correct metric measurements.
      Enable **Raster Images**.
      Enable **Import as block**.
      Click **OK**.

10.

   Set the View Scale to **1:1**.

11.   Launch the **Content Browser** from the Home
         ribbon.

12.  Locate the **Design Tool Catalog -Metric**.

13.  Scroll down and select **Walls**.

14.  Select **Brick**.

15.  Locate the Brick-090 Rigid-038 Air-050 Brick-090.

Right click and select **Insert into Drawing**.

16.  Go to the Render ribbon.

Select **Materials Browser**.

17.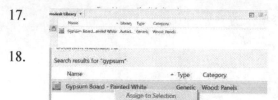

Locate the **Gypsum Board**.
Import into the current drawing.

18.

Highlight the Gypsum Board
material that is in the document.
Right click and select **Duplicate**.

19.

Rename the copied Gypsum Board
material:
**Gypsum Board - Painted Sage
Green.**

Double click on the material to edit.

20.

Change the Color to **RGB
171,171,155**.

*I use the ColorPicker website to
help you determine the RGB values
of paint colors.*

Right click on the image and select
Edit Image.

21.

Click on the Source file name.
Select the image file located in the
Lesson 3 download files named *SW
6178 Clary Sage*.
Close the dialog.

22.

The appearance has been edited.

Close the Materials Editor.

Close the Materials Browser.

23.

Go to the Manage ribbon.

Launch the **Style Manager**.

24.

Browse to **Wall Styles** under Architectural Objects.
You will see the wall style you just imported.

25.

Create a copy of the Brick-090 wall style.
Paste under Wall Styles.

26.

Rename the copied style:
Brick-090 Rigid-038 Air-050 Gypsum Wall Board.

27.

| General | Components | Materials | Endcaps / Opening Endcaps | Classifica |

Name:

Brick-090 Rigid-038 Air-050 Gypsum Wall Board

Description:

Gypsum Wall Board and 90mm Brick Wall with 38mm Rigid

Change the Description.

28.

Index	Name	Priority	Width	Edge Offset
1	Gypsum Wall Board	810	3.00	-5.00
2	Air Gap (Brick Separati...	805	50.00	38.00
3	Rigid Insulation	804	38.00	0.00
4	Brick Veneer (Structural)	800	90.00	38.00

Select the Components tab.

Set up the layers as shown.

The Brick Veneer should be on the exterior.

29.

Select the Materials tab.

Edit the Gypsum Wall Board.

30.

Select the Style Overrides.

31.

Set the Plan Linework color to **Green** on the Line/Color/Linetype tab.

32.

Set the Hatching to **PLAST** on the Hatching tab.

33.

Set the Render Material to the **Gypsum Wall Board Sage Green** on the Other tab.

Close the Display Properties dialog. Close the Styles Manager.

34.

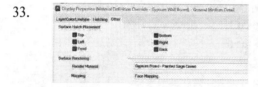

Select the **Wall** tool from the Home ribbon.

35.

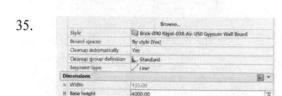

Set the Style to **Brick-090 Gypsum Wall Board**.
Set the Base Height to **4000**.

36.

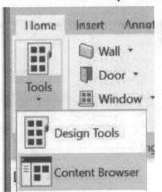

Draw the four walls that are the outline of the building.

37.

Launch the **Content Browser** from the Home ribbon.

38.

Locate the **Design Tool Catalog -Metric**.

39.

Locate the **Doors** category.

40.

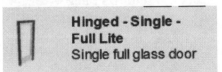

Locate the **Hinged – Single – Full Lite** door.

41.

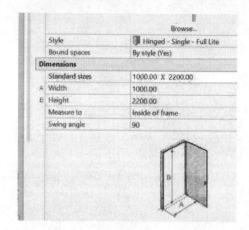

Right click and select **Insert into Drawing**.

42.

Set the Standard size to 1000 x 2200.

43.

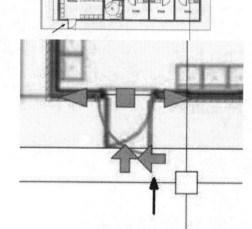

Place in the locations shown.

44.

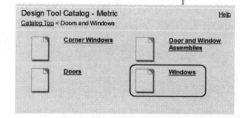

If required, select the placed door. Use the arrows to flip the door so it is oriented correctly.

45.

Return to the Design Tool Catalog – Metric.

Select the **Windows** category.

46.

Locate **the Casement – Single Casement window**.

47.

Right click and select **Insert into Drawing**.

48.

Set the size to **900.00 x 2000.00**.

49.

Place a window.
Right click and select Multiple.
Set the number of windows to 6.
Place the additional windows.

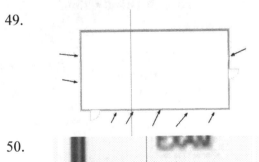

50.

You can use the arrows to orient the windows correctly.
The arrows should be located on the exterior side of the windows.
You can use the middle grip to position the window.
You can use the end grips to resize the window.

51.

Place windows around the building.

52.

Switch to a 3D view.

Save the *01 Core.dwg* and close.

Exercise 3-2:

Adding Interior Walls

Drawing Name: 01 Core.dwg
Estimated Time: 20 minutes

This exercise reinforces the following skills:

- ❑ Create Walls
- ❑ Insert Doors
- ❑ Content Browser

1.

Open the Project Browser.

2.

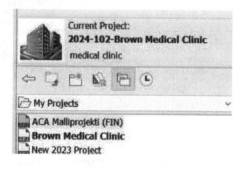

Set the **Brown Medical Clinic** Current.

The Brown Medical Clinic project was created using a metric project template. Placeholders were automatically generated.

3.

Switch to the Constructs tab.

Highlight **01 Core**.

Right click and select **Open**.

4.

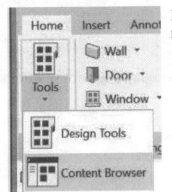

Launch the **Content Browser** from the Home ribbon.

5.

Locate the **Design Tool Catalog -Metric**.

6.

Scroll down and select **Walls**.

7. 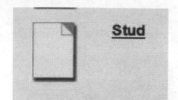   Select the **Stud** category.

8. 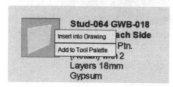   Locate the **Stud-064 GWB-018 2 Layers Each Side**.
Right click and select **Insert into Drawing**.

9. 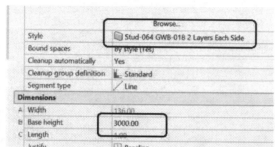   Verify that the Style is set correctly.
Set the Base Height to **3000**.

10. 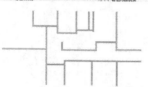   Trace over the PDF to place just the interior walls.

11.    You can freeze and thaw the PDF_Images to review your work.

Lock the PDF_Images layer to ensure that the PDF is not moved or accidentally deleted.

12. 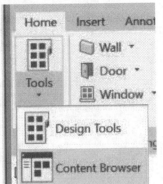 Launch the **Content Browser** from the Home ribbon.

13.  Locate the **Design Tool Catalog -Metric**.

14. 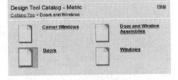 Locate the **Doors** category.

15. 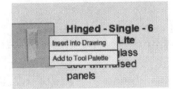 Locate the **Hinged – Single – - 6 Panel - Half Lite** door.
Right click and select **Insert into Drawing**.

16. 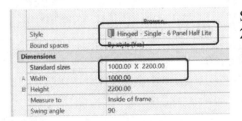 Set the Standard size to **1000 x 2200**.

17.

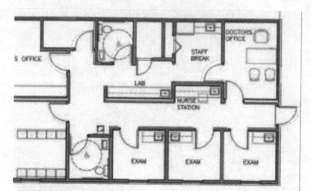

Use the grips and flip arrows to adjust the orientation, position, and size of the doors to match the PDF layout.

18.

Locate the **Design Tool Catalog -Metric**.

19.

Locate the **Doors** category.

20.

Locate the **Bifold – Single Louver** door.
Right click and select **Insert into Drawing**.

21.

Set the Standard size to **700 x 2200**.

Place in the Staff Break room.

22.  Locate the **Design Tool Catalog -Metric**.

23. 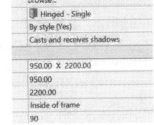 Locate the **Doors** category.

24.  Locate the **Hinged - Single** door. Right click and select **Insert into Drawing**.

25. 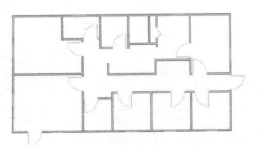 Set the Standard size to **950 x 2200**.

Place in the two bathrooms.

26. Save the file.

➤ If you draw a wall and the materials composing the wall are on the wrong side, you can reverse the direction of the wall. Simply select the wall, right click and select the Reverse option from the menu.

➤ To add a wall style to a drawing, you can import it or simply create the wall using the Design Tools.

➤ Many architects use external drawing references to organize their projects. That way, teams of architects can concentrate just on their portions of a building. External references also use fewer system resources.

➤ You can convert lines, arcs, circles, or polylines to walls. If you have created a floor plan in AutoCAD and want to convert it to 3D, open the floor plan drawing inside of AutoCAD Architecture. Use the Convert to Walls tool to transform your floor plan into walls.

➤ To create a freestanding door, click the ENTER key when prompted to pick a wall. You can then use the grips on the door entity to move and place the door wherever you like.

➤ To move a door along a wall, use Door→Reposition→Along Wall. Use the OSNAP From option to locate a door a specific distance from an adjoining wall.

Exercise 3-3:
Place a Grid

Drawing Name: grid.dwg
Estimated Time: 10 minutes

This exercise reinforces the following skills:

- ❑ Place a grid
- ❑ Create a new Construct

1.    Launch the US Metric version of ACA.

            This ensures that all the content will be metric.

2.   Launch the Project Navigator.
On the Constructs tab:
Highlight **Constructs**.
Right click and select
New→Category.

3.   Name the new category **General**.

4.   Highlight **General**.
Right click and select
New→Construct.

5.

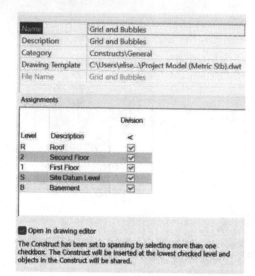

Change the Name to **Grid and Bubbles**.
In Description, type **Grid and Bubbles**.
Enable all the **Levels**.
Enable Open in drawing editor.

Click **OK**.

6.

Set the Scale to **1:1**.

7.

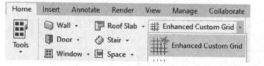

Click on **Enhanced Custom Grid** on the Home ribbon.

8.

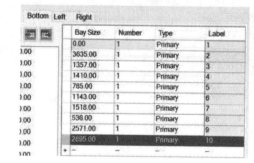

Select the **Bottom** tab.
Type in the following values:

3635
1357
1410
786
1143
1518
536
2571
2895

9.
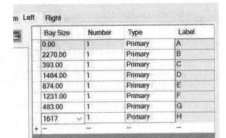

Switch to the Left tab.
Type in the following values:

2270
393
1463
874
1231
483
1617

10.
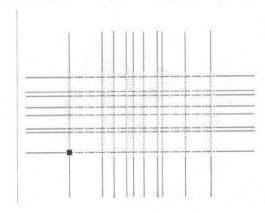

There is a preview window which shows the grids.
Click **OK** to place the grid.

11.
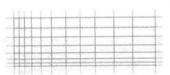

Type **0,0** as the insertion point.
Click **ENTER** to accept the rotation angle of 0.
The grid is placed.

12.
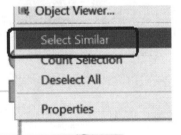

Select one of the column bubbles.
Right click and **Select Similar**.

13.

Bound spaces	By style (No)
Scale	
X	125.00
Y	125.00
Z	125.00
Location on column grid	

Change the Scale to **125** in the Properties palette.

Click ESC to release the selection.

14.

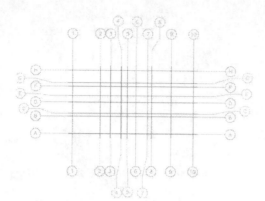

The grid updates.

You can select the overlapping column bubbles and move them to a new position.
This will create a copy, so you need to delete the original column bubble.

Save the file.

Exercise 3-4:
Combining Constructs

Drawing Name: Main Model,dwg
Estimated Time: 15 minutes

This exercise reinforces the following skills:

- ❑ Add Constructs to a 3D model
- ❑ Create a new Construct
- ❑ Visual Styles

1.

Launch the US Metric version of ACA.

This ensures that all the content will be metric.

2.

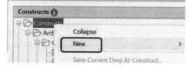

Launch the Project Navigator.
On the Constructs tab:
Highlight **Constructs**.
Right click and select **New→Category**.

3.

Name the new category **3D**.

4.

Highlight **3D**.
Right click and select
New→Construct.

5.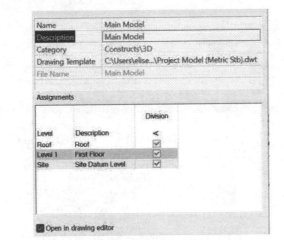

Change the Name to **Main Model**.
In Description, type **Main Model**.
Enable all the **Levels**.
Enable Open in drawing editor.

Click **OK**.

6.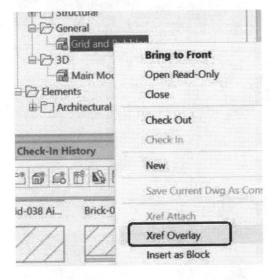

Highlight the **Grid and Bubbles** construct.
Right click and select **Xref Overlay**.

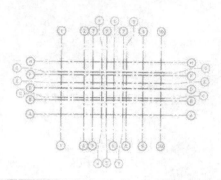

The Grid and Bubbles drawing is placed in the Main Model drawing.

7.

Highlight the **01 Core** construct.

Right click and select **Xref Overlay**.

8.

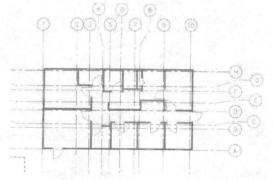

If necessary, reposition the 01 Core construct so it is aligned with the grid layout.

9.

[−][SW Isometric][2D Wireframe]

Switch to an isometric view.

10.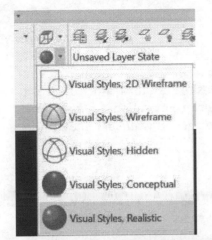

Set the Visual Style to Realistic.

Orbit around the model.

Check for any other errors and correct as needed.

11.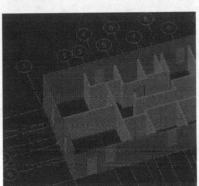

Save the drawing.

Curtain Walls

Curtain walls provide a grid or framework for inserting objects such as windows and doors. Curtain walls have many similarities to standard walls, such as baseline, roof line, and floor line, and they allow for interferences. You can insert doors, windows, and door/window assemblies into a curtain wall, just like standard walls, but the insertion process is different.

Curtain Wall Grids

Curtain walls are made up of one or more grids. Each grid in a curtain wall has either a horizontal division or a vertical division, but you can nest the grids to create a variety of patterns from simple to complex.

Elements of Grids

Grids are the foundation of curtain walls, curtain wall units, and door/window assemblies. Every grid has four element types:

- **Divisions:** Define the direction of the grid (horizontal or vertical) and the number of cells

- **Cell Infills:** Contain another grid, a panel infill, or an object such as a window or a door

- **Frames:** Define the edge around the outside of the primary grid and nested grids

- **Mullions:** Define the edges between the cells

Note: Division is an abstract element, in contrast to the other three element types that represent physical elements of the curtain wall.

Each element type is assigned a default definition that describes what elements of that type look like.

Element type	Default definitions
Divisions	Primary horizontal grid with a fixed cell dimension of 13' and secondary vertical grid with a fixed cell dimension of 3'
Cell Infills	Cells containing simple panels 2" thick
Frames	Left, right, top, and bottom outer edges of grid 3" wide and 3" deep
Mullions	Edges between cells 1" wide and 3" deep

Exercise 3-5:
Place a Curtain Wall

Drawing Name: 3D Model-Master_2
Estimated Time: 15 minutes

This exercise reinforces the following skills:

- ❑ Place a curtain wall
- ❑ External References
- ❑ Steering Wheel
- ❑ Views
- ❑ UCS

1.  Launch the ACA English (US Imperial) software.

2.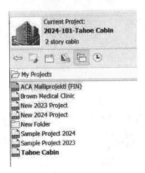

    Set the **Tahoe Cabin** project current.

3.

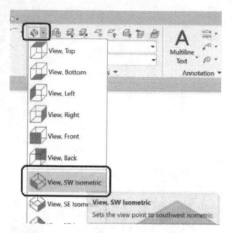

Open 3D Model-Master_2.dwg.

On the Home ribbon:
Switch to a **SW Isometric view**.

4.

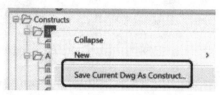

On the Project Navigator:
Open the **Constructs** tab.
Highlight **3D**.
Right click and select **Save Current Dwg As Construct**.

5.

Name	3D Model-Master_2
Description	3D Model-Master_2
Category	Constructs\3D
File Name	3D Model-Master_2

Assignments

Level	Description	Division 1
Roof	Roof	☑
Second Floor	Second Floor	☑
First Floor	First Floor	☑
Ground Level	Ground Level	☑
Foundation	foundation	☑

Enable all the levels.

Click **OK**.

6.

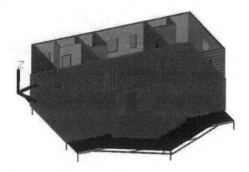

Select the lower exterior walls.

Edit Reference In-Place

On the ribbon, select **Edit Reference In-Place**.

7.

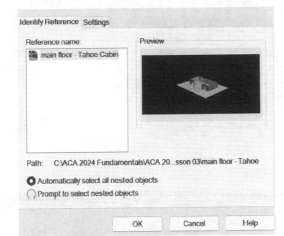

Click **OK**.

8.

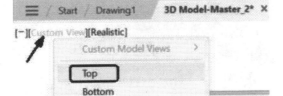

Switch to a **Top** view.

9.

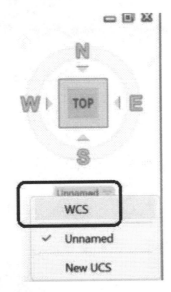

Under the Viewcube, select **WCS** to use the World UCS.

You will see the UCSICON adjust.

10.

Select the **Curtain Wall** tool on the Home ribbon.

11.

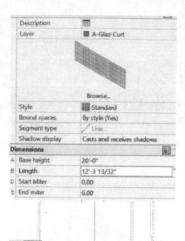

Set the Base Height to **20'-0"**.

12.

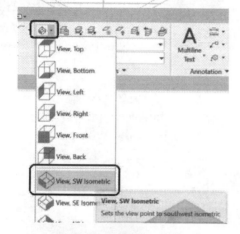

Draw two angled lines to place the curtain wall.

13.

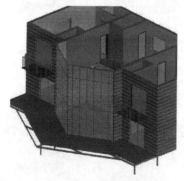

On the Home ribbon:
Switch to a **SW Isometric view**.

14.

Press the SHIFT key and the middle mouse button to orbit around the model and inspect the curtain wall.

15.

Switch the View ribbon.

Launch the **Full Navigation** Steering Wheel.

16. Use the different tools on the steering wheel to view the model.

When you are done, click the X located in the upper right corner of the steering wheel to close.

17.

Select **Save Changes** on the ribbon.

18.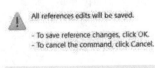

Click **OK**.

Exercise 3-6:
Add an Opening to a Wall

Drawing Name: 01 Core.dwg
Estimated Time: 10 minutes

This exercise reinforces the following skills:

❑   Add an opening to a wall

1.  Launch the US Metric version of ACA.

   This ensures that all the content will be metric.

2. Open the Project Browser.

3. 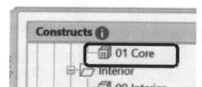 Set the **Brown Medical Clinic** Current.

   The Brown Medical Clinic project was created using a metric project template. Placeholders were automatically generated.

4. 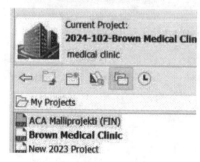 Launch the Project Navigator.

   Switch to the Constructs tab.

   Highlight **01 Core**.

   Right click and select **Open**.

5.

Switch to an Isometric view.

Select the wall located between the business office and the waiting area.

6.

Select **Opening** on the ribbon.

7.

Place the opening so it aligns with the opening shown in the PDF.

8. Click on the bottom arrow grip to raise the height of the opening.

9.

The opening should be adjusted to look like a window between the rooms.

10. Save and close the drawing.

Notes:

QUIZ 3

True or False

1. When you insert a PDF into a drawing, it cannot be converted to AutoCAD elements, like lines or text.

2. When you insert an image, it cannot be converted to AutoCAD elements, like lines or text.

3. The direction you place walls – clockwise or counter-clockwise – determines which side of the wall is oriented as exterior.

4. Curtain walls can only be linear, not arcs, in AutoCAD Architecture.

5. Grids can only be lines, not arcs, in AutoCAD Architecture.

6. To re-orient a door, use the Rotate command.

7. You can set cut planes for individual objects (such as windows), an object style (such as walls) or as the system default.

Multiple Choice

8. To change the hatch display of wall components in a wall style:

    A.   Modify the wall style
    B.   Change the visual style
    C.   Change the display style
    D.   Switch to a plan\top view

9. This tool on the status bar:

    A.   Sets the elevation of the active level
    B.   Controls the default cut plane of the view and the display range
    C.   Sets the distance between levels
    D.   Sets the elevation of the active plane

10. To assign a material to a door:

    A.   Modify the door component by editing the door style
    B.   Drag and drop the material from the Material Browser onto a door
    C.   Use the Display Manager
    D.   Define a new Visual Style

11. You want to create a wall that shows a specific paint color. Put the steps in the correct order.

    A. Place a wall.
    B. Open the Styles Manager.
    C. Import the desired material into the current drawing.
    D. Use the Materials Browser to locate a similar material.
    E. Duplicate the Material and redefine it with the correct color specification.
    F. Create a new wall style.
    G. Assign the desired material to a wall component.
    H. Set the Render material for the wall component to the desired material definition.

12. You want to place a door style that is not available on the Design Tools palette. You open the Design Tool Catalog and search for the desired door style. You locate the desired door style. Now what?

    A. Select the Door tool on the Home ribbon and use the Properties palette to select the desired door style.
    B. Add the Door Style from the Design Tool Catalog to the Design Tools palette.
    C. Right click on the desired door style in the catalog and select Insert into Drawing.
    D. Right click on the desired door style in the catalog and select Add to Styles Manager.

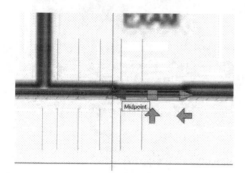

13. You select on a window that was placed in the model. The arrows can be used to:

    A. Change the window location/position.
    B. Change the window orientation.
    C. Change the window size.
    D. Change the window opening.

ANSWERS:

1) F; 2) T; 3) T; 4) F; 5) F; 6) F; 7) T; 8) A; 9) B; 10) A; 11) D, E, C, B, F, G, H, A; 12) C; 13) B

Lesson 4:
Space Planning

A residential structure is divided into three basic areas:

- ❑ Bedrooms: Used for sleeping and privacy
- ❑ Common Areas: Used for gathering and entertainment, such as family rooms and living rooms, and dining area
- ❑ Service Areas: Used to perform functions, such as the kitchen, laundry room, garage, and storage areas

When drawing your floor plan, you need to verify that enough space is provided to allow placement of items, such as beds, tables, entertainment equipment, cars, stoves, bathtubs, lavatories, etc.

AutoCAD Architecture comes with Design Content to allow designers to place furniture to test their space.

Exercise 4-1:
Creating AEC Content

Drawing Name: New
Estimated Time: 30 minutes

This lesson reinforces the following skills:

- ❑ Design Center
- ❑ AEC Content
- ❑ Content Browser
- ❑ Tool Palettes
- ❑ Customization

1.    Launch the ACA English (US Imperial) software.

2.

Launch the **Design Tools** from the Home ribbon.

3.

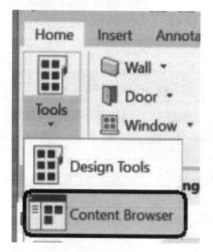

Launch the **Content Browser** from the Home ribbon.

4.

Right click on the Design Palette. Select **New Palette**.

5.

Rename the **Palette Furnishings and Fixtures**.

6.

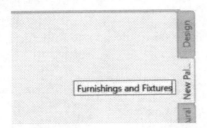

Locate the **Design Tool -Catalog – Imperial**.

We will be adding elements from the catalogs to the palettes. Feel free to select any elements you think might be useful to your work.

7.

Design Tool Catalog - Imperial
Catalog Top < Equipment < Residential

Browse to the Residential folder.

8.

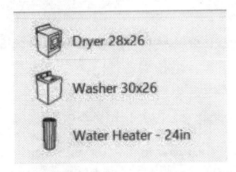

Add a Dryer, Washer, and Water Heater to the palette.

9.

Design Tool Catalog - Imperial
Catalog Top < Equipment < Office

Browse to the Office folder.

10.

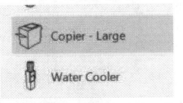

Add a Copier and Water Cooler to the palette.

11.

Design Tool Catalog - Imperial
Catalog Top < Equipment < Food Service < Refrigerator

Browse to the Refrigerator folder.

12.

Add a Refrigerator to the palette.

13. Catalog Top < General < People   Browse to the People folder.

14. Person   Add the Person to the palette.

15. Design Tool Catalog - Imperial
     Catalog Top < Mechanical < Plumbing Fixtures < Lavatory   Browse to the Lavatory folder.

16. Vanity
     Plumbing: Lavatory:
     Vanity w/ Lav [15400]   Add the Vanity to the palette.

17. Design Tool Catalog - Imperial
     Catalog Top < Mechanical < Plumbing Fixtures < Toilet   Browse to the Toilet folder.

18. Tank 2
     Plumbing: Toilet: Tank
     flush, floor mount WC
     [15400]   Add Tank 2 to the palette.

19. Design Tool Catalog - Imperial
     Catalog Top < Mechanical < Plumbing Fixtures < Shower   Browse to the Shower folder.

20. Tank 2
     HC Shower - Left
     HC Shower - Right   Add the Left and Right HC Shower elements to the palette.

21. Design Tool Catalog - Imperial
     Catalog Top < Mechanical < Plumbing Fixtures < Bath   Browse to the Bath folder.

22.		Add a couple of tubs to the palette.

HC Shower - Right

Tub 30x48

Tub 30x54

23.	**Design Tool Catalog - Imperial**	Browse to the Table folder.

Catalog Top < Furnishing < Furniture < Table

24.		Add a Night table, a Coffee Table, and a Conference table to the palette.

Night 18x18

Coffee Table

Conf 12ft - 12 Seat

25.	**Design Tool Catalog - Imperial**	Browse to the Bed folder.

Catalog Top < Furnishing < Furniture < Bed

26.		Add a King, Hospial and Queen Bed to the palette.

King

Hospital

Queen

27.	**Design Tool Catalog - Imperial**	Browse to the Desk folder.

Catalog Top < Furnishing < Furniture < Desk

28.		Add a desk to the palette.

42x18 Left

29.	**Design Tool Catalog - Imperial**	Browse to the Chair folder.

Catalog Top < Furnishing < Furniture < Chair

30.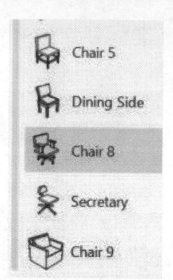

Select several chairs to add to the palette.

31.

Design Tool Catalog - Imperial
Catalog Top < Furnishing < Furniture < Sofa

Browse to the Sofa folder.

32.

Long

Modular 1 End

Modular 1 Mid

Modular 2 Mid

Modular 2 End

Add the modular sofa elements to the palette.

These will give you the most flexibility in placing a sofa.

33.

Design Tool Catalog - Imperial
Catalog Top < Equipment < Unit Kitchens

Browse to the Unit Kitchens folder.

34.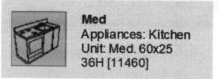

Med
Appliances: Kitchen
Unit: Med. 60x25
36H [11460]

Add the Medium Kitchen to the palette.

35.

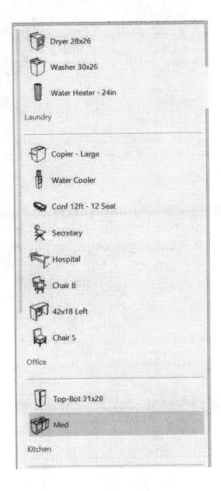

Add separators and text to organize the palette.

This will make it easier to locate the desired elements.

36.

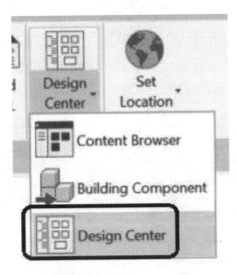

Go to the Insert ribbon.

Launch the Design Center.

37.

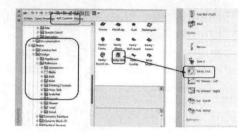

You can drag and drop content from the Design Center onto the tool palette.

Locate the Vanity Unit in the Basin folder under Metric/Design/Bathroom and place on the tool palette.

38.

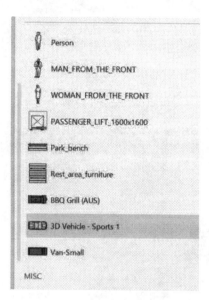

Browse through the Design Center and add any desired elements to the tool palette.

Exercise 4-2:

Inserting AEC Content

Drawing Name: second floor – interior – Tahoe Cabin.dwg
Estimated Time: 15 minutes

This lesson reinforces the following skills:

- ❑ Managing External Reference Files
- ❑ Inserting blocks

1.  Launch the ACA English (US Imperial) software.

2.  Set the **Tahoe Cabin** project current.

3. 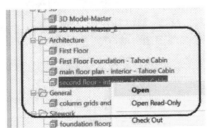 Launch the Project Navigator.

   Open the **second floor – interior – Tahoe Cabin** construct drawing.

4.

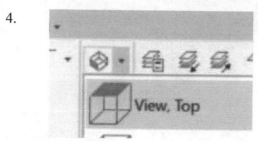

Switch to a **Top View**.

5.

Switch to a **2D Wireframe** display.

6.

Verify that the UCS is set to **WCS**.

7.

The imported PDF shows the rough location of the bedroom furniture and the bathroom fixtures.

If you do not see the PDF, thaw the **PDF_Images** layer.

8.

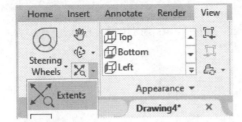

Zoom Extents.

You can do this by typing Z,E on the command line or using the tool on the View tab on the ribbon or double clicking the mouse wheel.

9.

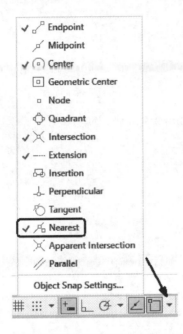

Enable the **Nearest** OSNAP to make it easier to place furniture.

10.

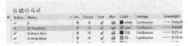

Create a new Layer called **Furniture**. Set the Layer **Current**.

11.

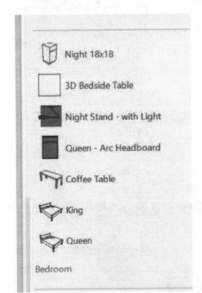

Use the elements placed on the Furnishing & Fixtures tool palette to populate the floor plan.

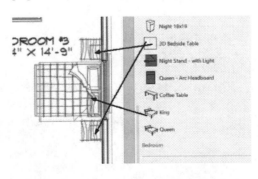

Simply drag from the palette to the desired location.

12.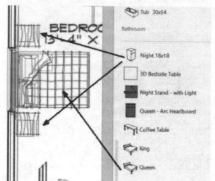

Place elements in Bedroom #2.

13.

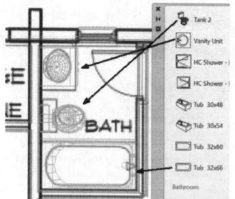

Place elements in the bathroom.

14.

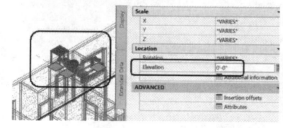

Switch to a 3D view and use 3D Orbit to view the furniture placement.

The bathroom fixtures were placed at the wrong elevation.

Select the blocks.
Right click and select **Properties**.
Change the Elevation to **0' 0"**.

15.

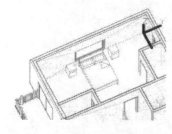

Verify the bedroom furniture looks OK.

Modify the position/elevation as needed.

16. 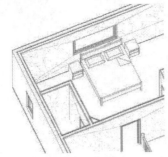 Save the *second floor – interior – Tahoe Cabin.dwg*

The Space Planning process is not just to ensure that the rooms can hold the necessary equipment but also requires the drafter to think about plumbing, wiring, and HVAC requirements based on where and how items are placed.

As an additional exercise, place towel bars, soap dishes and other items in the bathrooms.

Exercise 4-3:

Inserting Blocks

Drawing Name: main floor – interior – Tahoe Cabin.dwg
Estimated Time: 20 minutes

This lesson reinforces the following skills:

- ❑ Managing External Reference Files
- ❑ Inserting blocks

1.  Launch the ACA English (US Imperial) software.

2.

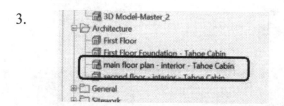

Set the **Tahoe Cabin** project current.

3.

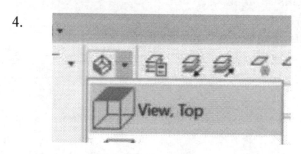

Launch the Project Navigator.

Open the **main floor – interior – Tahoe Cabin** construct drawing.

4.

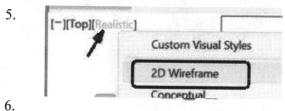

Switch to a **Top View**.

5.

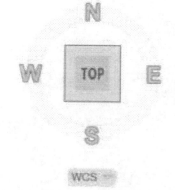

Switch to a **2D Wireframe** display.

6.

Verify that the UCS is set to **WCS**.

7.

The imported PDF shows the rough location of the bedroom furniture and the bathroom fixtures.

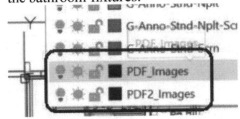

If you do not see the PDF, thaw the **PDF_Images** layer.

8.

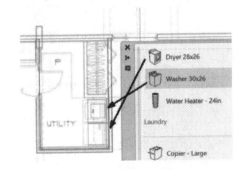

Create a new layer called **spaced planning**.

Set current.

9.

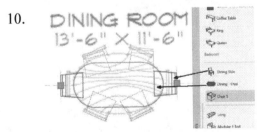

Place the washer and dryer in the utility room.

10.

Place a dining table and chairs in the dining room.

11.

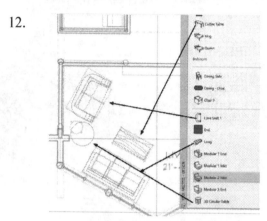

Add a queen bed and night stand in the Master Bedroom.

12.

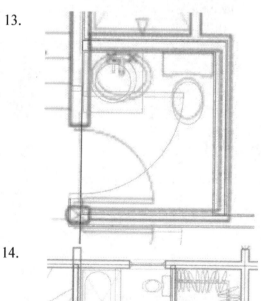

Add a loveseat, sofa, coffee table and end table to the living room.

13.

Add a toilet and vanity to the bathroom.

14.

Add a toilet, tub, toilet and two vanities to the master bath.

15.

Your drawing should look similar to this.

Save your work.

Designing a Kitchen

Drawing Name: main floor plan – interior – Tahoe Cabin
Estimated Time: 45 minutes

Common areas are the Living Room, Dining Room, and Family Room.

This lesson reinforces the following skills:

- ❑ Insert blocks
- ❑ Extrudes
- ❑ Boolean Operations – Subtract
- ❑ Materials
- ❑ Groups

1.

Launch the ACA English (US Imperial) software.

2.

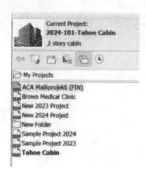

Set the **Tahoe Cabin** project current.

3.

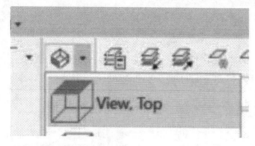

Launch the Project Navigator.

Open the **main floor – interior – Tahoe Cabin** construct drawing.

4.

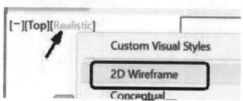

Switch to a **Top View**.

5.

Switch to a **2D Wireframe** display.

6.

Verify that the UCS is set to **WCS**.

7.

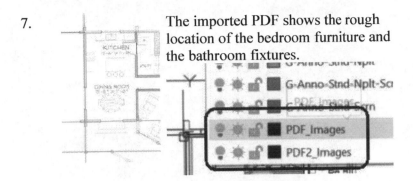

The imported PDF shows the rough location of the bedroom furniture and the bathroom fixtures.

If you do not see the PDF, thaw the **PDF_Images** layer.

8.

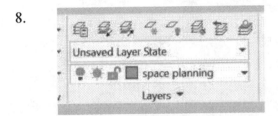

Set the layer called **spaced planning**. current.

9.

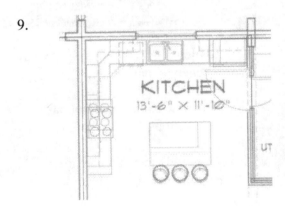

When organizing the kitchen, you will need to place base and wall cabinets.

Use the layout shown to place the kitchen elements.

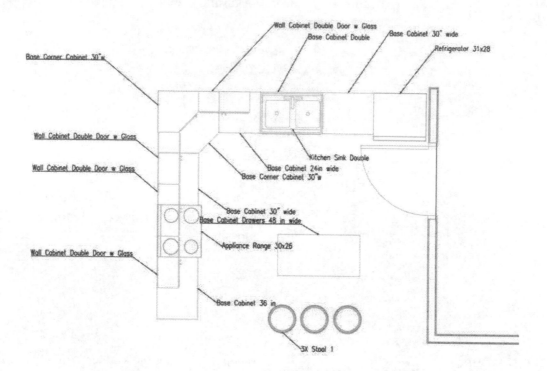

Wall Cabinet Double Door w Glass
Base Cabinet Double
Base Cabinet 30" wide
Refrigerator 31x28
Base Corner Cabinet 30"w
Wall Cabinet Double Door w Glass
Wall Cabinet Double Door w Glass
Kitchen Sink Double
Base Cabinet 24in wide
Base Corner Cabinet 30"w
Base Cabinet 30" wide
Base Cabinet Drawers 48 in wide
Wall Cabinet Double Door w Glass
Appliance Range 30x26
Base Cabinet 36 in
3X Stool 1

10.

Switch to a 3D view to inspect the kitchen layout.

The base cabinets do not have countertops.

You will have to create the countertops for the base cabinets.

11.

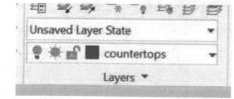

Unsaved Layer State

countertops

Layers ▼

Create a layer called **countertops** and set it current.

12.

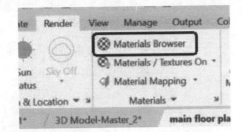

Switch to the Render ribbon.

Launch the **Materials Browser**.

13.

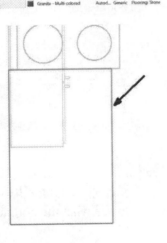

Locate the material you want to use for the countertops.

I selected **Granite – Gray Speckled**.

Add it to the document.

14.

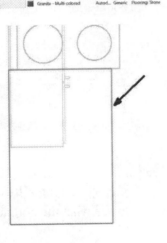

Switch back to a top view.
Set the WCS as current.
Turn off the PDF_Images layer.

Zoom into the lower left corner of the kitchen area.

Draw a rectangle aligned to the cabinet outline.

15. Type **EXTRUDE**.
Select the rectangle.
Press ENTER.
Set the Height to ¼".

16.

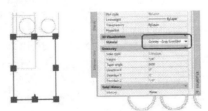

Select the 3D Solid that was just created.
On the Properties palette:
Assign the **Granite – Gray Speckled** material.

17. Switch to a 3D view.

You see the countertop on the floor instead of on top of the cabinet.

18. Use MOVE to move the countertop on top of the cabinet.

19. Switch to the Realistic Visual Style to see what the countertop looks like.

We want to group the cabinet with the countertop, so they move together.

20. Type **GROUP**.

Right click and select **Name**.

21. Type **36-in-cabinet-w-countertop**.

Note you need the dashes between the words or the entire name will not be used.

Select the cabinet and the countertop.

They are now grouped together.
Switch back to a top view.
Set the WCS as current.
Set the Visual Style to Wireframe.

We will create a countertop for the corner area.

22.

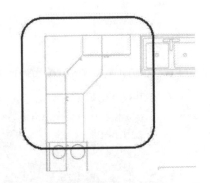

23.

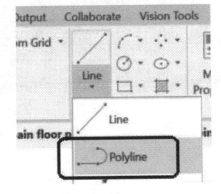

Select the **Polyline** tool.

24.

Draw a polygon aligned to the cabinet outlines.

Type **EXTRUDE**.
Select the polygon.
Press ENTER.
Set the Height to ¼".

25.

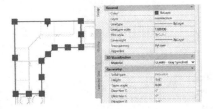

Select the 3D Solid that was just created.
On the Properties palette:
Assign the **Granite – Gray Speckled** material.

26.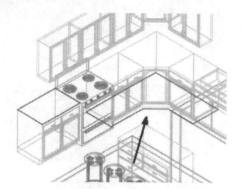

Switch to a 3D view.

You see the countertop on the floor instead of on top of the cabinet.

27.

Use the MOVE tool to move the countertop to the top of the cabinets.

28.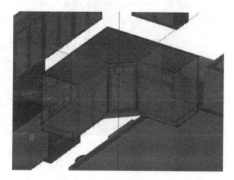

Group the countertop with the cabinets it sits on.
Name the group
Corner-cabinet-w-countertop.

29.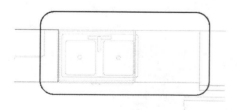

When you hover over the group, you should see the three cabinets and the countertop highlight.

30.

Switch back to a top view.
Set the **WCS** as current.
Set the Visual Style to **Wireframe**.

We will create a countertop for the sink and cabinet next the refrigerator.

31.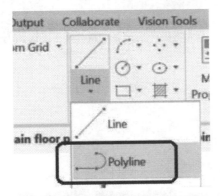

Select the **Polyline** tool.

32.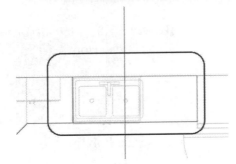

Draw a polygon aligned to the cabinet outlines.

33. Type **EXTRUDE**.
Select the outside polygon.
Select the sink cutout
Press ENTER.
Set the Height to ¼".

34.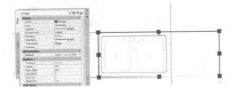

Select the 3D Solid that was just created.
On the Properties palette:
Assign the **Granite – Gray Speckled** material.

35.

Switch to a 3D view.

You see the countertop on the floor instead of on top of the cabinet.

Use the MOVE tool to move the countertop to the top of the cabinets.

36.

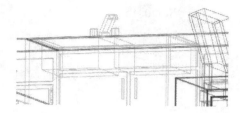

Position the sink so it is sitting on top of the countertop.

37.

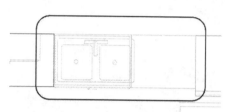

Switch back to a top view.
Set the **WCS** as current.
Set the Visual Style to **Wireframe**.

We will create a cutout for the sink in the countertop.

38.

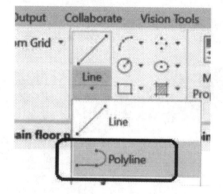

Select the **Polyline** tool.

39.

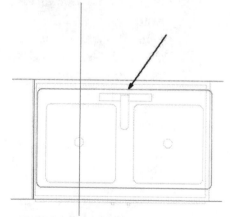

Draw a second polygon outlining the sink.
Add a 1" radius to the corners.
This will be the sink cutout in the countertop.

40.   Type **EXTRUDE**.
Select the sink cutout.
Press ENTER.
Set the Height to ¼".

41.

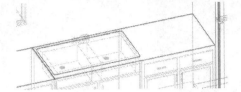

Switch to a 3D view.

You should see two 3d Solids – one for the countertop and one for the sink opening.

Verify that they are lying on top of each other.

42.

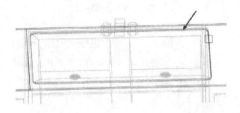

Type **SUBTRACT**.

Select the countertop.
Press ENTER.
Select the sink cutout.

Press ENTER.

43.

Switch to a Realistic Visual Style.

You now see the sink on top of the countertop placed properly.

44.

Create a group of the counter top, the sink, and the two cabinets.

45.

Switch back to a top view.
Set the **WCS** as current.
Set the Visual Style to **Wireframe**.

We will create a countertop for the kitchen island.

46.

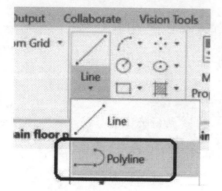

Select the **Polyline** tool.

47.

Draw a rectangle.

Extend one side of the rectangle down to create a sitting area for the stools.

48. Type **EXTRUDE**.
Select the rectangle.
Press ENTER.
Set the Height to ¼".

49.

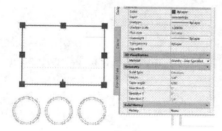

Select the 3D Solid that was just created.
On the Properties palette:
Assign the **Granite – Gray Speckled** material.

50.

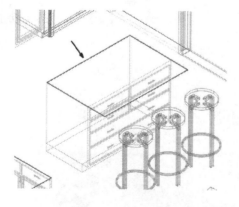

Move the countertop to the top of the kitchen island.

51.  Orbit around the 3D view to inspect your work and check for any errors.

52. Create a group for the kitchen island.
53. Save the file.

Tips Tricks

If the drawing preview in the Design Center is displayed as 2D, then it is a 2D drawing.

Extra:

Add towel bars, plants, mirrors, and other accessories to the floor plan using blocks from the design center or downloaded from the internet.

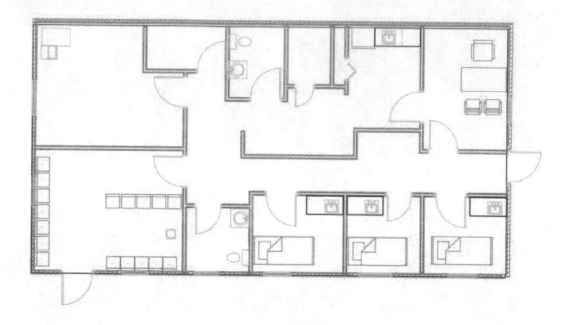

Exercise 4-5:
Modifying an Element

Drawing Name: 01 Core.dwg
Estimated Time: 15 minutes

This exercise reinforces the following skills:

- ❑   Content Browser
- ❑   External References

1.   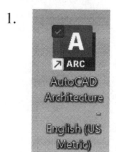   Launch the US Metric version of ACA.

This ensures that all the content will be metric.

2.   Open the Project Browser.

3.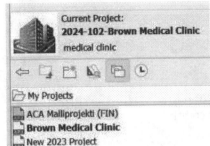

Set the **Brown Medical Clinic** Current.

The Brown Medical Clinic project was created using a metric project template. Placeholders were automatically generated.

4.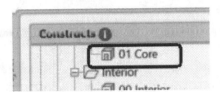

Launch the Project Navigator.

Switch to the Constructs tab.

Highlight **01 Core**.

Right click and select **Open**.

5.

Set the View Scale to **1:1**.

6.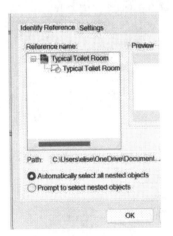

Select the cyan rectangle element that represents the Typical Toilet Room.
Click **Edit Reference In-Place** on the ribbon.

7.

Click **OK**.

8.

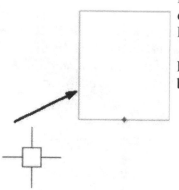

Locate the rectangle that is designated as the Typical Toilet Room.

Move it to overlay the top bathroom.

9.

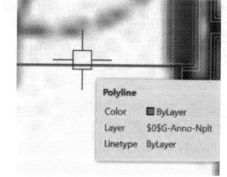

Scale the outline so it is inside the bathroom area.

10.

If you hover over the polyline outline, you will see that it is on a layer designated as NO PLOT.

11.

Thaw the PDF Images layer.

12.

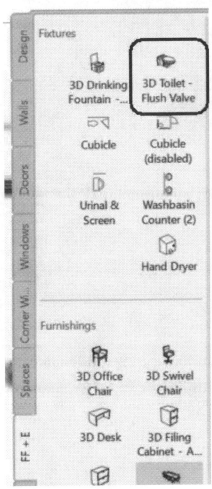

Launch the Design Tools palette.
Select the FF + E tab.

Locate the 3D Toilet – Flush Valve.

Drag and drop the toilet into the bathroom area.

Use the PDF to guide you in the placement of the toilet.

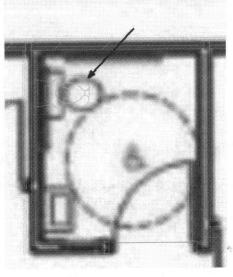

13.

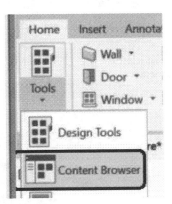

Launch the Content Browser.

You can use Ctl+4 as a shortcut key to launch the Content Browser.

14.

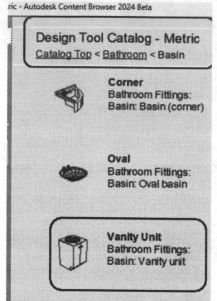

Open the Design Tool Catalog – Metric.

Locate the **Vanity Unit**.

15.

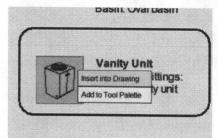

Right click and select **Insert into Drawing**.

16.

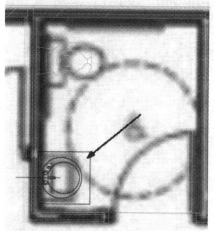

Place in the bathroom.

Right click and select ENTER to exit the command.

17.

Click **Save Changes** on the ribbon.

18.

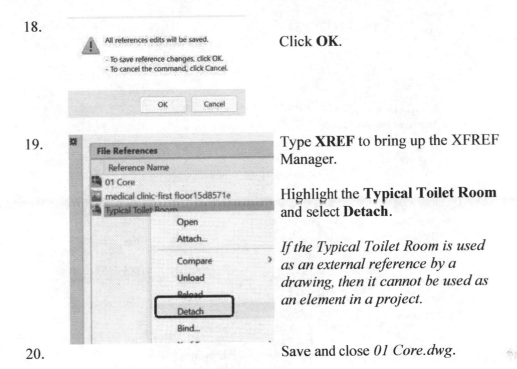

Click **OK**.

19.

Type **XREF** to bring up the XFREF Manager.

Highlight the **Typical Toilet Room** and select **Detach**.

If the Typical Toilet Room is used as an external reference by a drawing, then it cannot be used as an element in a project.

20.

Save and close *01 Core.dwg*.

Exercise 4-6:

Inserting an Element

Drawing Name: Main Model.dwg
Estimated Time: 30 minutes

This exercise reinforces the following skills:

- ❑ Content Browser
- ❑ External References
- ❑ Elements

1

Launch the US Metric version of ACA.

This ensures that all the content will be metric.

2.   Open the Project Browser.

3.   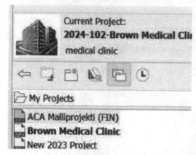   Set the **Brown Medical Clinic** Current.

The Brown Medical Clinic project was created using a metric project template. Placeholders were automatically generated.

4.   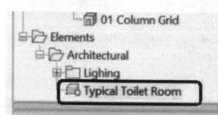   Launch the Project Navigator.

Switch to the Constructs tab.

Highlight **Typical Toilet Room** under Elements.

Right click and select **Open**.

5.   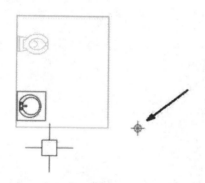   The magenta object represents the current basepoint.

Delete it.

6.   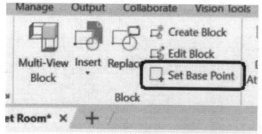   Open the Insert ribbon.

Select **Set Base Point**.

7.

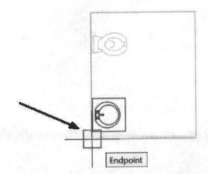

Select the lower left corner of the polyline.

Save the drawing and close.

8.

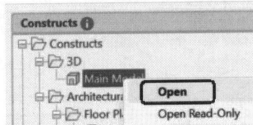

Launch the Project Navigator.

Switch to the Constructs tab.

Highlight **Main Model**.

Right click and select **Open**.

9.

Set the View Scale to **1:1**.

10.

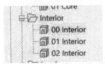

Look under Interior Constructs.
Delete 00 Interior and 02 Interior.

11.

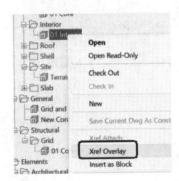

Highlight 01 Interior.

Right click and select **Xref Overlay**.

12.

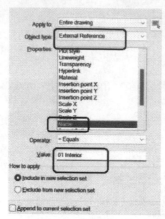

Type **QS** for Quick Select.
Set Object Type to **External Reference.**
Highlight **Name.**
Set Operator to **Equals.**
Set Value to **01 Interior.**
Click **OK.**

13.

Click **Edit Reference In-Place** on the ribbon.

14.

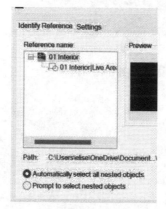

Click **OK.**

15.

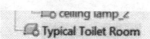

Locate the Typical Toilet Room under Elements.

Right click and select **Insert As Block.**

16.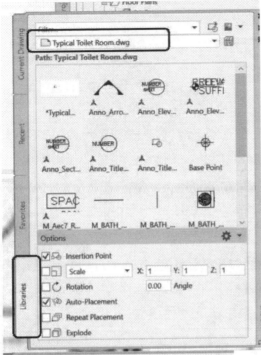

Open the Libraries tab.

Browse to the location where you are storing your project files.

Locate the *Typical Toilet Room.dwg.*

Enable **Insertion Point**.

17.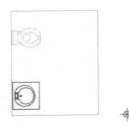

Right click on the block and select **Insert**.

18.

The bathroom comes in at a very large scale.

This will happen if the view scale was not set correctly.

Move the bathroom into the correct location on the floor plan.

19.

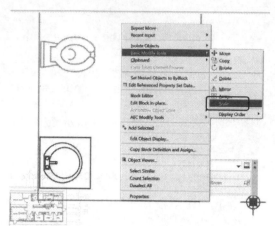

Select the block.
Right click and select **Basic Modify Tools→Scale**.

20.

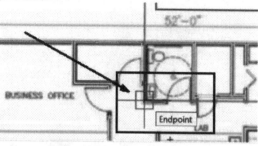

Click the lower left corner as the base point.

21.

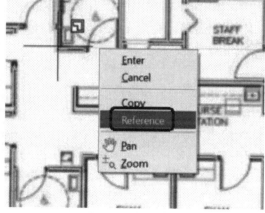

Right click and select **Reference**.

22.

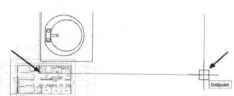

For the reference length, select the two end points of the bottom horizontal line of the block.

23.

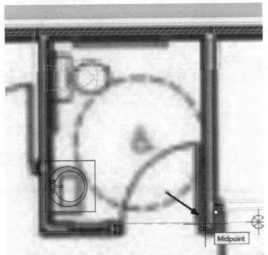

For the new length, click the inside face of the right-side wall of the bathroom.

24.

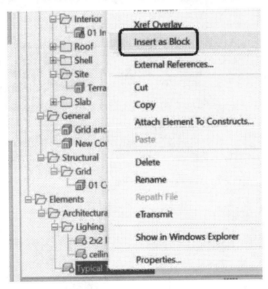

Select the block.
Right click and select **Properties**.
The scale is set to .99.

You can use this for other insertions.

25.

Locate the Typical Toilet Room under Elements.

Right click and select **Insert As Block**.

26.

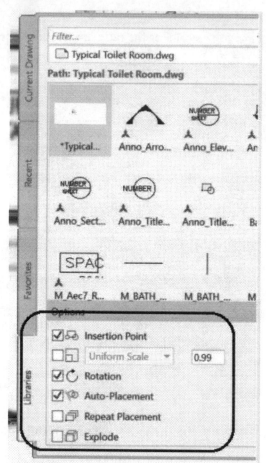

Open the Libraries tab.

Locate the *Typical Toilet Room.dwg*.

Enable **Insertion Point**.
Set the Scale to **0.99**, if needed.
Enable **Rotation**.
Enable **Auto-Placement**.

27.

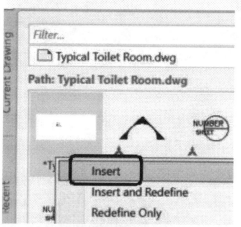

Right click on the block and select **Insert**.

28.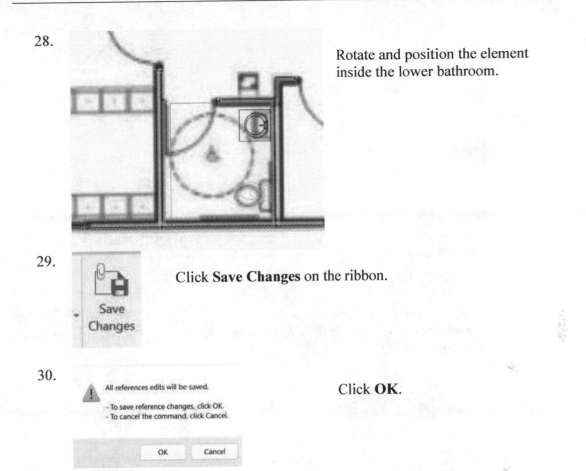

Rotate and position the element inside the lower bathroom.

29. Click **Save Changes** on the ribbon.

30. Click **OK**.

All references edits will be saved.

- To save reference changes, click OK.
- To cancel the command, click Cancel.

OK   Cancel

31. Save and close the *Main Model.dwg*.

Extra: Using what you have learned, edit the 01 Interior construct to include the furniture shown in the PDF. Can you create casework in the business office, lab, and nurse station areas?

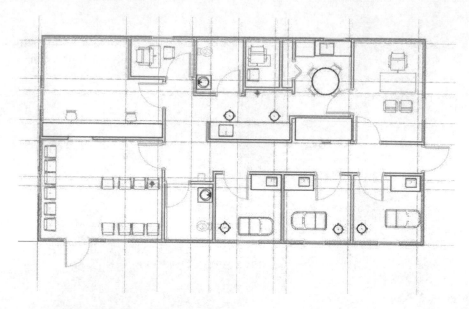

I have included blocks and textures in the exercise files to help you plan the interior floor plan for the medical clinic.

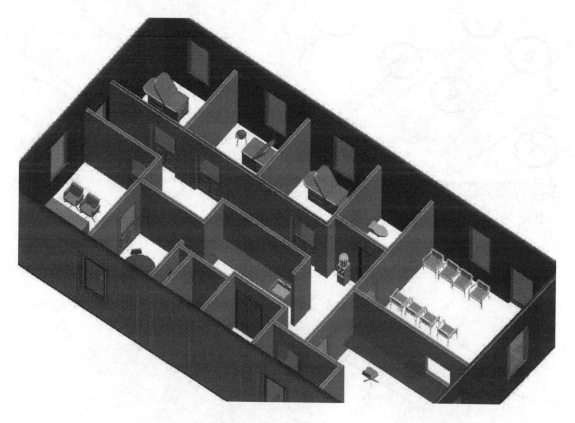

QUIZ 4

True or False

1. Custom content can be located in any subdirectory and still function properly.

2. The sole purpose of the Space Planning process is to arrange furniture in a floor plan.

3. The Design Center only has 3D objects stored in the Content area because ADT is strictly a 3D software.

4. Appliances are automatically placed on the APPLIANCE layer.

5. When you place a wall cabinet, it is automatically placed at the specified height.

6. You can create tools on a tool palette by dragging and dropping the objects from the Design Center onto the palette.

Multiple Choice

7. A residential structure is divided into:

    A. Four basic areas
    B. Three basic areas
    C. Two basic areas
    D. One basic area

8. Kitchen cabinets are located in the _____ subfolder.

    A. Casework
    B. Cabinets
    C. Bookcases
    D. Furniture

9. Select the area type that is NOT part of a private residence:

    A. Bedrooms
    B. Common Areas
    C. Service Areas
    D. Public Areas

10. To set the layer properties of a tool on a tool palette:

    A. Use the Layer Manager
    B. Select the tool, right click and select Properties
    C. Launch the Properties dialog
    D. All of the above

11. Vehicles placed from the Design Center are automatically placed on this layer:

    A. A-Site-Vhcl
    B. A-Vhcl
    C. C-Site Vhcl
    D. None of the above

12. To combine elements so that they act as a single object, use this command:

    A. GROUP
    B. SUBTRACT
    C. ADD
    D. INCLUDE

13. To create a 3DSolid, use this command:

    A. SOLID
    B. EXTRUDE
    C. MASS
    D. 3DSOLID

ANSWERS:

1) T; 2) F; 3) F; 4) F; 5) T; 6) T; 7) B; 8) A; 9) D; 10) B; 11) C; 12) A; 13) B

Lesson 5:
Roofs

Roofs can be created with single or double slopes, with or without gable ends, and with or without overhangs. Once you input all your roof settings, you simply pick the points around the perimeter of the building to define your roof outline. If you make an error, you can easily modify or redefine your roof.

You need to pick three points before the roof will begin to preview in your graphics window. There is no limit to the number of points to select to define the perimeter.

To create a gable roof, uncheck the gable box in the Roof dialog. Pick the two end points for the sloped portion of the roof. Turn on the Gable box. Pick the end point for the gable side. Turn off the Gable box. Pick the end point for the sloped side. Turn the Gable box on. Pick the viewport and then Click ENTER. A Gable cannot be defined with more than three consecutive edges.

Roofs can be created using two methods: ROOFADD places a roof based on points selected or ROOFCONVERT which converts a closed polyline or closed walls to develop a roof.

> ➢ If you opt to use ROOFCONVERT and use existing closed walls, be sure that the walls are intersecting properly. If your walls are not properly cleaned up with each other, the roof conversion is unpredictable.
> ➢ The Plate Height of a roof should be set equal to the Wall Height.
> ➢ You can create a gable on a roof by gripping any ridgeline point and stretching it past the roof edge. You cannot make a gable into a hip using grips.

Shape – Select the Shape option on the command line by typing 'S.'	Single Slope – Extends a roof plane at an angle from the Plate Height.	 end elevation view
	Double Slope – Includes a single slope and adds another slope, which begins at the intersection of the first slope and the height specified for the first slope.	 end elevation view
Gable – Select the Gable option on the command line by typing 'G.'	If this is enabled, turns off the slope of the roof place. To create a gable edge, select Gable prior to identifying the first corner of the gable end. Turn off gable to continue to create the roof.	 gable roof end
Plate Height – Set the Plate Height on the command line by typing 'PH.'	Specify the top plate from which the roof plane is projected. The height is relative to the XY plane with a Z coordinate of 0.	
Rise – Set the Rise on the command line by typing 'PR.'	Sets the angle of the roof based on a run value of 12.	A rise value of 5 creates a 5/12 roof, which forms a slope angle of 22.62 degrees.
Slope – Set the Slope on the command line by typing 'PS.'	Angle of the roof rise from the horizontal.	If slope angles are entered, then the rise will automatically be calculated.

Upper Height – Set the Upper Height on the command line by typing 'UH.'	This is only available if a Double Slope roof is being created. This is the height where the second slope will start.	
Rise (upper) – Set the Upper Rise on the command line by typing 'UR.'	This is only available if a Double Slope roof is being created. This is the slope angle for the second slope.	A rise value of 5 creates a 5/12 roof, which forms a slope angle of 22.62 degrees.
Slope (upper) – Set the Upper Slope on the command line by typing 'US.'	This is only available If a Double Slope roof is being created. Defines the slope angle for the second slope.	If an upper rise value is set, this is automatically calculated.
Overhang – To enable on the command line, type 'O.' To set the value of the Overhang, type 'V.'	If enabled, extends the roofline down from the plate height by the value set.	

	The Floating Viewer opens a viewer window displaying a preview of the roof.
	The match button allows you to select an existing roof to match its properties.
	The properties button opens the Roof Properties dialog.
	The Undo button allows you to undo the last roof operation. You can step back as many operations as you like up to the start.
	Opens the Roof Help file.

Exercise 5-1:

Creating a Roof using Existing Walls

Drawing Name: roofs1.dwg
Estimated Time: 10 minutes

This exercise reinforces the following skills:

- Roof
- Roof Properties
- Visual Styles

1.  Open *roofs1.dwg*.

2.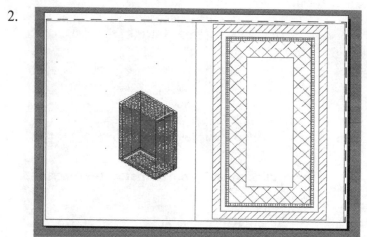

   Select the Work icon. Activate the right viewport that shows the top or plan view.

3. Select the **Roof** tool from the Home ribbon.

4.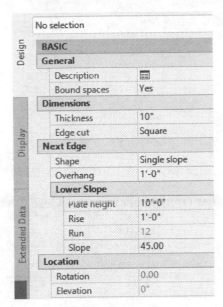

Expand the Dimensions section.
Set the Thickness to **10″ [254.00 mm]**.
Set the Shape to **Single slope**.
Set the Overhang to **1′-0″ [609.6 mm]**.

Dimensions	
Thickness	254.00
Edge cut	Square
Shape	Single slope
Overhang	609.60
	🔲 Edges/Faces

5.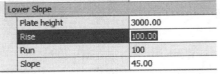

Expand the Lower Slope section.
Set the Plate height to **10′-0″ [3000 mm]**.

The plate height determines the level where the roof rests.

6.

Lower Slope	
Plate height	3000.00
Rise	100.00
Run	100
Slope	45.00

Set the Rise to **1′-0″ [100 mm]**.

7.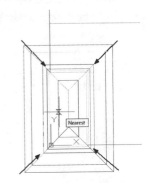

Pick the corners indicated to place the roof.

Click **Enter**.

8. 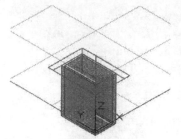   Activate the left 3D viewport.

9. 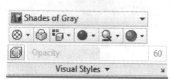   Activate the View ribbon.

Select the **Shades of Gray** under Visual Styles.

10.    Save as *ex5-1.dwg*.

In order to edit a roof, you have to convert it to Roof Slabs.

Roof Slabs

A roof slab is a single face of a roof. Roof slab elements differ from roof elements in that each roof slab is independent of other roof slabs. When you use multiple roof slabs to model a roof, you have more flexibility as you can control the shape and appearance of each slab or section of the roof.

While roof slabs are independent, they can interact. You can use one roof slab as a trim or extend boundary. You can miter roof slabs together. You can cut holes in roof slabs and add or subtract mass elements.

You can use grips to modify the shapes of roof slabs. You can use basic modify tools, like move or rotate, to reposition roof slabs.

Layer States

Layer states are used to save configurations of layer properties and their state (ON/OFF/FROZEN/LOCKED). For example, you can set a layer to be a different color depending on the state. Instead of constantly resetting which layers are on or off, you can save your settings in layer states and apply them to layouts.

Layer States are saved in the drawing. To share them across drawings, you need to export them. Each layer state has its own LAS file. To export a layer state, select it in the Layer States Manager and choose Export.

To import a layer state, open the Layer States Manager and click the Import tool. Then select the *.las file you wish to import.

You can save layer states in your drawing template.

Exercise 5-2:

Create Layer Filters and Layer States

Drawing Name: 3D Model-Master_3.dwg
Estimated Time: 15 minutes

This exercise reinforces the following skills:

- ❑ Layers
- ❑ Layer Filters
- ❑ Layer States

1.

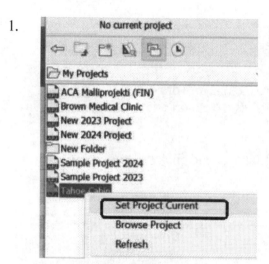

Launch the Project Browser.
Set the **Tahoe Cabin project** current.

2.

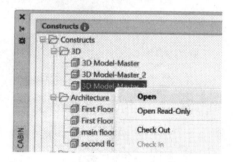

Launch the Project Navigator.

Open the *3D Model-Master_3.dwg*.

You may need to redefine the path of your external references.

3.

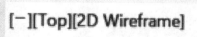

Set the View to **Top/2D Wireframe**.

4. 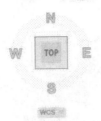   Set the UCS to **WCS**.

5.    Launch the **Layer Manager**.

6.    Highlight All.

Right click and select **New Properties Filter**.

7. 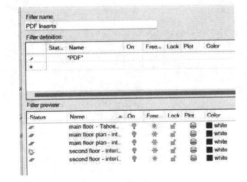   Name the Filter:
PDF Inserts.
Under Name:
Type ***PDF***.

All the layers with PDF in the name are displayed.

Press **OK**.

8.    Freeze all the PDF layers.

9.

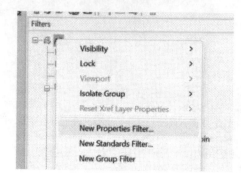

Highlight All.

Right click and select **New Properties Filter**.

10.

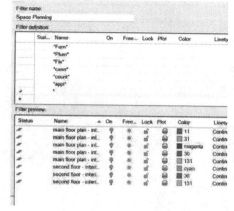

Name the Filter:
Space Planning.
Under Name:
Type ***Furn***.
Plum
Fix
case
count
appl

All the layers that meet the filter are displayed.

Press **OK**.

11.

Freeze all the space planning layers.

12.

Select the **Layer States Manager**.

13.

Select **New**.

14.

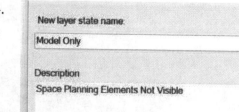

Type **Model Only** for the layer state name.

In the Description field:
Type **Space Planning Elements Not Visible**.

Click **OK**.

15.

Click **Close**.

16.

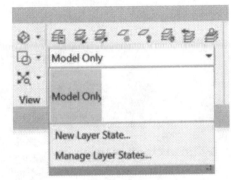

You can see the Layer State now in the drop-down list on the Layers panel.

Close the Layer Manager.

17.

Save the *3D Model-Master_3.dwg.*

Exercise 5-3:
Roof Slabs

Drawing Name: 3D Model-Master_3.dwg , roof_slabs – Tahoe Cabin.dwg
Estimated Time: 15 minutes

This exercise reinforces the following skills:

- ❑ AddRoof
- ❑ Project Navigator
- ❑ Convert to Roof Slabs
- ❑ Modify Roof Lines
- ❑ Trim Roof Slabs
- ❑ Roof Properties

1.

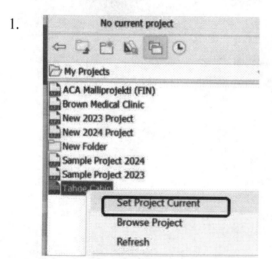

Launch the Project Browser.
Set the **Tahoe Cabin project** current.

2. Open *roof_slabs- Tahoe Cabin.dwg.*

3.

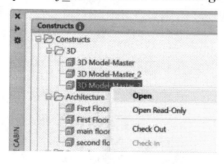

Launch the Project Navigator.

Open the *3D Model-Master_3.dwg.*

You may need to redefine the path of your external references.

4.

Switch to the *roof_slabs- Tahoe Cabin.dwg.*

5.

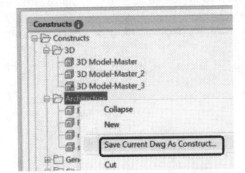

Right click on Architectural.

Select **Save Current Dwg As Construct**.

6.

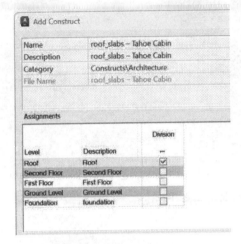

Enable **Roof**.

Click **OK**.

7.

| roof_slabs – Tahoe Cabin* | **3D Model-Master_3** × | + |

Switch back to the *3D Model-Master_3.dwg* tab.

8.

Highlight the Construct:
roof_slabs- Tahoe Cabin.dwg.

Right click and select **Xref Overlay**.

9.

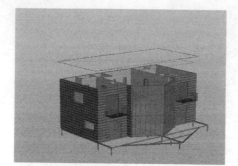

The roof slabs construct has some polylines.

Select it and move it into the correct position.

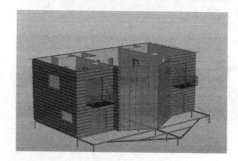

It should sit right on top of the building.

Select the roof slabs construct.
Right click.

10.

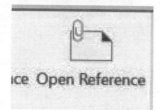

Select **Open Reference** from the ribbon.

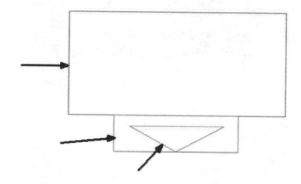

There are three
polygons in the
drawing:
A large rectangle,
a small rectangle
and a triangle.

These will be used
to create three
different roofs.

11.

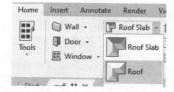

Select the **Roof** tool from the Home ribbon.

12.

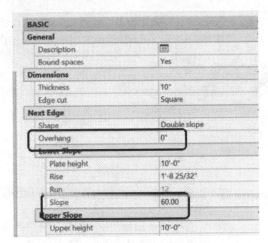

Set the Overhang to **0"**.
Set the Slop to **60**.

13.

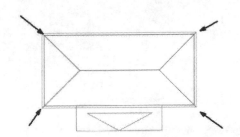

Select the four corners on main building to place a roof.

Click ENTER to complete the command.

14.

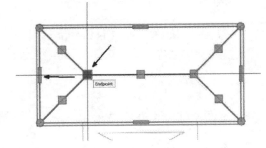

Select the grip indicated in red and shift the vertex so it overlies the midpoint on the left.

This eliminates the slope on the left side of the roof.

The roof now looks like this.

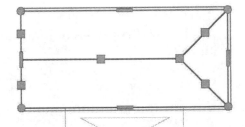

15.

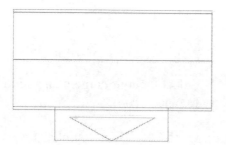

Select the grip indicated in red and shift the vertex so it overlies the midpoint on the right.

This eliminates the slope on the right side of the roof.

16.

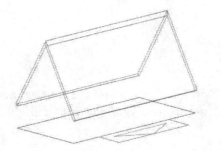

Release the grips and the roof should appear as shown.

17.

[−][SW Isometric][2D Wireframe]

Switch to a 3D view so you can see what the roof looks like.

Notice that the roof is elevated above the polylines.

18.

Return to a top view.

Launch the Design Tools.

Right click on the **Roof** tool and select **Apply Tool Properties to Linework and Walls**.

Select the cyan triangle.

When prompted to erase linework, select No.

19.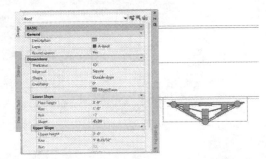

The roof is created.

Change the Slope to **45** on the Properties palette.

20.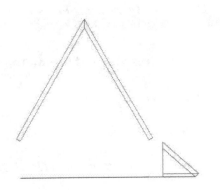

Select the center grip of the triangular roof.
Place that grip on the middle top grip.

The roof should look like this.

21.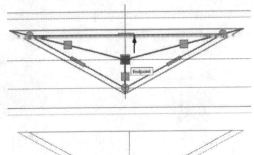

[−][Left][2D Wireframe]

Switch to a LEFT view.

Move the A-frame roof on the left down to the polyline.

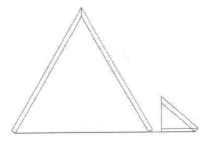

You roofs should look like this.

22.

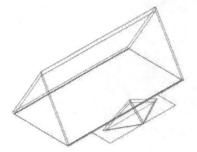

[−][SW Isometric][2D Wireframe]

Switch to a SW Isometric view.

23.

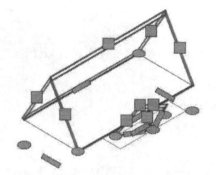

Select the two roofs.

Convert

Select **Convert** on the ribbon.

This converts the roofs to roof slabs.

24.

Roof Slab Style:	Standard	∨
☐ Erase Layout Geometry		
OK	Cancel	Help

Disable **Erase Layout Geometry.**

Click **OK**.

Click **ESC** to release the selection.

Select the left rectangular side of the smaller roof.

25.

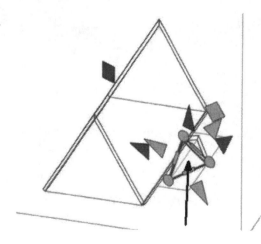

Extend

Select **Extend** on the ribbon.

26.

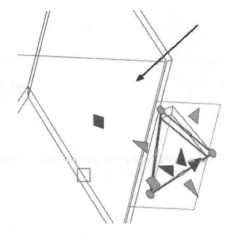

You are prompted to select a boundary to extend to.

Select the A-frame slab closest to the small roof.

27.

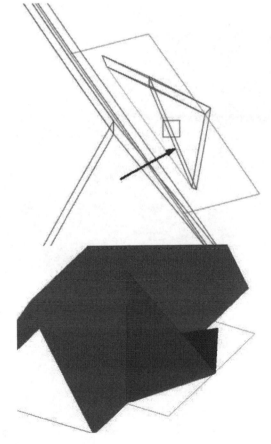

Select the top edge of the slab to extend. Select the bottom edge of the slab to extend.

I switched to a Shaded view so you can see what the extended roof looks like.

28.

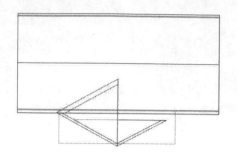

[-][Top][2D Wireframe]

Switch to a **Top View/2D Wireframe**.

29.

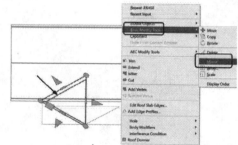

Select the right small roof slab and delete it.

30.

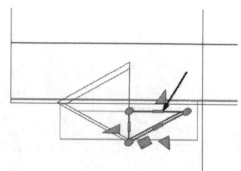

Select the remaining small roof slab.

Right click and select **Basic Modify Tools→Mirror**.

31.

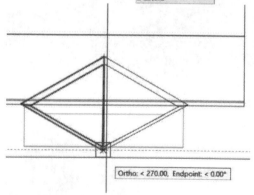

Mirror to the other side.

32.
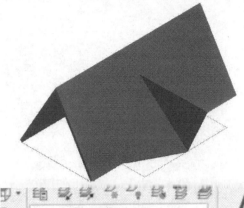
Switch to an Isometric view so you can see the roof.

33.
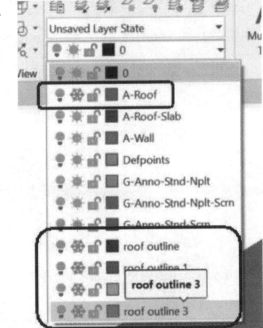
Freeze the following layers:

A-Roof
Roof outline
roof outline_1
Roof outline_2
Roof outline_3

34. Save the file and close.

35.
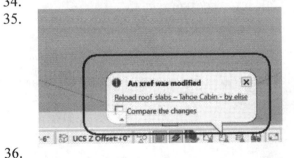
Disable **Compare the changes**.

Click **Reload roof slabs – Tahoe cabin** to see the changes.

36. Save and close the *3D Model-Master-3.dwg* file.

Attach Walls to Roofs

Drawing Name: 3D Model-Master_3
Estimated Time: 15 minutes

This exercise reinforces the following skills:

- ❑ Attach walls to roofs
- ❑ Modify Roof Line

1.
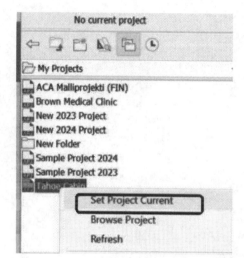
Launch the Project Browser.
Set the **Tahoe Cabin project** current.

2.
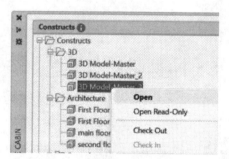
Launch the Project Navigator.

Open the *3D Model-Master_3.dwg.*

You may need to redefine the path of your external references.

3.
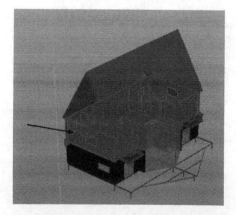
Select the exterior wall of the second floor.

4.  Select **Edit Reference In-Place** from the ribbon.

5. 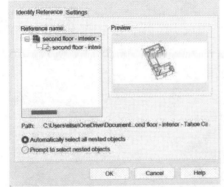 Select **OK**.

6. 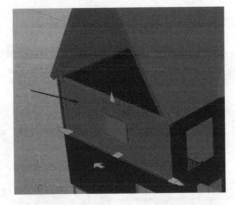 Select the wall.

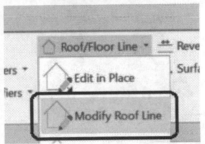

On the ribbon: select **Modify Roof Line**.
Right click and select Auto project.

7. 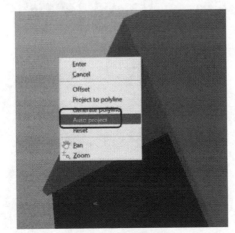 Select the roof.

The wall adjusts to meet the roof.

Orbit around to the other side of the building model.

8.

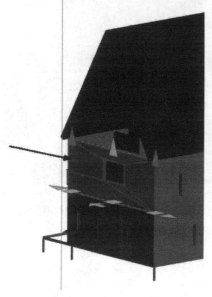

Select the wall.

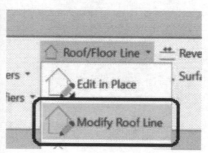

On the ribbon: select **Modify Roof Line**.

9.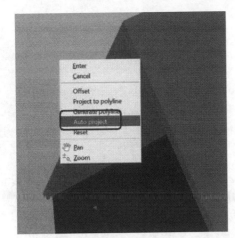

Right click and select Auto project.

Select the roof.

10.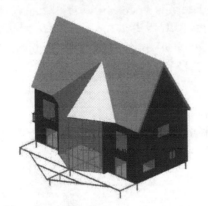

The wall now meets the roof.

Orbit the model around to inspect the walls and roof.

11.

Select **Save Changes** on the ribbon.

12.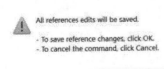

Click **OK**.

13. Save the file.

Rendering Materials in ACA

Materials represent substances such as steel, concrete, cotton and glass. They can be applied to a 3D model to give the objects a realistic appearance. Materials are useful for illustrating plans, sections, elevations and renderings in the design process. Materials also provide a way to manage the display properties of object styles.

The use of materials allows you to more realistically display objects. You need to define the display of a material such as brick or glass only once in the drawing or the drawing template and then assign it to the component of an object where you want the material to display. You typically assign materials to components in the style of an object such as the brick in a wall style. Then whenever you add a wall of that style to your drawing, the brick of that wall displays consistently. Defining materials in an object style can provide control for the display of objects across the whole project. When the characteristics of a material change, you change them just once in the material definition and all objects that use that material are updated. With the Material tool you can apply a material to a single instance of the object.

You can take advantage of Visual Styles, Rendering Materials, Lights, and Cameras in AutoCAD® Architecture. Materials provide the ability to assign surface hatches to objects. Surface hatches can be displayed in model, elevation, and section views. This is helpful to clearly illustrate sections and elevations.

Exercise 5-5:
Creating Materials

Drawing Name: 3D Model-Master_3.dwg
Estimated Time: 30 minutes

This exercise reinforces the following skills:

- ❑ Design Palette
- ❑ Materials
- ❑ Style Manager

1.
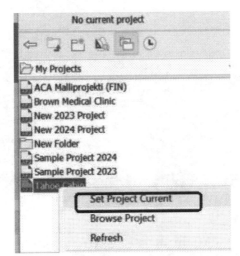
Launch the Project Browser.
Set the **Tahoe Cabin project** current.

2.
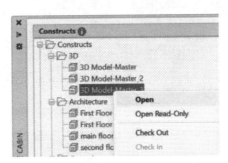
Launch the Project Navigator.

Open the *3D Model-Master_3.dwg.*

You may need to redefine the path of your external references.

3.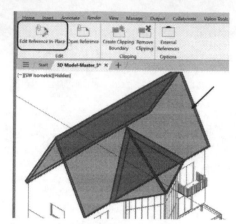

Select the roof.
On the ribbon:
Select **Edit Reference In-Place**.

4.

Click **OK**.

5. [−][SW Isometric][Hidden]

Switch to an **SW Isometric Hidden** view.

6.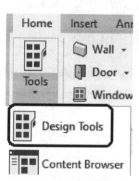

Launch the Design Tools.

7.

Right click on the title bar of the Design Tools and enable **All Palettes**.

8.  Right click on the tabs of the Design Tools and enable **Materials**.

9.  Locate the **Asphalt Shingles** material.

10. 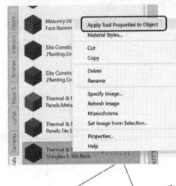 Right click on the **Asphalt Shingles**.

Select **Apply Tool Properties to Object.**

11. 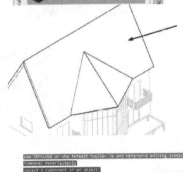 Select the roof.

You are prompted if you are modifying the style or just the object.

Click ENTER to modify the Style.

 The roof material updates.

12.

Type **MATERIALS**.

This brings up ACA's Materials Browser.

13.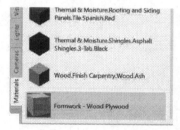

Type **plywood** in the search field.

14.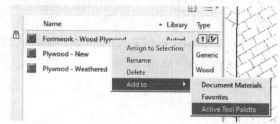

Highlight the **Formwork – Wood Plywood.**

Right click and select **Add to→Active Tool Palette.**

15.

The material appears at the bottom of the Materials palette.

16.

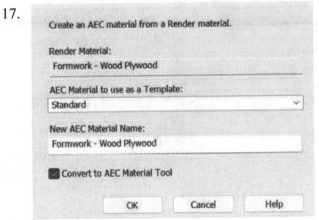

Highlight the **Formwork – Wood Plywood.**

Right click and select **Convert to AEC Material.**

17.

Create an AEC material from a Render material.

Render Material:
Formwork - Wood Plywood

AEC Material to use as a Template:
Standard

New AEC Material Name:
Formwork - Wood Plywood

☑ Convert to AEC Material Tool

| OK | Cancel | Help |

Set the AEC Material to use as a Template to **Standard.**

Click **OK.**

18.

Tool Style Update - Save Drawing   ✕

The style for this tool has not been saved with the drawing.

To ensure that the tool will work properly, the style must be saved first. You should save this drawing now.

Close

Click **Close.**

Save the drawing file.

19.

rubber

Type **rubber** in the search field.

20. 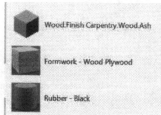 Highlight the **Rubber-Black.**

Right click and select **Add to→Active Tool Palette.**

21. The material appears at the bottom of the Materials palette.

22. 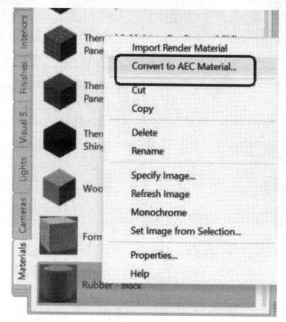 Highlight the **Rubber - Black.**

Right click and select **Convert to AEC Material.**

23.

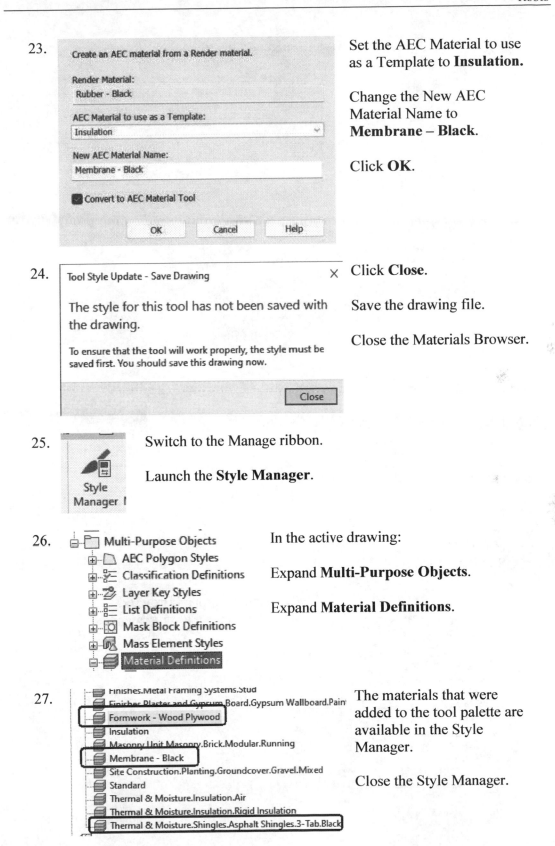

Set the AEC Material to use as a Template to **Insulation.**

Change the New AEC Material Name to **Membrane – Black**.

Click **OK.**

24.

Tool Style Update - Save Drawing   ✕

The style for this tool has not been saved with the drawing.

To ensure that the tool will work properly, the style must be saved first. You should save this drawing now.

Close

Click **Close.**

Save the drawing file.

Close the Materials Browser.

25.

Style Manager

Switch to the Manage ribbon.

Launch the **Style Manager**.

26.

Multi-Purpose Objects
- AEC Polygon Styles
- Classification Definitions
- Layer Key Styles
- List Definitions
- Mask Block Definitions
- Mass Element Styles
- Material Definitions

In the active drawing:

Expand **Multi-Purpose Objects**.

Expand **Material Definitions**.

27.

Finishes.Metal Framing Systems.Stud
Finishes.Plaster and Gypsum Board.Gypsum Wallboard.Pain
Formwork - Wood Plywood
Insulation
Masonry Unit Masonry.Brick.Modular.Running
Membrane - Black
Site Construction.Planting.Groundcover.Gravel.Mixed
Standard
Thermal & Moisture.Insulation.Air
Thermal & Moisture.Insulation.Rigid Insulation
Thermal & Moisture.Shingles.Asphalt Shingles.3-Tab.Black

The materials that were added to the tool palette are available in the Style Manager.

Close the Style Manager.

28. 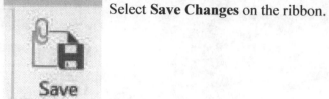 Select **Save Changes** on the ribbon.

29. 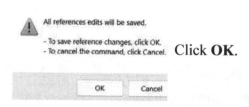 Click **OK**.

All references edits will be saved.

- To save reference changes, click OK.
- To cancel the command, click Cancel.

OK   Cancel

30. Save the file.

Exercise 5-6:

Roof Slab Styles

Drawing Name: 3D Model-Master_3.dwg
Estimated Time: 10 minutes

This exercise reinforces the following skills:

- Roof Slab Styles
- Design Palette
- Materials
- Style Manager

1. 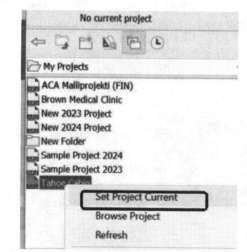 Launch the Project Browser.
Set the **Tahoe Cabin project** current.

2.

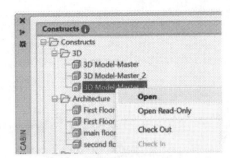

Launch the Project Navigator.

Open the *3D Model-Master_3.dwg*.

You may need to redefine the path of your external references.

3.

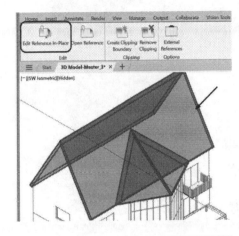

Select the roof.
On the ribbon:
Select **Edit Reference In-Place**.

4.

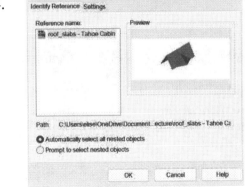

Click **OK**.

5.

[−][SW Isometric][Hidden]

Switch to an **SW Isometric Hidden** view.

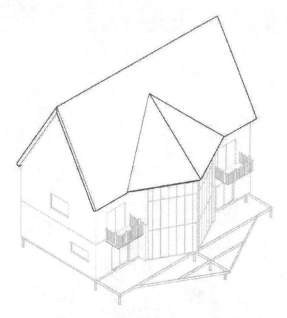

6. Activate the Manage ribbon.

    Select the **Style Manager**.

7. Locate the Roof Slab Styles under Architectural Objects.

    There are two Standard styles.
    The 0Standard style is used by the external reference.

8. Right click on the Roof Slab Style and select **New**.

9. Rename the style **Asphalt Shingles**.

10.

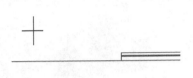

Define three components:
- Asphalt Shingles with a thickness of ½" and Thickness Offset of 0.
- Membrane with a Thickness of 1/8" and Thickness offset of ½".
- Plywood Sheathing with a thickness of ¼" and Thickness offset of 5/8".

11.

Use the preview to check the placement of each component.

12.

Highlight the **Asphalt Shingles** component.

Select the Asphalt Shingles material from the drop-down list.

This material was added to the drawing when it was applied to the rood slab.

13.

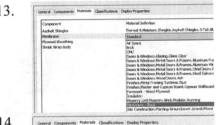

Highlight the **Membrane** component.

Select the **Membrane - Black** material from the drop-down list.

14.

Highlight the **Plywood Sheathing** component.

Select the **Formwork – Wood Plywood** material from the drop-down list.

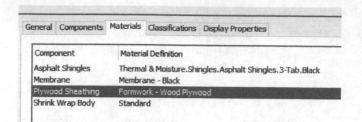

The assigned materials should look like this.

Click **OK**.
Close the Style Manager.

15.

Select **Save Changes** on the ribbon.

16.

Click **OK**.

17.

Save the file.

Exercise 5-7:

Applying a Roof Slab Style

Drawing Name: 3D Model-Master_3.dwg
Estimated Time: 5 minutes

This exercise reinforces the following skills:

- ❑ Roof Slab Styles
- ❑ Select Similar
- ❑ Properties
- ❑ Edit External Reference In-Place

1.

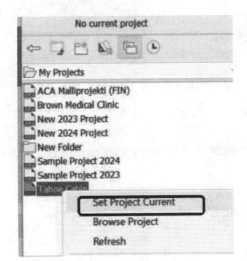

Launch the Project Browser.
Set the **Tahoe Cabin project** current.

2.

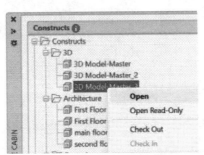

Launch the Project Navigator.

Open the *3D Model-Master_3.dwg*.

You may need to redefine the path of your external references.

3.

Select the roof.
On the ribbon:
Select **Edit Reference In-Place**.

4.

Click **OK**.

5.

[−][SW Isometric][Hi

Switch to an **SW Isometric Hidden** view.

6.

Select a roof slab.
Right click and select **Select Similar**.

7.

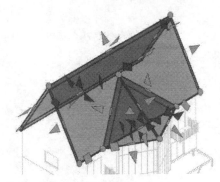

All the roof slabs are selected.

Right click and select **Properties**.

8.

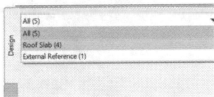

Select **Roof Slab** from the selection drop-down.

9.

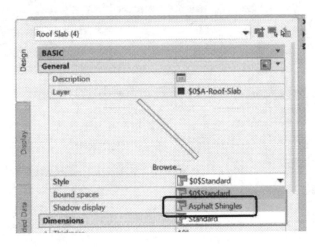

On the Properties palette,
set the Style to **Asphalt Shingles**.

Click **ESC** to release the selection.

10.

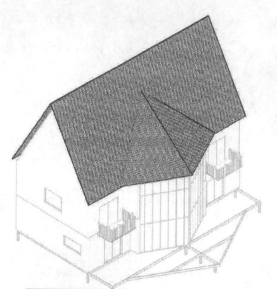

The roof appearance updates.

11.

Select **Save Changes** on the ribbon.

12.

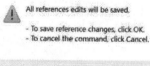

Click **OK**.

13.   Save the file.

Extra:

Open the Brown Medical Clinic project.
Create a Layer filter and Layer State to control the visibility of the furniture and the fixtures.

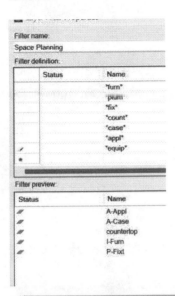

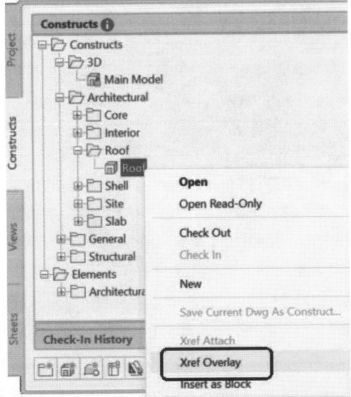

Add the roof construct to the Main Model as an Xref Overlay.

Edit the Roof external reference in-place.
Hint: Use QSELECT to select an external reference using the Name "Roof".
Add a roof.

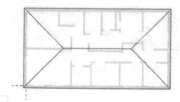

Create a roof slab style using Spanish tiles.
Convert the roof to roof slabs.
Assign the Spanish Tile roof style to the roof slabs.

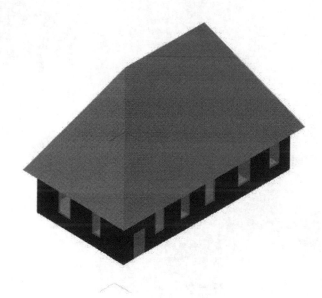

QUIZ 5

True or False

1. Roof slabs are a single face of a roof, not an entire roof.

2. The thickness of a roof slab can be modified for an individual element without affecting the roof slab style.

3. If you change the slope of a roof slab, it rotates around the slope pivot point.

4. Once you place a roof or roof slab, it cannot be modified.

5. You can use the TRIM and EXTEND tools to modify roof slabs.

Multiple Choice

6. Materials are displayed when the view setting is: (select all that apply)

    A. wireframe
    B. Shaded
    C. Hidden
    D. Rendered

7. Roof slab styles control the following properties: (select all that apply)

    A. Slab Thickness
    B. Slab Elevation
    C. Physical components (concrete, metal, membrane)
    D. Materials

8. Identify the tool shown.

    A. Add Roof
    B. Add Roof Slab
    C. Add Gable
    D. Modify Roof

9. To change the elevation of a slab:

    A. Modify the slab style.
    B. Use Properties to change the Elevation.
    C. Set layer properties.
    D. All of the above.

10. Roof slabs can be created: (select all that apply)

     A. Independently (from scratch)
     B. From Existing Roofs
     C. From Walls
     D. From Polylines

ANSWERS:

1) T; 2) T; 3) T; 4) F; 5) T; 6) A, B, D; 7) A, C, D; 8) B; 9) B; 10) A, B, C, D

Lesson 6:
Floors, Ceilings and Spaces

Slabs are used to model floors and other flat surfaces. You start by defining a slab style which includes the components and materials which are used to construct the floor system.

You can create a floor slab by drawing a polyline and applying the style to the polyline or by selecting the slab tool and then drawing the boundary to be used by the slab.

Ceiling grids can be used to define drop ceilings.

You can also use spaces to define floors and ceilings. You can use the Space/Zone Manager to apply different styles to floors and ceilings.

There are two reasons why you want to include floors and ceilings in your building model. 1) for your construction documentation and 2) for any renderings.

Exercise 6-1:

Creating Slab Styles

Drawing Name: slab_styles.dwg
Estimated Time: 30 minutes

This exercise reinforces the following skills:

- ❏   Creating Styles
- ❏   Copying Styles
- ❏   Renaming Styles
- ❏   Use of Materials

1. 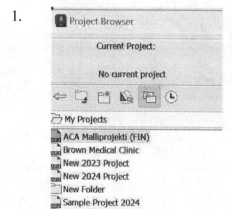   Open the *slab_styles.dwg*.
    Close any current projects.

2.    Activate the Manage ribbon.

    Select **Style Manager**.

3.    Browse to the Slab Styles
    category.

    Highlight Slab Styles.

    Right click and select **New**.

4.   Rename **Carpet**.

5.

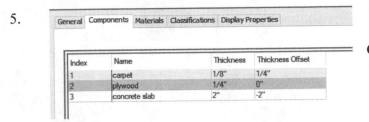

Create three components:

- Carpet
- Plywood
- Concrete Slab

Set the carpet at 1/8" and ¼" Thickness Offset.
Set the plywood at ¼" and 0" Thickness Offset.
Set the concrete slab at 2" and -2" Thickness Offset.

6.

Use the preview window to verify that the thicknesses and offsets are correct.

7.

Create new materials for each of the components.

I have pre-loaded several materials in the exercise files. Some of the materials use image files that are part of the exercise files.

8.

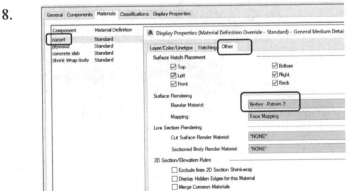

Edit the carpet material definition to use the **Berber – Pattern 2** render material.

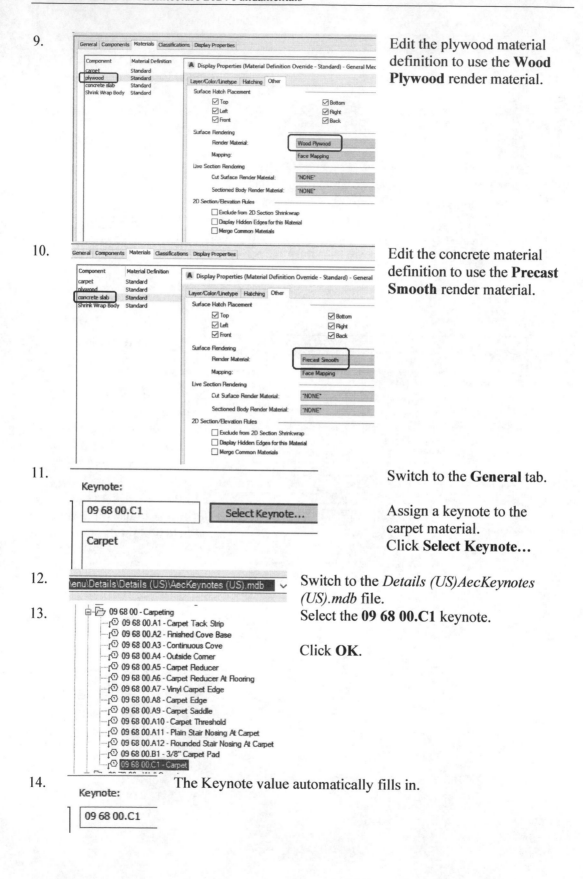

9.

Edit the plywood material definition to use the **Wood Plywood** render material.

10.

Edit the concrete material definition to use the **Precast Smooth** render material.

11.

Switch to the **General** tab.

Assign a keynote to the carpet material.
Click **Select Keynote…**

12.

Switch to the *Details (US)\AecKeynotes (US).mdb* file.

13.

Select the **09 68 00.C1** keynote.

Click **OK**.

14.

The Keynote value automatically fills in.

15. Click **Apply** to save your work.

16. Highlight the Carpet slab style.

    Right click and select **Copy**.

17. Highlight the Slab Styles.

    Right click and select **Paste**.

18. Rename the slab style **Concrete**.

19. Switch to the Components tab.

    Highlight the carpet component.
    Select **Remove**.

20. Highlight the plywood component.
    Select **Remove**.

21. Click **Apply** to save your work.

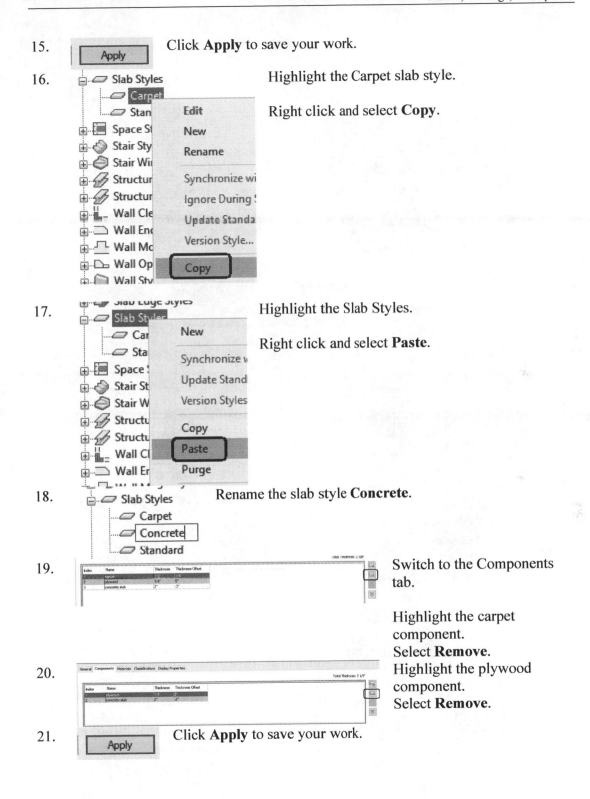

22.

Highlight the Carpet slab style.

Right click and select **Copy**.

23.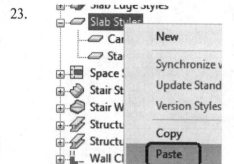

Highlight the Slab Styles.

Right click and select **Paste**.

24.

Rename **Oak Floor**.

25.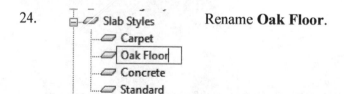

Switch to the Components tab.

Change the name of the carpet component to **oak flooring.**

26.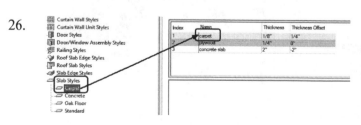

Click **Apply.**
Verify that the carpet component is still the top component for the carpet slab style.

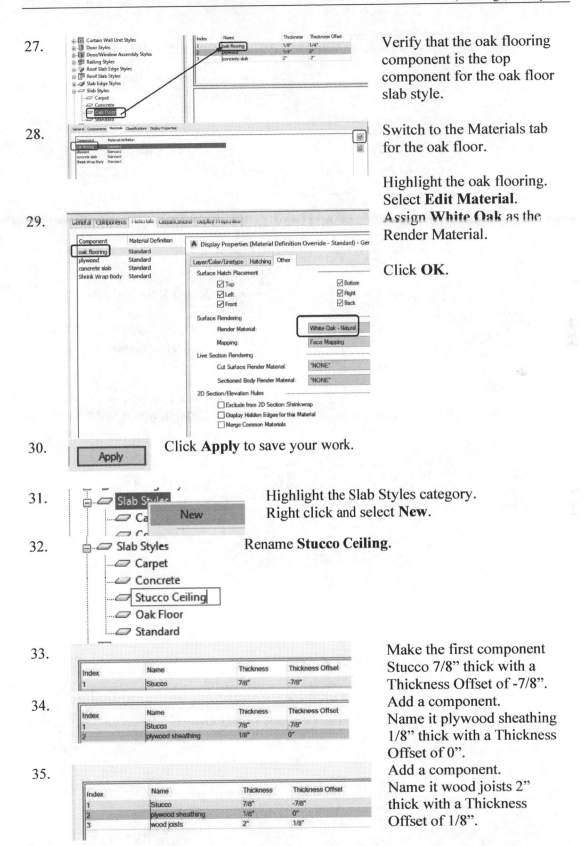

27. Verify that the oak flooring component is the top component for the oak floor slab style.

28. Switch to the Materials tab for the oak floor.

Highlight the oak flooring. Select **Edit Material**. Assign **White Oak** as the Render Material.

29. Click **OK**.

30. Click **Apply** to save your work.

31. Highlight the Slab Styles category. Right click and select **New**.

32. Rename **Stucco Ceiling**.

33. Make the first component Stucco 7/8" thick with a Thickness Offset of -7/8".

34. Add a component. Name it plywood sheathing 1/8" thick with a Thickness Offset of 0".

35. Add a component. Name it wood joists 2" thick with a Thickness Offset of 1/8".

36.

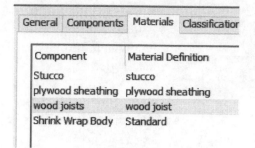

Switch to the Materials tab.

Create new materials for the components:

- Stucco
- Plywood sheathing
- Wood joist

37.

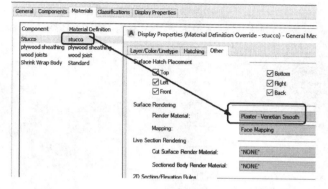

Highlight the stucco component.

Select **Edit Material**.

Set the Render Material to **Plaster – Venetian Smooth**.

Click **OK** twice to close the dialog boxes.
Switch to the Display Properties tab.

38.

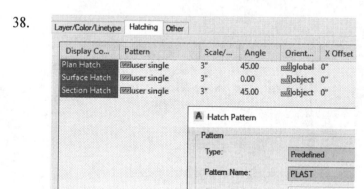

Set the Hatching to **PLAST**.

Click **OK**.

39.

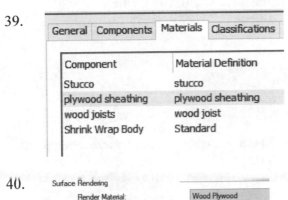

Highlight the plywood sheathing component.

Select **Edit Material**.

40.

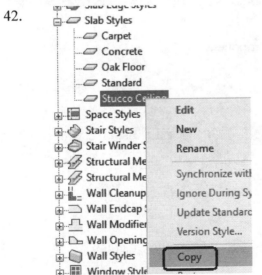

Select **Edit Display Properties**.
Select the Other tab.

Set the Render Material to **Wood Plywood**.
Click **OK** twice.

41.

Apply

Click **Apply** to save your work.

42.

Highlight the **Stucco Ceiling** slab style.

Right click and select **Copy**.

43.

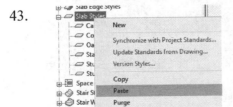

Highlight the Slab Styles.

Right click and select **Paste**.

44. 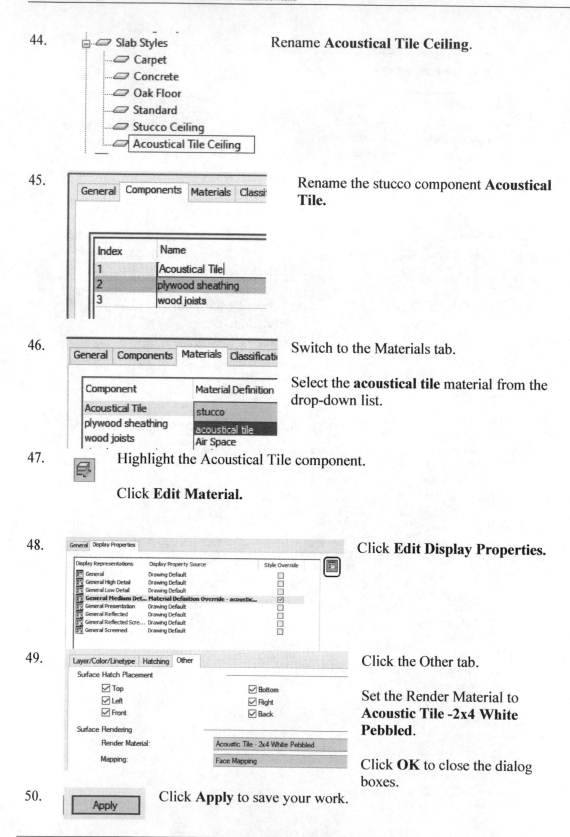 Rename **Acoustical Tile Ceiling**.

45. Rename the stucco component **Acoustical Tile.**

46. Switch to the Materials tab.

    Select the **acoustical tile** material from the drop-down list.

47. Highlight the Acoustical Tile component.

    Click **Edit Material.**

48. Click **Edit Display Properties.**

49. Click the Other tab.

    Set the Render Material to **Acoustic Tile -2x4 White Pebbled.**

    Click **OK** to close the dialog boxes.

50. Click **Apply** to save your work.

51.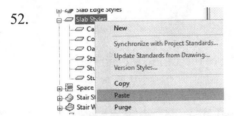

Highlight the **Oak Floor** slab style.

Right click and select **Copy**.

52.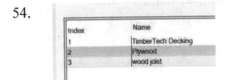

Highlight the Slab Styles.

Right click and select **Paste**.

53.

Rename **TimberTech Decking - Mahogany**.

54.

Index	Name
1	TimberTech Decking
2	Plywood
3	wood joist

Rename the stucco component
TimberTech Decking.

55.

New Name: TimberTech Decking - Mahogony

OK   Cancel

Switch to the Materials tab.
Create a new material called **TimberTech Decking - Mahogany**
Assign the material to the TimberTech Decking component.

56.

Highlight the TimberTech Decking component.

Click **Edit Material.**

57.

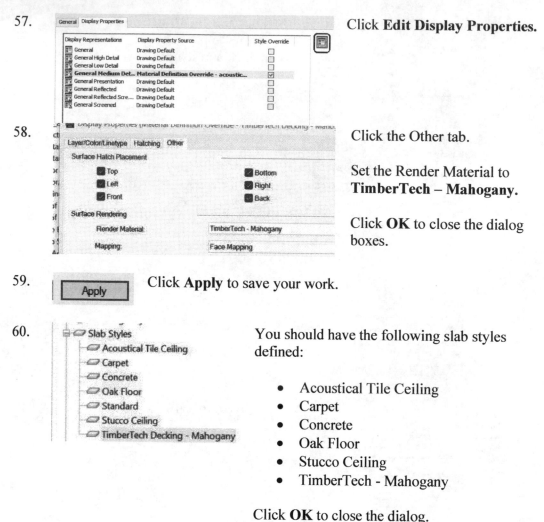

Click **Edit Display Properties.**

58.

Click the Other tab.

Set the Render Material to **TimberTech – Mahogany.**

Click **OK** to close the dialog boxes.

59.

Click **Apply** to save your work.

60.

You should have the following slab styles defined:

- Acoustical Tile Ceiling
- Carpet
- Concrete
- Oak Floor
- Stucco Ceiling
- TimberTech - Mahogany

Click **OK** to close the dialog.

61. Save as *slab_styles_2.dwg*.

Exercise 6-2:

Create Slab from Linework

Drawing Name: 3D Model-Master_4.dwg, deck outline.dwg
Estimated Time:105 minutes

This exercise reinforces the following skills:

- ❏ Creating Slabs
- ❏ Project Navigator
- ❏ External References
- ❏ Constructs
- ❏ Polylines

1. 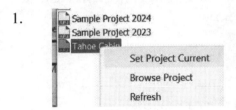 Set the **Tahoe Cabin** project current.

2. Open *3D Model_Master_4.dwg*.

   In the Project Navigator:
   Go to the Constructs tab.
   Highlight the 3D category.
   Right click and select **Save Current Dwg As Construct**.

3. Open *deck outline.dwg*.
   In the Project Navigator:
   Go to the Constructs tab.
   Highlight the Architecture category.
   Right click and select **Save Current Dwg As Construct**.

4.  Enable **First Floor**.

   Click **OK**.

5.

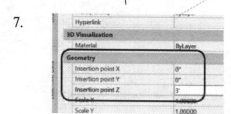

Highlight the deck outline construct.

Right click and select **Xref Overlay**.

6.

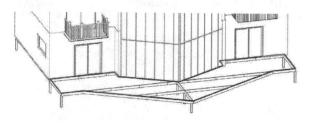

Switch to a 3D view.

You see the deck outline is located below the first floor.

7.

Select the deck outline external reference.
Right click and select **Properties**.
Change the Z value to **3'-0"**.

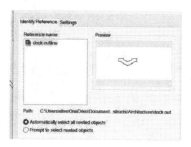

The deck outline is now sitting on top of the framing.

8.

Edit Reference In-Place

Select the deck outline external reference.

Select **Edit Referenced In-Place** from the ribbon.

9.

Click **OK**.

10.

Click the Home ribbon.

Launch the **Design Tools**.

11.

Locate the Slab tool.

Right click and select **Apply Tool Properties to →Linework and Walls.**

12.

Select the deck outline polyline.

Click **ENTER**.

13.
`⌐▾ SLABTOOLTOLINEWORK Erase layout geometry? [Yes No] <No>:`

Select **No**.

14.
`⌐▾ SLABTOOLTOLINEWORK Creation mode [Direct Projected] <Projected>:`

Select **Direct**.

15.
`SLABTOOLTOLINEWORK Specify slab justification [Top Center Bottom Slopeline] <Bottom>:`

Set the Justification to **Top**.

16.

Select the midpoint of the top of the boundary as the pivot point for the slab.

The slab will be horizontal, so it doesn't really need a pivot point.

17.

Set the slab style to **TimberTech-Decking**.

Note that the slope is set to 0.0.

18.

Switch to an isometric view to inspect your slab floor.

Set the view style to **Realistic.**

19.

Select **Save Changes** to save the changes to the external reference file.

20.

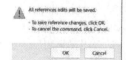

Click **OK**.

21.

Save as *3D Model-Master_4.dwg.*

Views

Typically, you create view drawings and model space views after you have defined levels, divisions, and the basic constructs. For example, you could create the following series of views for individual floors: First Floor Plan, Second Floor Plan, Basement Plan, Roof Plan. You could also create a series of views based on plan type, such as: Plan View, Model View, Reflected View. Another approach could be by function; for example: Walls, Framing, Furniture. Only constructs can be referenced into a view drawing. To see elements in a view, they must be referenced by a construct.

After the structure of the building project is defined, and constructs are assigned to levels and divisions, you can start to create view drawings. A view drawing references a number of constructs to present a specific view of the building project. To create a view drawing, you first decide which portion of the building you wish to look at and which type of view to generate. You could, for example, create a first-floor reflected ceiling plan or a second-floor framing plan, or create a composite view of all floors in the building. View drawings automatically reference the appropriate constructs according to their level/division assignments within the building. For example, to create a floor plan of the west wing of the second floor, you would create a view that references all constructs assigned to the second floor and the west wing. This would also include a curtain wall spanning the first through fifth floors.

Types of View Drawings

There are three different types of view drawings in the Drawing Management feature of the Project Navigator:

- **General view drawing**: A general view drawing contains referenced constructs from the project, representing a specific view on the building model. General view drawings are based on the general view template defined in the project settings.

  You can reference a view drawing in a sheet. When you reference the view drawing into a sheet, a sheet view is created that contains the view drawing reference.

- **Detail view drawing**: A detail view drawing contains one or more model space views that show a defined portion of the detail drawing in the level of detail you specify. A model space view containing a detail can be associated with a callout. Detail view drawings are based on the detail view template defined in the project settings.

- **Section/Elevation view drawing**: A section/elevation view drawing contains one or more model space views, each showing a defined portion of the section/elevation view drawing. A model space view containing a section or elevation can be associated with a callout. Section/Elevation view drawings are based on the section/elevation view template defined in the project settings.

Model Space Views

A model space view is a portion of a view drawing that may be displayed in its own paper space viewport on a sheet. Model space views are an evolution of the Named Views concept of AutoCAD. Unlike Named Views, a model space view has a defined boundary. When a model space view is placed onto a sheet, a sheet view is created. A view drawing can contain any number of model space views.

Exercise 6-3:

Create View Categories and Views

Drawing Name: create_slab2.dwg
Estimated Time: 2 minutes

This exercise reinforces the following skills:

- ❏ Views
- ❏ Project Navigator
- ❏ Layer Filters
- ❏ Layer Properties
- ❏ Style Manager
- ❏ Slab Styles

1.

   The Tahoe Cabin Project should be active.

   Launch the Project Navigator.

   Select the Views tab.

   Create additional view categories.
   Under Architectural:
   Elevations
   Plans
   Reflected Ceiling Plans
   Under Views:
   Details
   Sections
   Model Space

2.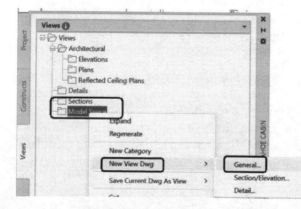

   Highlight the **Model Space** category.

   Right click and select **New View Dwg→General**.

3. Type **First Floor** in the Name field.
Click in the Description field and type **First Floor**.
Enable. **Open in drawing editor**.

Click **Next**.

4. Select the **First Floor** level.

Click **Next**.

5. Disable the 3D category.
Disable the Sitework category.
Under Architecture:
Disable deck outline.
Disable First Floor Foundation – Tahoe cabin.
Enable First Floor.
Enable main floor – Tahoe cabin.
Enable main floor plan – interior – Tahoe cabin.
Under General:
Enable the column grids and bubbles.
Click Finish.

6. If you see an error message, it means you need to repath some of the files.

7.

Set the view scale to **1'-0" = 1'-0"**.

8.

Launch the **Layer Manager**.

9.

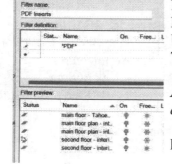

Highlight All.

Right click and select **New Properties Filter**.

10.

Name the Filter:
PDF Inserts.
Under Name:
Type ***PDF***.

All the layers with PDF in the name are displayed.

Press **OK**.

11.

Freeze all the PDF layers.

12.

Highlight **All**.

Right click and select **New Properties Filter**.

13.

Name the Filter:
Space Planning.
Under Name:
Type ***Furn***.
Plum
Fix
case
count
appl

All the layers that meet the filter are displayed.

Press **OK**.

14.

Freeze all the space planning layers.

15.

Select the **Layer States Manager**.

16.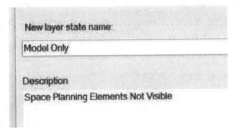

Select **New**.

17.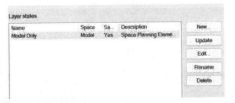

Type **Model Only** for the layer state name.

In the Description field:
Type **Space Planning Elements Not Visible**.

Click **OK**.

18.

Click **Close**.

19.

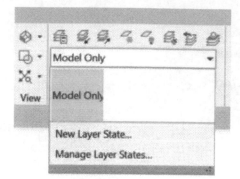

You can see the Layer State now in the drop-down list on the Layers panel.

Close the Layer Manager.

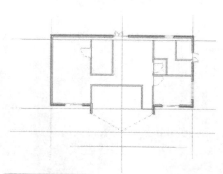

You see the view updates with just the floor plan.

20.

Switch to the Manage ribbon.
Open the **Style Manager**.

21.

Open the *slab_styles_2.dwg*.

You should see several slab styles.

Copy and paste the slab styles into the *First Floor.dwg*.

22. Enable **Leave Existing**.

    Click **OK.**

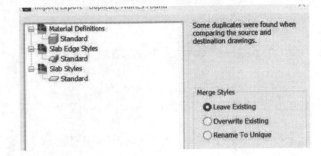

23. The slab styles are now available in the First Floor.dwg.

    Close the Style Manager.

24.   Save the *First Floor.dwg*.

Exercise 6-4:

Create Slab from Boundary

Drawing Name: First Floor.dwg
Estimated Time: 15 minutes

This exercise reinforces the following skills:

- ❑ Saved Views
- ❑ View Cut Height
- ❑ Create Slab

1.

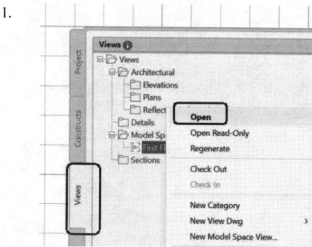

Go to the Views tab of the Project Navigator.

Right click on the **First Floor** view.

Select **Open**.

2.

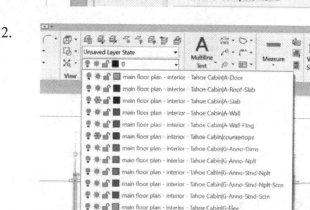

Reset the layer filter to show all the layers.

Thaw the **PDF_Images** layer.

3.

Zoom into the Master Bedroom area.

4.

Change the Cut Height to adjust the view depth.

Set the Value to **3'-0"**.

5.

Specify cut plane heights:

Display Above Range: 10'-0"

Cut Height: 3'-0"

Display Below Range: 0"

Calculate Cut Plane from Current Project: [Calculate]

[OK] [Cancel]

Set the Display Above Range to **10'-0"**.
Set the Cut Height to **3'**.
Set the Display Below Range to **0"**.

Click **OK**.

6.

Select the **Slab** tool from the Home ribbon.

We will add a carpet slab to this room.

7.

Set the Style to **Carpet**.

8.

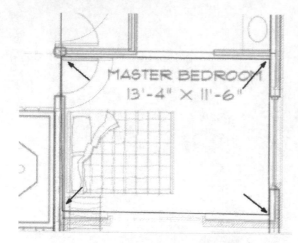

You are going to trace the boundary for the floor to be placed.

Start at the upper left corner of the room.

Trace around the room.

At the last segment, right click and select **Close**.

Click ENTER when the boundary is completed.

9.

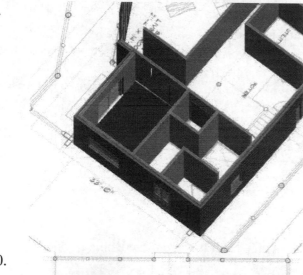

~][Custom View][Realistic]

Switch to an isometric view and inspect the floor you just placed in the bedroom area.

10.

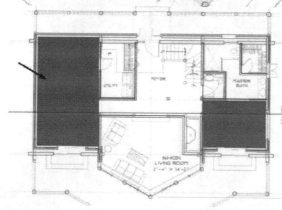

Add a tile slab to the kitchen and dining room area.

11.

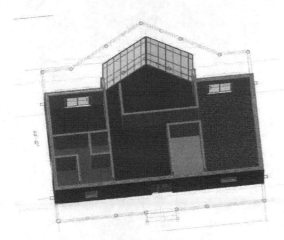

Place slabs in the remaining rooms using appropriate slab styles.

Save the file.

Exercise 6-5:
Create a Reflected Ceiling Plan View

Drawing Name: First Floor RCP.dwg
Estimated Time: 25 minutes

This exercise reinforces the following skills:

- Views
- Project Navigator
- Layer Filters
- Layer States
- Style Manager
- Slab Styles

1.

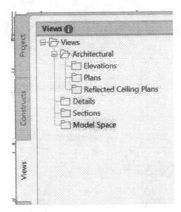

The Tahoe Cabin Project should be active.

Launch the Project Navigator.

Select the Views tab.

2.

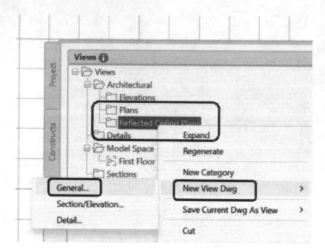

Highlight the **Reflected Ceiling Plans** category.

Right click and select **New View Dwg→General**.

3.

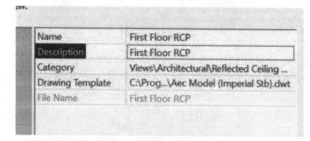

Type **First Floor RCP** in the Name field.
Click in the Description field and type **First Floor RCP**.
Enable. **Open in drawing editor**.

Click **Next**.

4.

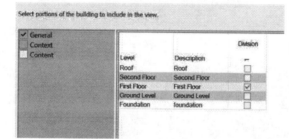

Select the **First Floor** level.

Click **Next**.

5.

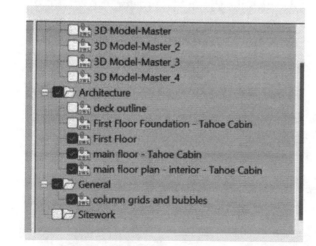

Disable the 3D category.
Disable the Sitework category.
Under Architecture:
Disable deck outline.
Disable First Floor Foundation – Tahoe cabin.
Enable First Floor.
Enable main floor – Tahoe cabin.
Enable main floor plan – interior – Tahoe cabin.
Under General:
Enable the column grids and bubbles.
Click **Finish**.

6.
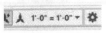

Could not reference file
'C:\Users\elise\OneDrive\Documen...\My
Projects\Tahoe
Cabin\Constructs\Architecture\First
Floor.dwg' in
'C:\Users\elise\OneDrive\Documen...\My
Projects\Tahoe Cabin\Views\Model
Space\First Floor.dwg'.

If you see an error message, it
means you need to repath some
of the files.

Zoom all to see the drawing.

7.
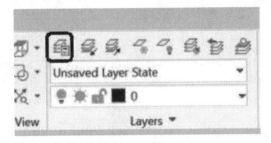

Set the view scale to **1'-0" = 1'-0"**.

8.
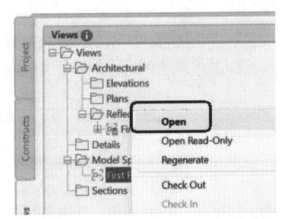

Open the **First Floor** view.

9.

Launch the **Layer Manager**.

10.
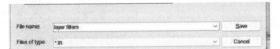

Click the **Save** icon.

11.

Browse to your work folder.
Name the file *layer filters*.
Click **Save**.

12.

Current layer: 0 : 0

Open the **Layer States Manager**.

13.

Highlight the **Model** Layer State.

Click **Export**.

14.

Browse to your work folder.
Name the file *Model Only.las*.
Click **Save**.

15. Close all the dialogs.
Close the *First Floor.dwg*.
Set the *First Floor RCP.dwg* active.

16.

Launch the **Layer Manager**.

17.

Select **Load Filter Groups**.

18.

Note, that this action will remove any
Layer Filter assignments that already
exist.

Do you wish to continue?

Click **Yes**.

19.

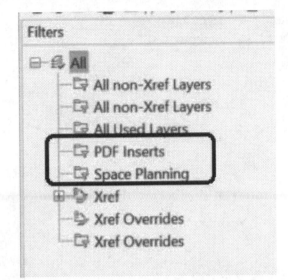

The layer filters are now available in this drawing.

Open the **Layer States Manager**.

20.

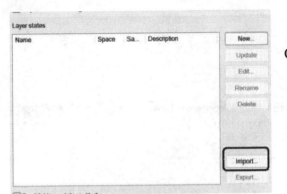

Click **Import**.

21.

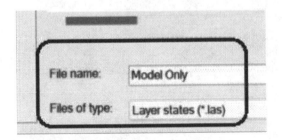

Set the Files of type: to ***.las**.

Select the *Model Only* filc.

Click **Open**.

22. The layers states in C:\ACA 2024 Fundamentals\ACA 2024 Fundamentals exercises\ACA 2024 exercises reorganized\Lesson 06\completed exercises\Model Only.las were successfully imported. Do you want to restore any layer states?

Click **Restore states**.

Close the Layer Manager.

23.

Switch to the Manage ribbon.
Open the **Style Manager**.

24.

Open the *slab_styles_2.dwg*.

You should see several slab styles.

Copy and paste the slab styles into the *First Floor RCP.dwg*.

25. Enable **Leave Existing**.

Click **OK**.

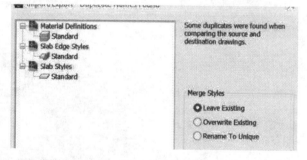

26. *The slab styles are now available in the First Floor RCP.dwg.*

Close the Style Manager.

27. Save and close the *First Floor RCP.dwg*.

Ceiling Grids

Grids are AEC objects on which you can anchor other objects, such as columns, and constrain their locations. Grids are useful in the design and documentation phases of a project. A ceiling grid allows you to mount light fixtures to a ceiling.

The Design Tools palette contains tools which quickly place ceiling grids with a specific ceiling grid style and other predefined properties. You can use the default settings of the tool, or you can change its properties. You can also use ceiling grid tools to convert linework to ceiling grids and to apply the settings of a ceiling grid tool to existing ceiling grids.

You can change existing ceiling grids in different ways. You can change the overall dimensions of the grid, the number and position of grid lines, and the location of the grid within the drawing. You can also specify a clipping boundary for the grid and use that to mask the grid or insert a hole into the grid.

Exercise 6-6:
Create Ceiling Grid

Drawing Name:	First Floor - RCP
Estimated Time:	25 minutes

This exercise reinforces the following skills:

- ❑ Named Views
- ❑ Ceiling Grid
- ❑ Properties
- ❑ Set Boundary

1. 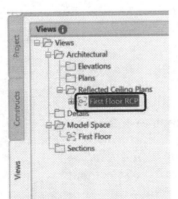   Open the **First Floor RCP** view.

2. 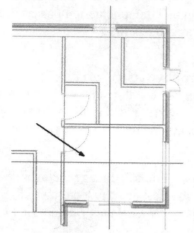 Switch to the Home ribbon.
Select **Ceiling Grid**.

3. Select the inside corner of the bedroom.

Click **ENTER** to accept the default rotation of 0.00.

Click **ESC** to exit the command.

4.  Switch to an isometric view.
The ceiling grid was placed at the 0 elevation.

Select the ceiling grid.

5. Right click and select **Properties.**

6.  Change the Elevation to **8' 0"**.

The ceiling grid position will adjust.
Click **ESC** to release the selection.

Location	
Rotation	0.00
Elevation	8'-0"

7.

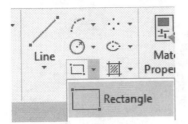

Switch back to the **top** view.

You won't see the ceiling grid because it is above the cut plane.

Select the **Rectangle** tool from the Draw panel on the Home ribbon.

Place a rectangle in the bedroom

8.

9.

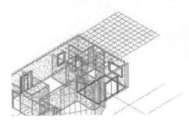

Select the rectangle.

Set the Global Width to ¼"

This just makes it a little easier to see.

Click **ESC** to release the selection.

10.

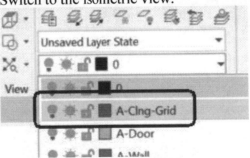

Switch to the isometric view.

Turn ON the **A-Clng-Grid** layer.

11.

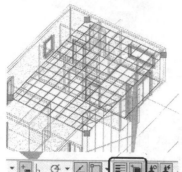

Move the ceiling grid so that it is located inside the bedroom.

12.

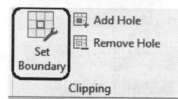

Enable **Lineweight**.
Enable **Selection Cycling**.

13.

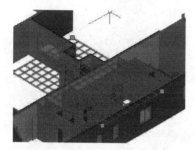

Select the ceiling grid.

14.

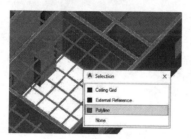

Select **Set Boundary** from the Clipping panel.

Note: You won't see this tool unless the ceiling grid is selected.

15.

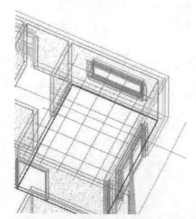

Click near the bottom edge of a wall.

The Selection Cycling dialog should list the polyline that is the rectangle.

Click on the Polyline in the list to select.

16.

The ceiling grid boundary is modified to fit inside the room.

Click ESC to exit the command.

17.
[−][Top][2D Wireframe]

Switch back to a top view.

18.
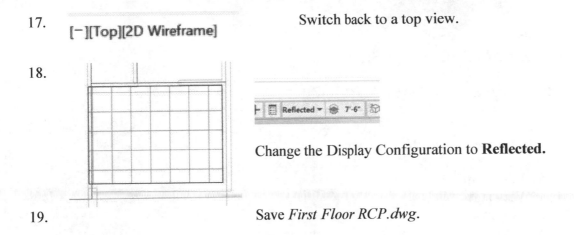

Change the Display Configuration to **Reflected.**

19.
Save *First Floor RCP.dwg.*

Lighting

When there are no lights in a scene, the scene is rendered with default lighting. Default lighting is derived from one or two distant light sources that follow the viewpoint as you orbit around the model. All faces in the model are illuminated so that they are visually discernible. You can adjust the exposure of the rendered image, but you do not need to create or place lights yourself.

When you place user-defined lights or enable sunlight, you can optionally disable default lighting. Default lighting is set per viewport, and it is recommended to disable default lighting when user-defined lights are placed in a scene.

Light fixtures are placed for three reasons – 1) for rendering 2) to create lighting fixture schedules, so the construction documents can detail which light fixtures are placed where; and 3) for electrical wiring drawings.

Exercise 6-7:

Add Lights to Ceiling Grid

Drawing Name: First Floor RCP.dwg
Estimated Time: 10 minutes

This exercise reinforces the following skills:

- ❏ Reflected Ceiling Plan Display
- ❏ Viewports
- ❏ Design Center
- ❏ Insert Block

1.

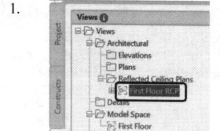

Open the **First Floor RCP** view.

2.

On the status bar:

Notice that the Display Configuration is set to **Reflected.**

3.

Switch to the View ribbon.

Launch the **Design Center**.

4.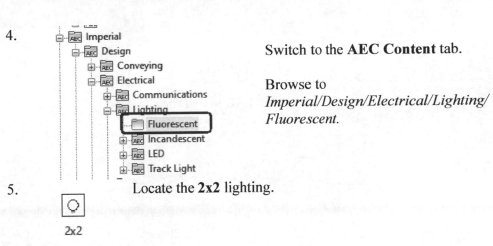

Switch to the **AEC Content** tab.

Browse to
Imperial/Design/Electrical/Lighting/Fluorescent.

5. Locate the **2x2** lighting.

2x2

6.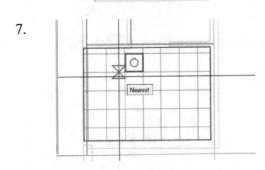

Select and right click.

Select **Insert**.

7.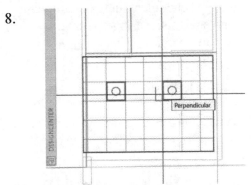

Select the ceiling grid when prompted to select the layout node.

Then select the grid intersection to use to place the lighting fixture.

8.

Move the cursor over and select another location for the lighting fixture.

9.

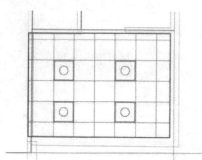

Place four lighting fixtures.

Click **ESC** to exit the command.

10.

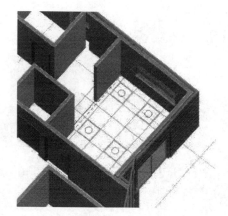

Switch to an isometric view.

The lighting fixtures are displayed in the ceiling grid.

Save *First Floor RCP.dwg*.

Spaces

Spaces are 2-dimensional or 3-dimensional style-based architectural objects that contain spatial information about a building, including floor area, wall area, volume, surface information, and more. Spaces can be used for organizing schedules, such as statements of probable construction cost, energy requirements and analysis, leasing documents, operating costs and fixtures, and lists of furniture and equipment.

Spaces can have a number of different data types attached to them:

- Geometric data belonging to the space: This is data inherent to the space like its height or width. This data can be displayed in the Properties palette or used in schedule tables and space tags.

- Modified and calculated data derived from the space: this is data derived from the geometric data by applying calculation modifier styles, boundary offset calculation formulas, or formula properties. Schedule properties exist for values created by calculation modifiers, boundary offsets, and formulas, so that they can be displayed in the Properties palette and used in a schedule table, too.

- Data generated by a geometric decomposition of the space area.

- Space surface properties: Data attached to the surfaces of a space.

- User-defined information sets: you can define any set of relevant property set data for spaces, like for example, floor finish, ceiling material etc. This data can be displayed in the Properties palette and used in schedule tables and space tags.

Exercise 6-8:
Add a Space

Drawing Name: First Floor.dwg
Estimated Time: 5 minutes

This exercise reinforces the following skills:

- Named Views
- Space

1. 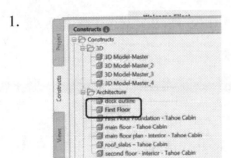 Open the **First Floor** construct.

2.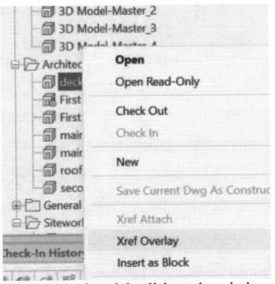

   It should appear like this.

   If it doesn't, then right click on the missing constructs and select **Xref Overlay**.

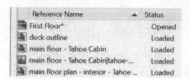

The First Floor construct should have the following xref overlays:

- Deck outline
- Main floor
- Main floor plan – interior

3.

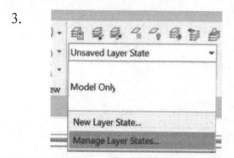

Import the **Model** layer state into the drawing.

Restore the **Model Only** layer state.

4.

Select the **Space** tool from the Build panel on the Home ribbon.

5.

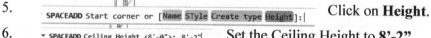

Click on **Height**.

6.

SPACEADD Ceiling Height <8'-0">: 8'-2"

Set the Ceiling Height to **8'-2"**.

7.

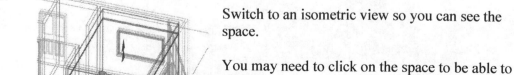

Draw a rectangle in the master bedroom by selecting the opposing corners.

Click **ESC** to exit the command.

8.

Switch to an isometric view so you can see the space.

You may need to click on the space to be able to see it.

9.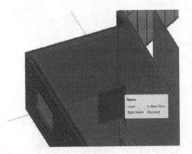

Rotate the view so you can see the underside of the space and the model.

Save *First Floor.dwg*.

Ceilings and Floors

You can use the Space/Zone Manager to modify the style that is applied to the floor or ceiling of a space. This will automatically update in any existing schedules.

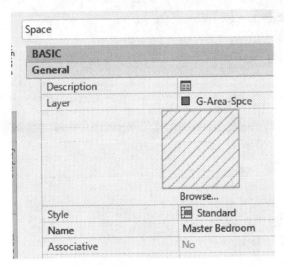

It helps if you name the space assigned to each room before you start applying different styles to ensure that you apply the correct styles to each space.

You can create space styles that use specific floor and ceiling materials to make it more efficient when defining your spaces.

Exercise 6-9:
Add a Ceiling and Floor to a Space

Drawing Name: First Floor.dwg
Estimated Time: 10 minutes

This exercise reinforces the following skills:

- ❑ Add Slab
- ❑ Slab Styles
- ❑ Properties

1.
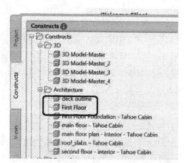
Open the **First Floor** construct.

2. -][SW Isometric][Hidden Switch to a **SW Isometric** view.

3.

Switch to the Manage ribbon.
Open the **Style Manager**.

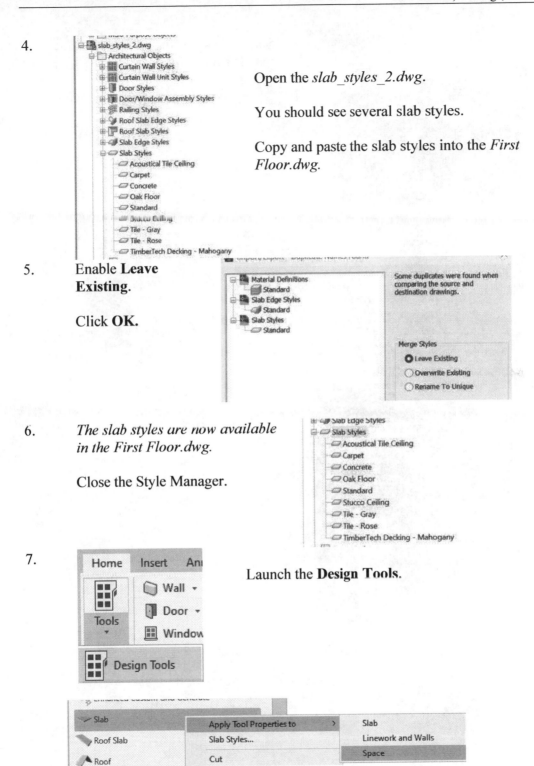

4. Open the *slab_styles_2.dwg*.

You should see several slab styles.

Copy and paste the slab styles into the *First Floor.dwg*.

5. Enable **Leave Existing**.

Click **OK**.

6. *The slab styles are now available in the First Floor.dwg.*

Close the Style Manager.

7. Launch the **Design Tools**.

8. Locate the Slab tool.
Right click and select **Apply Tool Properties to→ Space.**

9.  Select the space that is located in the master bedroom.

Click **ENTER** to complete selection.

10. 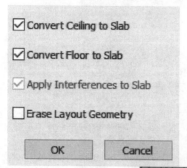 Enable **Convert Ceiling to Slab**.
Enable **Convert Floor to Slab**.

Click **OK**.

11.  *The two slabs appear.*

Click **ESC** to release the selection.

12. [−][SW Isometric][2| Switch to a **2D Wireframe** display.

13.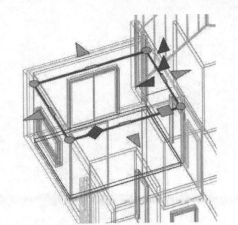

Select the ceiling slab.

14.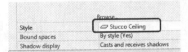

Set the Style to **Stucco Ceiling**.

Click **ESC** to release the selection.

15.

Orbit the view so you can select the floor slab.

16.

Set the Style to **Oak Floor**.

Click **ESC** to release the selection.

17.

Save as *First Floor.dwg*.

Exercise 6-10:
Add a Lighting Fixture to a Ceiling

Drawing Name: First Floor.dwg
Estimated Time: 5 minutes

This exercise reinforces the following skills:

❑   Inserting Blocks
❑   Properties

1. 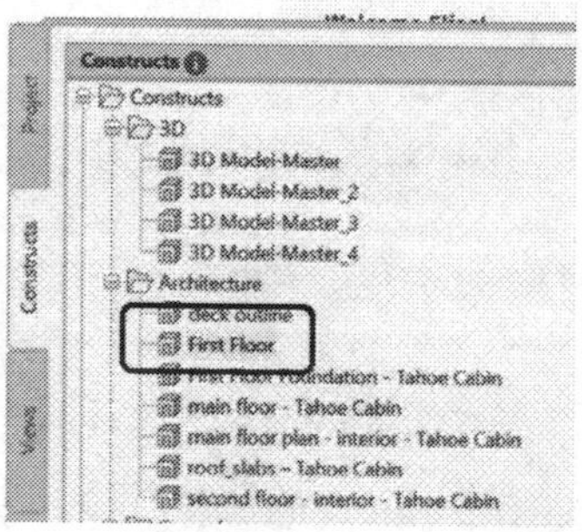   Open the **First Floor** construct.

2.   [–][Top][Real   Set the view to **Top Realistic**.

3. 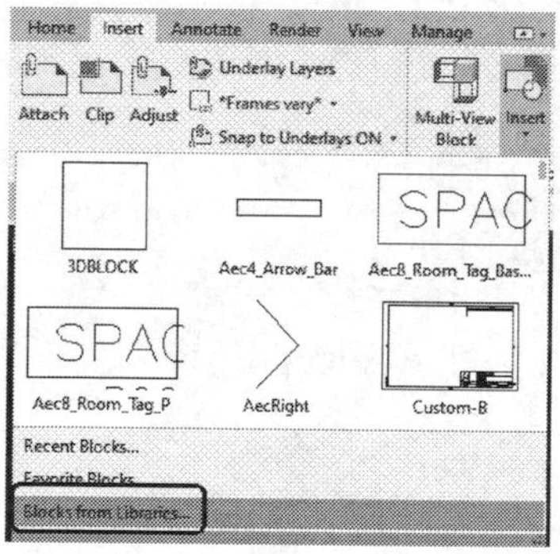   Switch to the Insert ribbon.

Select **Insert →Blocks from Libraries**.

4.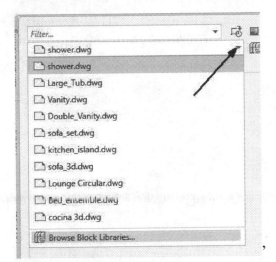

Select **Browse Block Libraries** from the drop-down list.

5.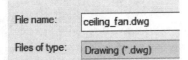

Locate the *ceiling_fan.dwg* file located in the downloaded exercises.

Click **Open**.

6.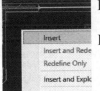

Highlight the file.

Right click and select **Insert**.

7.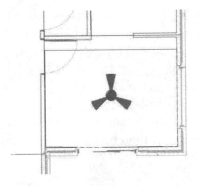

Place in the center of the master bedroom.

8.

Material	ByLayer
Geometry	
Position X	996'-7 13/32"
Position Y	-91'-4 13/32"
Position Z	7'-4"
Scale X	1.00000
Scale Y	1.00000
Scale Z	1.00000

Select the fan and set the Position Z to **7'-4"**.

It took me some trial and error to figure out the correct Z value to place the fan.

9. 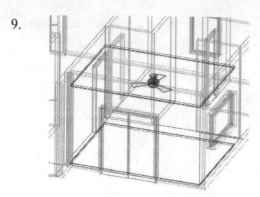 Switch to an isometric 2D Wireframe view to inspect the placement of the ceiling fan.

10. Save the file.

Stairs

Stairs are AEC objects that use flights of treads and risers to accommodate vertical circulation. Stairs also interact with railing objects. You can control the style of the stair, the shape of the landing, the type of treads, and the height and width of the stair run. By default, ACA will comply with standard business codes regarding the width of the tread and riser height. An error will display if you attempt to enter values that violate those standards.

You can modify the edges of stairs so that they are not parallel, and ACA allows curved stairs.

Landings can also be non-rectangular. Stairs allow the use of nearly arbitrary profiles for the edges of flights and landings. In addition, railings and stringers can be anchored to stairs and can follow the edges of flights and landings. You can create custom stairs from linework or profiles to model different conditions as well.

Materials can be assigned to a stair. These materials are displayed in the Realistic visual style, or when rendered. Materials have specific settings for the physical components of a stair, such as risers, nosing, and treads. You can include the material properties in any schedules.

Stair styles allow you to predefine materials for the stair. A stair style allows you to specify the dimensions, landing extensions, components, and display properties of the stair.

You can create a stair tool from any stair style. You can drag the style from the Style Manager onto a tool palette. You can then specify default settings for any stair created from the tool.

Exercise 6-11:
Placing a Straight Stair

Drawing Name: 3D Model-Master_5.dwg
Estimated Time: 20 minutes

This exercise reinforces the following skills:

- ❏ Stair
- ❏ Project Navigator
- ❏ Constructs
- ❏ External References
- ❏ Slabs
- ❏ Slab Styles
- ❏ Stairs

1.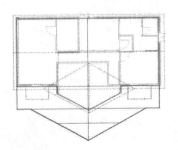

   Open *3D Model-Master_5.dwg*

   In the Project Navigator:
   Select the Constructs tab.
   Highlight 3D.
   Right click and select **Save Current Dwg As Construct**.

2.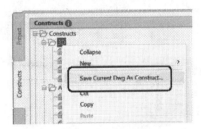

   Enable all the levels.

   Click **OK**.

3.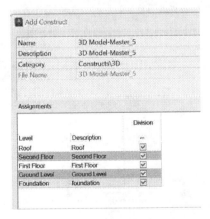

   Set the view to **Top 2D Wireframe**.

   [–][Top][2D Wireframe]

6-51

4.
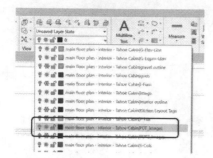

Thaw the **main floor plan – interior – Tahoe Cabin|PDF Images** layer.

This will allow you to see where stairs should be placed.

5.

Set **deck outline** as the current layer.

6.
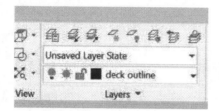

Draw a rectangle at the top of the building.

7.
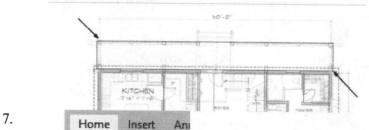

Launch the **Design Tools**.

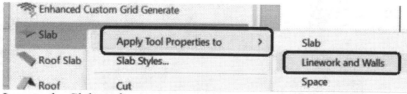

8. Locate the Slab tool.
Right click and select **Apply Tool Properties to→ Linework and Walls**.

9. Select the rectangle.
Click ENTER.

10. Click **Yes** to erase the layout geometry.

11. Click ENTER to accept **Projected**.

12.   Set the Base Height to **3'-0"**.

13.   Set the Justification to **Bottom**.

14.

Set the top midpoint as the pivot point.

15.

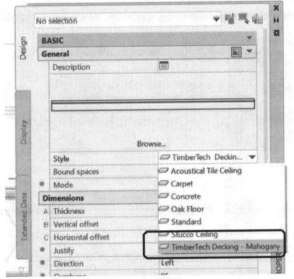

Set the Style to **TimberTech Decking – Mahogany**.

16.

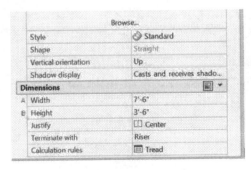

Select the **Stair** tool on the Home ribbon.

17.

On the Properties palette:

Set the Shape to **Straight**.
Set the Width to **7'-6"**.
Set the Height to **3'-6"**.
Set Justify to **Center**.

18.

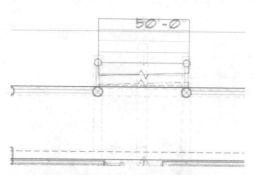

Select a point above the building model and draw the stairs straight down.

When all the risers are shown, click to end the stair run.

Click ENTER to exit the command.
You can move them next to the deck after the stairs are placed.

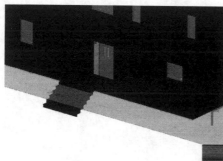

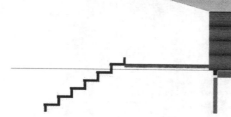

19.

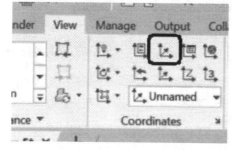

Switch to a 3D view to inspect your stairs.

Switch to a left view.

20.

If you zoom in, you can see the stairs do not quite meet the floor slab for the deck.

21.

Switch to the View ribbon.

Select the **UCS** tool.

Select the **View** option.

22.

The UCS should update to an XY orientation.

23.   Adjust the position of the stairs to meet the deck using the MOVE tool.

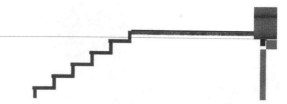

24.

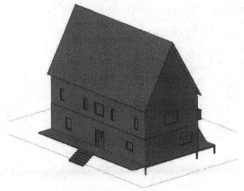

Switch to an isometric view to inspect the stairs and verify they are placed correctly.

25.

Save the file.

Railings

Railings are objects that interact with stairs and other objects. You can add railings to existing stairs, or you can create freestanding railings.

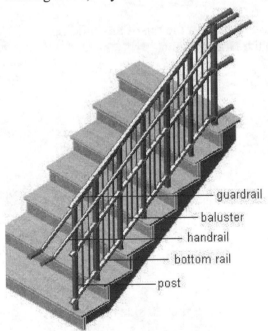

Railings can have guardrails, handrails, posts, balusters, and one or more bottom rails. Additionally, you can add custom blocks to railings.

A railing style can control the properties for all railings that use that style, rather than changing properties for each railing in the drawing. Various styles exist in the templates for common railing configurations, such as guardrail pipe, handrail grip, and handrail round.

Within the railing style, you can specify the rail locations and height, post locations and intervals, components and profiles, extensions, material assignments, and display properties of the railing.

Exercise 6-12:
Placing a Railing on a Stair

Drawing Name: 3D Model-Master_5.dwg
Estimated Time: 5 minutes

This exercise reinforces the following skills:

❑   railing

1.

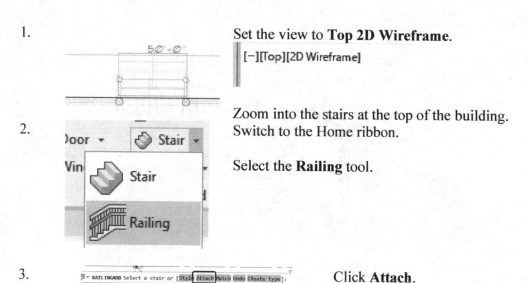

Set the view to **Top 2D Wireframe**.

[−][Top][2D Wireframe]

2.

Zoom into the stairs at the top of the building. Switch to the Home ribbon.

Select the **Railing** tool.

3.

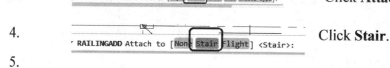

Click **Attach**.

4.

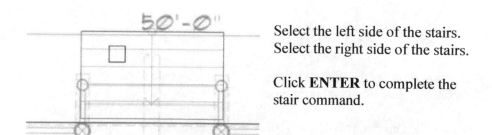

Click **Stair**.

5.

Select the left side of the stairs. Select the right side of the stairs.

Click **ENTER** to complete the stair command.

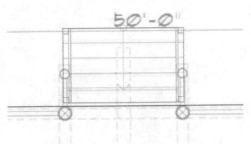

Railings should be visible on the left and right side of the stairs.

6.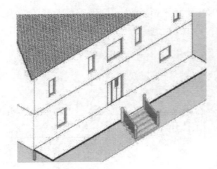

Set the view to **NE Isometric Hidden**.

[−][NE Isometric][Hidden]

Inspect the railing placed on the stair.

7.

Save the file.

Exercise 6-13:
Placing a Railing on Slab

Drawing Name: 3D Model-Master_5.dwg
Estimated Time: 5 minutes

This exercise reinforces the following skills:

❑   railing

1.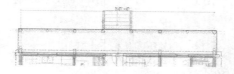

   Set the view to **Top 2D Wireframe**.

   [−][Top][2D Wireframe]

   Zoom into the deck at the top of the building.

2.

   Switch to the Home ribbon.

   Select the **Railing** tool.

3.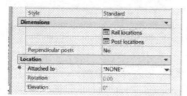

   On the Properties palette:

   Change Attached to: **None**.

4.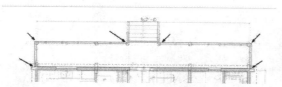

   Sketch a path for the railing following the outer edge of the deck.

   There should be one railing on the left and one on the right.

5.

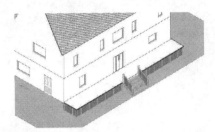

Set the view to **NE Isometric Hidden**.

[−][NE Isometric][Hidden]

Inspect the railing placed on the deck.
Notice the railing is at the wrong elevation – it
is sitting below the deck.

Select the two railings.

6.

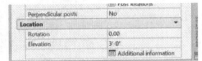

On the Properties palette:

Change the Elevation to **3'-0"**.

Release the selection.

7.

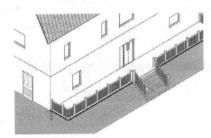

The railing now looks correct.

Save the file.

Exercise 6-14:
Placing a Stair with a Landing

Drawing Name: 3D Model-Master_5.dwg
Estimated Time: 15 minutes

This exercise reinforces the following skills:

- ❑ Stair
- ❑ Project Navigator
- ❑ Constructs
- ❑ External References
- ❑ Stairs
- ❑ Railings

1.

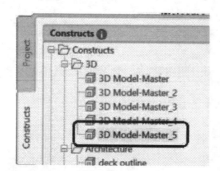

Open Construct:
3D Model-Master_5.dwg

2.

Set the view to **Top 2D Wireframe**.

[−][Top][2D Wireframe]

Switch to the View ribbon.

Activate the **Entry Foyer** view.

3.

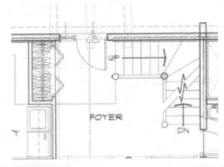
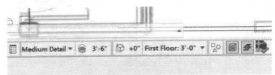

Set the elevation to **First Floor: 3'-0"**.
The view should show the Foyer area. The PDF layer can be thawed so that you can see where to locate the stairs
It can be easier to create stairs outside of the building model and then move them into the correct location.

4.

Launch the **Layer Manager**.

5.

Create a layer called **stair outline**.
Assign it the color **red**.

6.

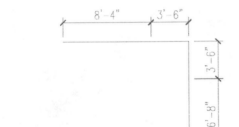

Draw four lines as shown.

These lines will act as guidelines to place the stairs. The 3'-6" lines are used to define the landing.
The 8'4" line is the first stair run.
The 6'-8" line is the second stair run.

7.

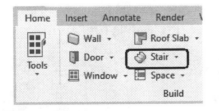

Select the **Stair** tool on the Home ribbon.

8.

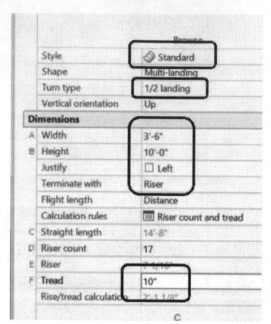

On the Properties palette:

Set the Shape to **Multi-Landing**.
Set the Turn type to **1/2 landing**.
Set the Width to **3'-6"**.
Set the Height to **10'-0"**.
Set Justify to **Left**.
Set the Tread to **10"**.

9.

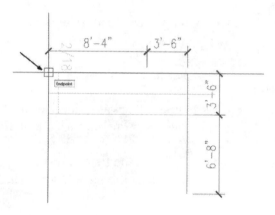

Start at the left endpoint of the guidelines.

10.

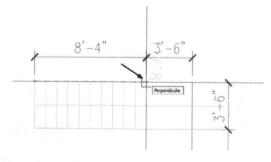

Select a point to indicate the first landing corner.

11.

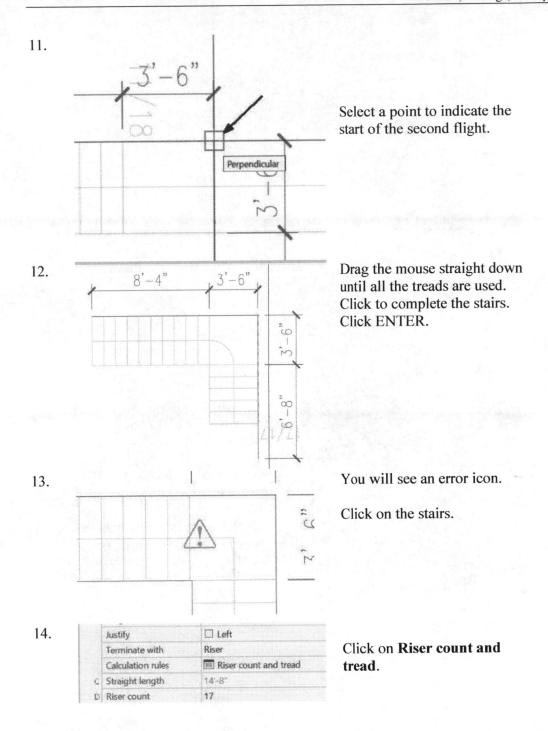

Select a point to indicate the start of the second flight.

12.

Drag the mouse straight down until all the treads are used. Click to complete the stairs. Click ENTER.

13.

You will see an error icon.

Click on the stairs.

14.

Justify	☐ Left
Terminate with	Riser
Calculation rules	🔲 Riser count and tread
C Straight length	14'-8"
D Riser count	17

Click on **Riser count and tread**.

15.

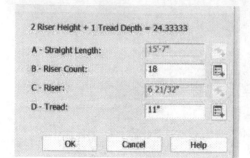

Change the Riser Count to **18**.
Change the Tread to **11"**.

Click **OK**.

16.

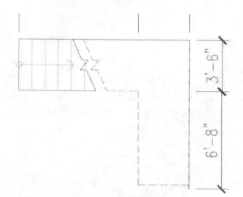

The error icon goes away.

Move the stair into the entry foyer.

17.

Switch to an isometric 2D wireframe view to see the stairs.

Set the stairs elevation to **3'-0"**, if necessary.

Switch to an isometric view to see the stairs and railing now look OK.

18.

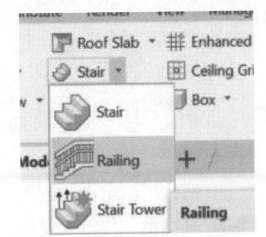

Switch back to a Top view.

Select the **Railing** tool from the Home ribbon.

Select **Attach**.
Select **Flight**.

19.

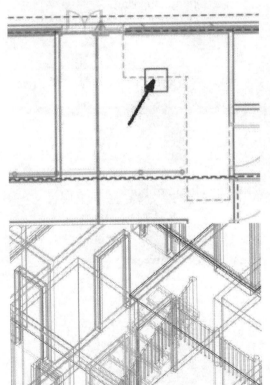

Select the outside of the stair to place the railing.

Click **ENTER**.

Repeat to add to the second flight.

20.

Switch to an isometric view to inspect the railings.

We need to make some corrections to the Building Model.

These will be done in a future exercise.

Save the model.

Exercise 6-15:
Create a Stair and Railing Style

Drawing Name: 3D Model-Master_5.dwg
Estimated Time: 30 minutes

This exercise reinforces the following skills:

- ❑ Stair
- ❑ Style Browser
- ❑ Materials
- ❑ Style Manager
- ❑ Railings

1.

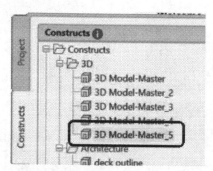

Open Construct:
3D Model-Master_5.dwg

2.

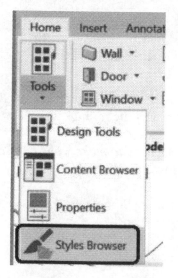

Go to the Home ribbon.

Under Tools:

Select **Styles Browser**.

3.

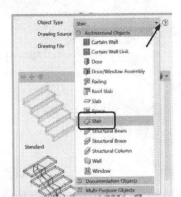

Under Object Type:

Select **Stair**.

4.

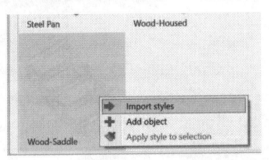

Locate the Wood-Saddle stair.

Right click and select **Import styles**.

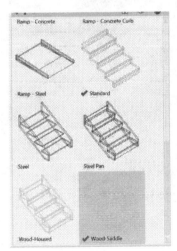

A green check is displayed next to the style.

A green check indicates the style exists in the active drawing.

5.

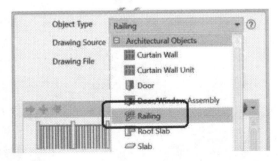

Select **Railing** under Object Type.

6.

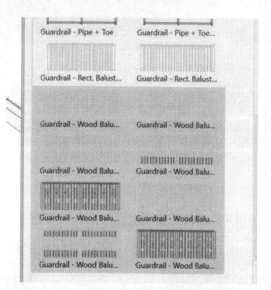

Hold down the CTL key and select all the Guardrail – Wood Baluster styles.

Right click and select **Import styles**.

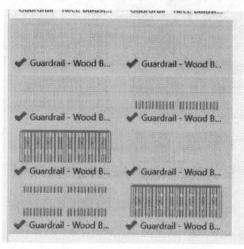

Green checks indicate the styles now are available in the active drawing.

7.   Close the Style Browser.

8.

Go to the Manage ribbon.

Select **Style Manager**.

9.

Locate the **Wood-Saddle** stair style.

Right click and select **Copy**.
Paste the copied style into the 3D Model-Master_5.dwg.

10.

Rename the copy **TimberTech Stairs**.

11.

Select the Materials tab.

Assign the TimberTech material to all the components.

Click **Apply**.

12.

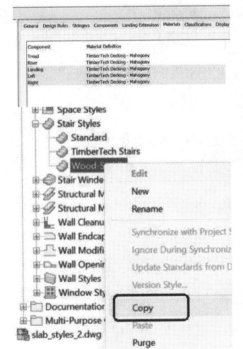

Locate the **Wood-Saddle** stair style.

Right click and select **Copy**.
Paste the copied style into the 3D Model-Master_5.dwg.

13.

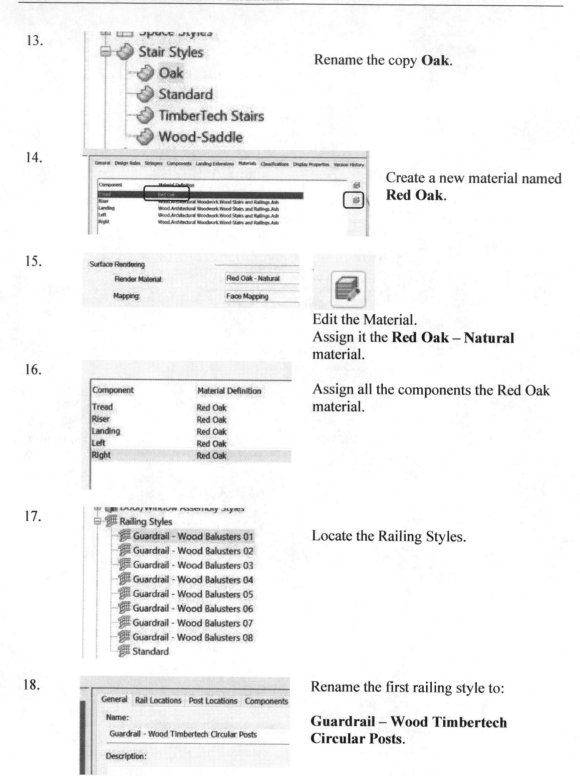

Rename the copy **Oak**.

14.

Create a new material named **Red Oak**.

15.

Edit the Material.
Assign it the **Red Oak – Natural** material.

16.

Assign all the components the Red Oak material.

17.

Locate the Railing Styles.

18.

Rename the first railing style to:

Guardrail – Wood Timbertech Circular Posts.

19.

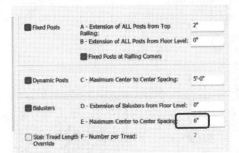

Select the Post Locations tab.

Set the value for E to **6"**.

20.

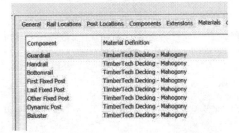

Select the Components tab.

Change the Profile Name to ***circular*** for all the components.

21.

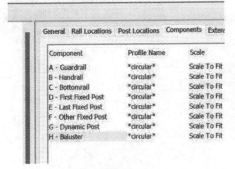

Select the Materials tab.

Assign the **TimberTech** material to all the components.

Click **Apply**.

22.

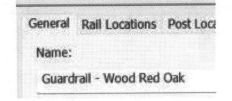

Rename the 02 railing style to:

Guardrail – Wood Red Oak

23.

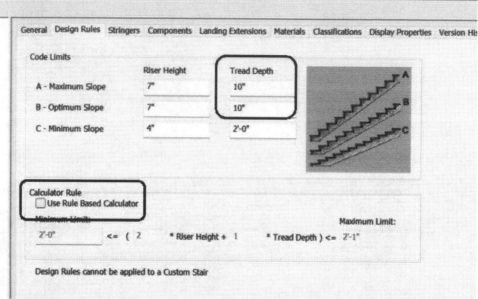

Select the Design Rules tab.
Change the Tread Depth to **10"**.
Set the Optimum Slope to **10"**.
Disable **Use Rule Based Calculator**.

24.

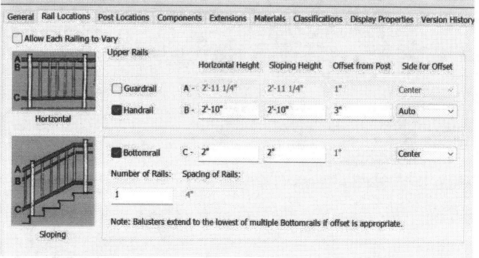

Select the Rail Locations tab.
Disable **Guardrail**.
Enable **Handrail**.
Enable **Bottomrail**.

25.

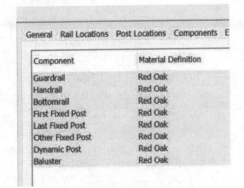

Select the Components tab.

Set the Handrail to **HR_GRIP**.
This profile is available in the Content Browser.
Set the Bottomrail to **Rectangular**.
Set the remaining components to **Baluster 01**.

26.

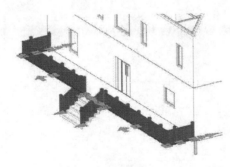

Select the Materials tab.

Set all the Components to **Red Oak**.

Click **OK**.

27.

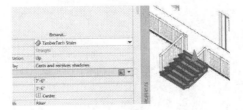

Select the railings at the front of the building model.

28.

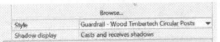

On the Properties palette:
Assign the TimberTech railing style to the selected railings.

Click ESC to release the selection.

29.

Select the stairs and assign the TimberTech Stairs style.

Click ESC to release the selection.

30.

[−][Custom View][Realistic]

Change the display to Realistic to see how the new styles appear.

31.

	Browse..	
Style	⬦ Oak	
Shape	Multi-landing	
Turn type	1/2 landing	
Vertical orientation	Up	
Shadow display	Casts and receives shadows	
Dimensions		
A	Width	3'-6"
B	Height	10'-0"
	Justify	☐ Left
	Terminate with	Riser
	Calculation rules	▦ Riser count and tread
C	Straight length	14'-2"
D	Riser count	18
E	Riser	6 21/32"

For the stair in the entry way:

Set the Style to **Oak**.
Set the Height to **10'-0".**
Set the Riser Count to **18**.

If needed, try to create the stair outside of the building and then move it into the correct location.

Save the file.

Exercise 6-16:
Add a Hole to a Slab

Drawing Name: 3D Model-Master_5.dwg
Estimated Time: 20 minutes

This exercise reinforces the following skills:

- ❑ Modifying External References
- ❑ Add a Hole to a Slab

1.

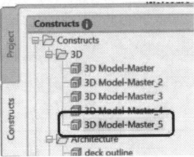

Open Construct:
3D Model-Master_5.dwg

2.

Select the slab on the second floor.

On the ribbon:

Select **Edit Reference In-Place**.

3.

Click **OK**.

4. Switch to a **Top|2D Wireframe** view.

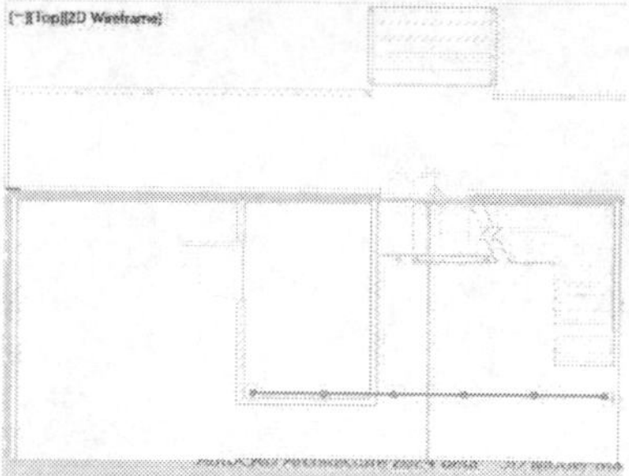

5. Create a layer called **stair outline**.
Set the layer current.

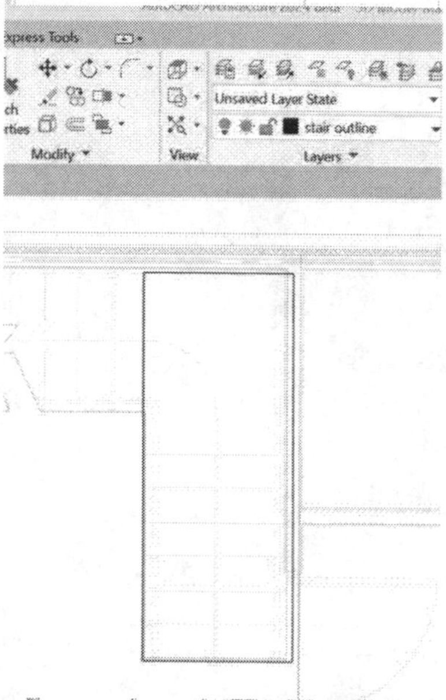

Draw a rectangle around the second/top flight of stairs.

6. Switch back to a 3D view.

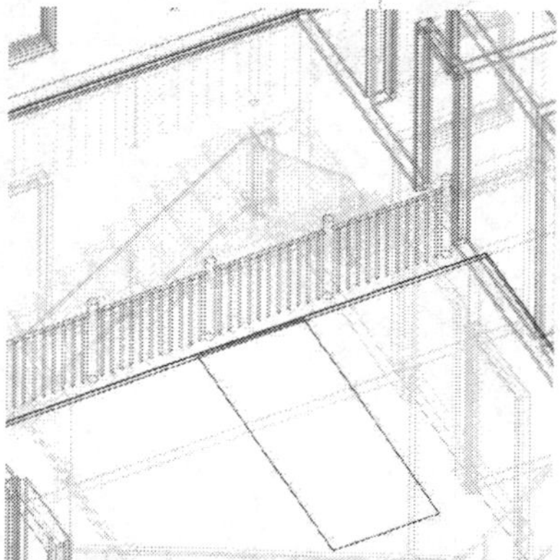

Orient the view so you see the floor slab on the second level, the rectangle and the interior stairs.

7. Select the floor slab.

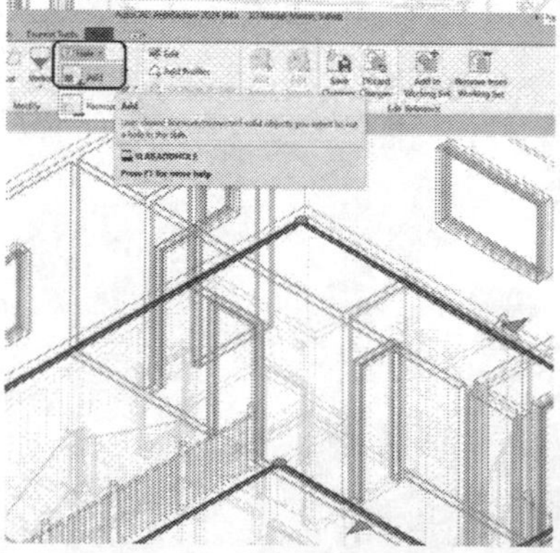

On the ribbon:

Select **Hole→Add**.

Select the rectangle.

Click **Yes** to erase the layout geometry.

A hole is created in the floor slab.

8.

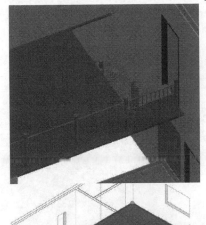

Switch to a Realistic display.

You can now see the upper flight of stairs.

9.

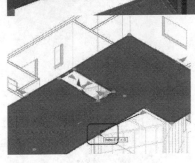

Select the floor slab.

Use the grip to extend the slab edge to allow more space for the door.

Click ESC to release the selection.

10.

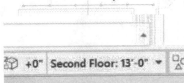

Move the railing to the new edge of the floor slab.

11.

Set the Level to **Second Floor**.

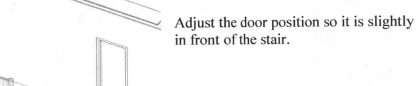

12.

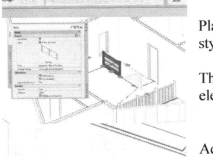

Place a railing using the Wood Red Oak style to the left of the stair.

The railing should be placed at an elevation of 13'-0".

13.

Adjust the door position so it is slightly in front of the stair.

14.

Select **Save Changes** on the ribbon to save the changes to the external reference.

15.

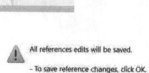

All references edits will be saved.

- To save reference changes, click OK.
- To cancel the command, click Cancel.

OK   Cancel

Click **OK**.

16.

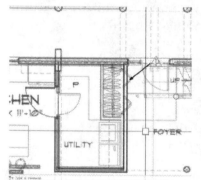

Switch to a Top|2D Wireframe view.

Select the main floor external reference.

Edit Reference In-Place

On the ribbon:

Select **Edit Reference In-Place**.

17.

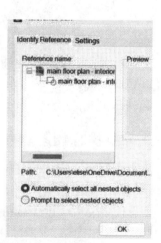

Identify Reference Settings

Reference name: Preview

main floor plan - interior
main floor plan - inte

Path: C:\Users\elise\OneDrive\Document...

● Automatically select all nested objects
○ Prompt to select nested objects

OK

Select **OK**.

18.

Shift the interior wall to the left so we can adjust the location of the front door.

19.

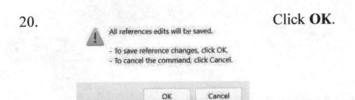

Select **Save Changes** on the ribbon to save the changes to the external reference.

20.

Click **OK**.

All references edits will be saved.

- To save reference changes, click OK.
- To cancel the command, click Cancel.

OK   Cancel

21.

Click on the front door.

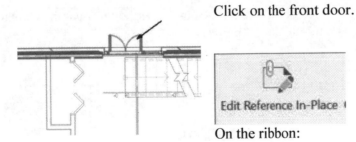

On the ribbon:

Select **Edit Reference In-Place**.

22.

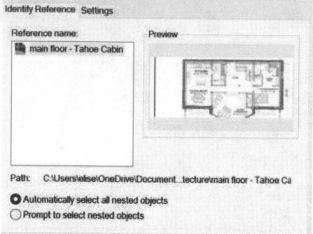

Select **OK**.

23.

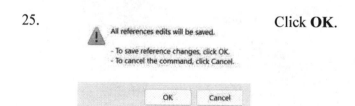

Shift the front door to the left.

24.

Select **Save Changes** on the ribbon to save the changes to the external reference.

25.

⚠ All references edits will be saved.

- To save reference changes, click OK.
- To cancel the command, click Cancel.

OK   Cancel

Click **OK**.

26.

Save the file.

Extra: *Add a ceiling to the second level. Add a chandelier to the foyer area. I have included two chandelier drawings in the exercise folder for you to choose from.*

QUIZ 6

True or False

1. You can create a floor or ceiling slab by converting a space.

2. Once you place a ceiling grid, it can't be modified.

3. When you place a lighting fixture on a ceiling grid, you cannot snap to the grid.

4. Railings are automatically placed on both sides of a selected stair.

5. You can assign materials to different components of a railing.

Multiple Choice

6. Slabs can be used to model: (select all that apply)

   A. Roofs
   B. Floors
   C. Ceilings
   D. Walls

7. Ceiling grids can be created using: (select all that apply)

   A. Linework
   B. Walls
   C. Roof Slabs
   D. Arcs

8. Space Styles: (select all that apply)

   A. Control the hatching displayed
   B. Control the name displayed when spaces are tagged
   C. Cannot be modified
   D. Have materials assigned

9. Railings can have the following: (select all that apply)

   A. Guardrails.
   B. Handrails.
   C. Posts.
   D. Balusters.

10. The first point selected when placing a set of stairs is:

    A. The foot/bottom of the stairs
    B. The head/top of the stairs
    C. The center point of the stairs
    D. Depends on the property settings

11. The Styles Browser can be used to:

    A. Create styles
    B. Modify existing styles
    C. Import styles into the current drawing
    D. Delete styles

ANSWERS:

1) T; 2) F; 3) F; 4) F; 5) T; 6) D; 7) A, B, C; 8) A, B; 9) A, B, C, D; 10) A; 11) C

Lesson 7:
Views, Schedules, and Legends

Before we can create our construction drawings, we need to create layouts or views to present the design. AutoCAD Architecture is similar to AutoCAD in that the user can work in Model and Paper Space. Model Space is where we create the 3D model of our house. Paper Space is where we create or set up various views that can be used in our construction drawings. In each view, we can control what we see by turning off layers, zooming, panning, etc.

To understand paper space and model space, imagine a cardboard box. Inside the cardboard box is Model Space. This is where your 3D model is located. On each side of the cardboard box, tear out a small rectangular window. The windows are your viewports. You can look through the viewports to see your model. To reach inside the window so you can move your model around or modify it, you double-click inside the viewport. If your hand is not reaching through any of the windows and you are just looking from the outside, then you are in Paper Space or Layout mode.
You can create an elevation in your current drawing by first drawing an elevation line and mark, and then creating a 2D or 3D elevation based on that line. You can control the size and shape of the elevation that is generated. Unless you explode the elevation that you create, the elevation remains linked to the building model that you used to create it. Because of this link between the elevation and the building model, any changes to the building model can be made in the elevation as well.

When you create a 2D elevation, the elevation is created with hidden and overlapping lines removed. You can edit the 2D elevation that you created by changing its display properties. The 2D Section/Elevation style allows you to add your own display components to the display representation of the elevation and create rules that assign different parts of the elevation to different display components. You can control the visibility, layer, color, linetype, lineweight, and linetype scale of each component. You can also use the line work editing commands to assign individual lines in your 2D elevation to display components and merge geometry into your 2D elevation.

After you create a 2D elevation, you can use the AutoCAD BHATCH and AutoCAD DIMLINEAR commands to hatch and dimension the 2D elevation.

AutoCAD Architecture comes with standard title block templates. They are located in the Templates subdirectory under *Program Data \ Autodesk\ ACA 2024 \enu \Template*.

Many companies will place their templates on the network so all users can access the same templates. In order to make this work properly, set the Options so that the templates folder is pointed to the correct path.

You use the Project Navigator to create additional drawing files with the desired views for your model. The views are created on the Views tab of the Project Navigator. You then create a sheet set which gathers together all the necessary drawing files that are pertinent to your project.

Previously, you would use external references, which would be external drawing files that would be linked to a master drawing. You would then create several layout sheets in your master drawing that would show the various views. Some users placed all their data in a single drawing and then used layers to organize the data.

This shift in the way of organizing your drawings will mean that you need to have a better understanding of how to manage all the drawings. It also means you can leverage the drawings so you can reuse the same drawing in more than one sheet set.

You can create five different types of views using the Project Navigator:

- ❑ Model Space View – a portion that is displayed in its own viewport. This view can have a distinct name, display configuration, description, layer snapshot and drawing scale.
- ❑ Detail View – displays a small section of the model, i.e. a wall section, plumbing, or foundation. This type of view is usually associated with a callout. It can be placed in your current active drawing or in a new drawing.
- ❑ Section View – displays a building section, usually an interior view. This type of view is usually associated with a callout. It can be placed in your current active drawing or in a new drawing.
- ❑ Elevation View – displays a building elevation, usually an exterior view. This type of view is usually associated with a callout. It can be placed in your current active drawing or in a new drawing.
- ❑ Sheet View – this type of view is created when a model space view is dropped onto a layout sheet.

Note: Some classes have difficulty using the Project Navigator because they do not use the same workstation each class or the drawings are stored on a network server. In those cases, the links can be lost and the students get frustrated trying to get the correct results.

I have created an archive file at the end of most of the exercises to help students keep up. Simply select the zipped archive file for the exercise BEFORE the exercise you will be doing, extract it, redefine the paths to the different external references and go from there.

When creating elevation and section views as new drawings to be added as views to the Project Navigator, several of the views were not generated correctly. This is a bug in the software and not user error. If this happens to you, add the view in the current drawing and then use WBLOCK to create the external view drawing, then add it to the Project Navigator as a view.

As we start building our layout sheets and views, we will be adding schedule tags to identify different building elements.

Schedule tags can be project-based or standard tags. Schedule tags are linked to a property in a property set. When you anchor or tag a building element, the value of the property displays in the tag. The value displayed depends on which property is defined by the attribute used by the schedule tag. ACA comes with several pre-defined schedule tags, but you will probably want to create your own custom tags to display the desired property data.

Property sets are specified using property set definitions. A property set definition is a documentation object that is tracked with an object or object style. Each property has a name, description, data type, data format, and default value.

To access the property sets, open the Style Manager. Highlight the building element you wish to define and then select the Property Sets button. You can add custom property sets in addition to using the default property sets included in each element style.

To determine what properties are available for your schedules, select the element to be included in the schedule, right click and select Properties.

Tags and schedules use project information. If you do not have a current project defined, you may see some error messages when you tag elements.

A Property Set Definition defines the bits of data that we want to be able to show in a schedule. In the schedule table style definition, these properties can become columns. When the properties are attached to an object, the schedule table that is based on that schedule table style will display those bits of data.

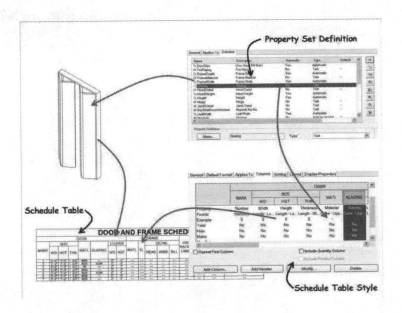

Elevation Views

Elevations of the building models are created by using the Elevation tool on the Annotate ribbon. You first draw an elevation line and mark to indicate the direction of the elevation, selecting the scope/depth of the view, and then placing a 2D or 3D elevation based on the elevation line. You can control the size and shape of any elevation that you create. If you modify the building model, you can update the elevation view so it shows the changes. 2D elevations are created with hidden and overlapping lines removed. You can control the display of 2D elevations by modifying or creating elevation styles and modifying the display properties of the 2D elevation.

We can also create live 3D sections to see the model interiors.

Exercise 7-1:

Creating Elevation Views

Drawing Name: elevation.dwg
Estimated Time: 15 minutes

This exercise reinforces the following skills:

   ❑ Creating an Elevation View
   ❑ Project Navigator

1.

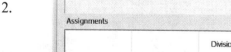

Open *3D Model-Master_6.dwg*.

In the Project Navigator:

Save the **Current Dwg As Construct**.

2.

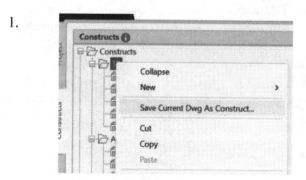

Enable all the levels.

Click **OK**.

3. Activate the **Top** view by clicking the top plane on the view cube.
You can also use the shortcut menu in the upper left corner of the display window to switch views.

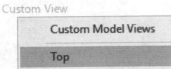

4. 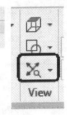 Verify using the View cube that North is oriented on top.

5. [−][Top][2D Wireframe]   Set the view to **2D Wireframe** mode by selecting the View Display in the upper left corner of the window.

6.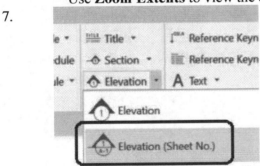

Use **Zoom Extents** to view the entire model.

7. Activate the **Annotate** ribbon.

Select the **Elevation (Sheet No.)** tool on the Callouts panel.

8. 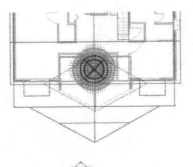 Place the elevation mark below the model.
Use your cursor to orient the arrow toward the building model.

9.

Set your view name to **South Elevation**.

Enable **Generate Section/Elevation**.

Enable **Place Titlemark**.

Set the Scale to **1/8″ = 1′-0″**.

Click the **New View Drawing** button.

If the view does not generate properly for you, use the Current Drawing option, then use WBLOCK to create a new drawing using the view.

10.

Type **South Elevation** for the Name.
Type **Exterior South Elevation** for the Description.
Click **Next**.

11.

Enable all the levels.

Click **Next**.

12.

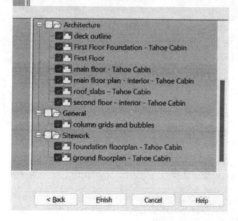

Click **Finish**.

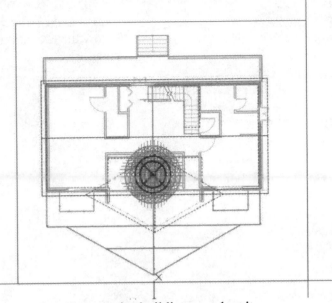

13.   Window around the entire building to select it.
        Select the upper left corner above the building and the lower right
        corner below the building.

14.   When prompted for the inscrtion point, type **0,0,0**.

15.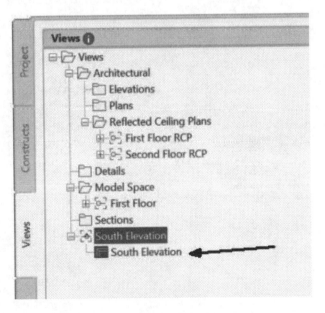

Go to the Project Navigator.

Locate the **South Elevation**
view on the Views tab.

16.

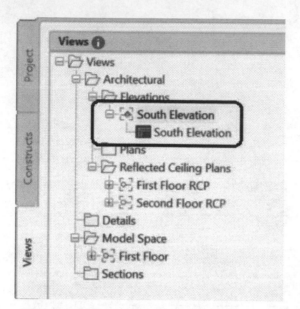

Drag and drop the view below the Elevations category.

17.

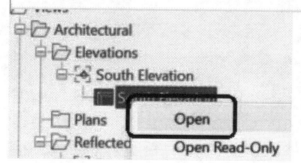

Click **Repath project now**.

Remember that you should not use the File Explorer to move, delete, rename, or copy files in a project. Only use the Project Navigator.

18.

Right click on the South Elevation.

Select **Open**.

19.

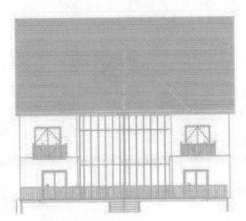

You should see the elevation.

Save and close all the drawings.

Exercise 7-2:

Creating a 3D Section View

Drawing Name: 3D Model-Master_6.dwg
Estimated Time: 15 minutes

This exercise reinforces the following skills:

- ❏ Creating an Elevation View
- ❏ Enable and Disable Live Section
- ❏ Project Navigator

1.

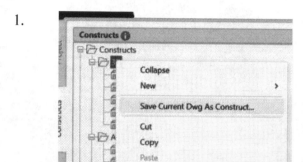

Open *3D Model-Master_6.dwg construct.*

2.

[−][Top][2D Wireframe]

Switch to a Top|2D Wireframe view.

Zoom into the 3D model.

3.

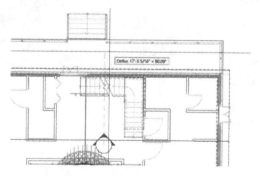

Select the Elevation tool.

Place the elevation marker below the stairs and orient towards the stairs.

Window around the stairs.

4.

Type **Stairs Elevation**.

Click **New View Drawing**.

If the view does not generate properly for you, use the Current Drawing option, then use WBLOCK to create a new drawing using the view.

5.

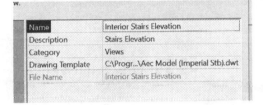

Type **Interior Stairs Elevation** for the Name.
Type **Stairs Elevation** for the Description.

Click **Next**.

6.

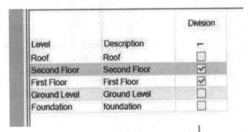

Enable the **First Floor** and **Second Floor** only.

Click **Next**.

7.

Click **Finish**.

Window around the interior stairs only to include them in the view.

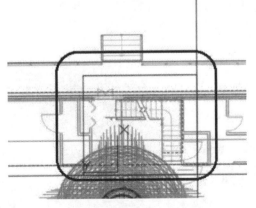

8.

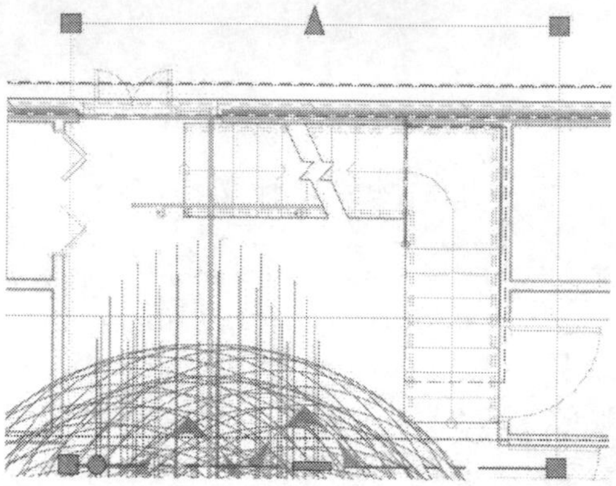

Select the section line.

9.

Enable
Live Section

Click **Enable Live Section** on the ribbon.

10.

[−][SW Isometric][Hidden]

Switch to a **SW Isometric|Hidden** view.

11.

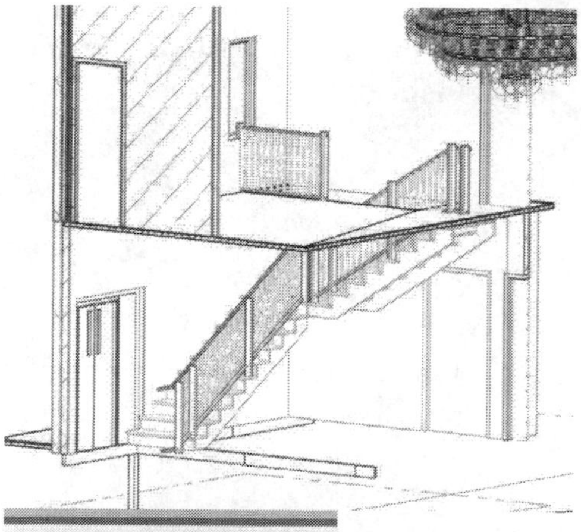

You see a 3D view of the interior stairs.

12.

-][Top][2D Wireframe]

Disable
Live Section

Return to a Top|2D Wireframe view.

Select the elevation line.

Select **Disable Live Section**.

13.

Go to the Project Navigator.

Select the Views tab.

Right click on the **Stairs Elevation** view and select **Open**.

14.

A view of the stairs is created.

Save and close all the drawings.

Exercise 7-3:
Creating a Detail View

Drawing Name: column detail.dwg
Estimated Time: 5 minutes

This exercise reinforces the following skills:

- Creating a DetailView
- Project Navigator

1. Open *column detail.dwg*.

2. Go to the Project Navigator.
   Open the Views tab.
   Highlight **Details**.
   Right click and select **Save Current Dwg As View→Detail**.

3. Type **column detail** for the Name.
   Type **column detail – Foundation** for the description.
   Click **Next**.

4. *This is a detail view, so do not select any of the levels.*

   Click **Next**.

5.

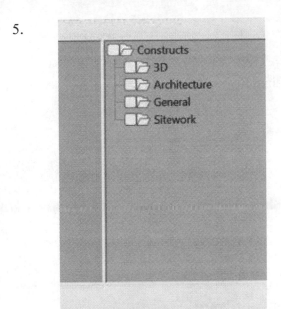

Do not select any of the constructs.

Click **Finish**.

6.

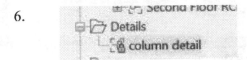

The view is now listed under the Details category.

Save and close all drawings.

Wall Sections

Most building documents include section cuts for each wall style used in the building model. The section cut shows the walls from the roof down through the foundation. Imagine if you could cut through a house like a cake and take a look at a cross section of the layers, that's what a wall section is. It shows the structure of the wall with specific indication of materials and measurements. It doesn't show every wall, it is just a general guide for the builder and the county planning office, so they understand how the structure is designed.

If you define the components for each wall style with the proper assigned materials, you can detail the walls in your model easily.

Exercise 7-4:

Creating a Wall Section

Drawing Name: main floor – Tahoe Cabin.dwg
Estimated Time: 10 minutes

This exercise reinforces the following skills:

- ❑   Wall Section
- ❑   Views

1.
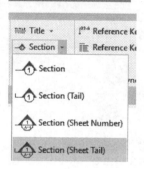
Open *main floor – Tahoe Cabin* from the Constructs tab on the Project Navigator.

2.

Activate the Model layout tab.

3.
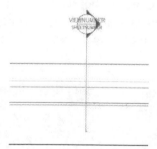
Activate the **Annotate** ribbon.

Select the **Section (Sheet Tail)** tool.

4.
Draw the section through the north wall..

Select a point below the wall.
Select a point above the wall.
Right click and select ENTER.
Left click to the right of the section line to indicate the section depth.

If the block is small, check the properties and verify that the scale is set to 1.

5.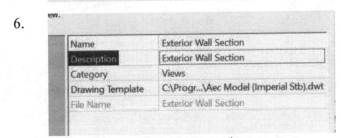

Type Exterior Wall Section in the View Name field.

Enable Generate Section/Elevation.

Enable Place Title mark.
Set the Scale to 1/8″ = 1′-0″.

Click **New View Drawing**.

6.

Name	Exterior Wall Section
Description	Exterior Wall Section
Category	Views
Drawing Template	C:\Progr...\Aec Model (Imperial Stb).dwt
File Name	Exterior Wall Section

Type **Exterior Wall Section** for the Name.
Type **Exterior Wall Section** for the description.
Click **Next**.

7.

Level	Description	DIVISION
Roof	Roof	☐
Second Floor	Second Floor	☐
First Floor	First Floor	☑
Ground Level	Ground Level	☐
Foundation	foundation	☐

Enable **First Floor**.

Click **Next**.

8.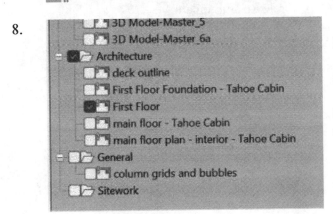

Enable **First Floor**

Click **Finish**.

Click outside the building model to create the view.

9.

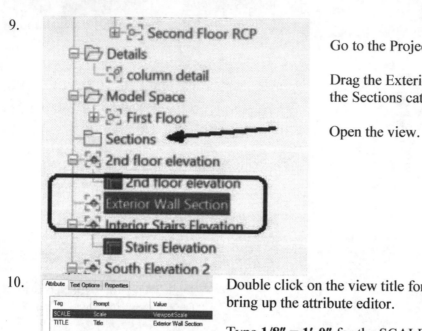

Go to the Project Navigator.

Drag the Exterior Wall Section to the Sections category.

Open the view.

10. Double click on the view title for the wall section to bring up the attribute editor.

Type **1/8″ = 1′-0″** for the SCALE.

Type **Exterior Wall Section** for the TITLE.

Click ESC to release and update.

11. Select the View title bubble.

Type **S1.1** for the NUMBER.
Click **OK**.

Click **ESC** to release and update.

12. Save and close all the drawings.

EXTRA: Create a section view for the interior wall.

Exercise 7-5:
Add Keynotes

Drawing Name: Exterior Wall Section.dwg
Estimated Time: 25 minutes

This exercise reinforces the following skills:

- ❏ Keynote Database
- ❏ Adding Keynotes

1.

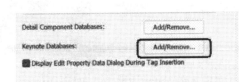

Open *Exterior Wall Section.dwg*.

2.   *Verify that the keynote database is set properly in Options.*

Type **OPTIONS** to launch the Options dialog.

Select the AEC Content tab.

Select the Add/Remove button next to **Keynote Databases**.

3.   Verify that the paths and file names are correct.

Click **OK**.

Many companies set up their own keynote files.

Close the Options dialog.

4.

Index	Name	Priority	Width
1	Logs	810	6"
2	Air Gap	700	1"
3	Rigid Insulation	600	1 1/2"
4	Stud	500	4"
5	GWB	1200	5/8"

If you go to the Style Manager for the main floor drawing and check the wall style, you can verify the components used in the wall.

5. 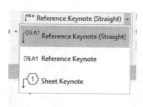 Activate the Annotate ribbon.

Select the **Reference Keynote (Straight)** tool on the Keynoting panel.

6. Select the outside component of the wall section.

The Keynote database dialog window will open.

7. 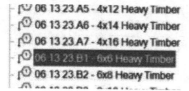 Select the *AecKeynotes (US).mdb*.

8. 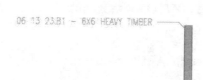 Locate the keynote for the **06 13 23 B1 – 6x6 Heavy Timber**.

Highlight and Click **OK**.

9. Select the outside component to start the leader line.
Left click to locate the text placement.
Click ENTER to complete the command.

10.  Select the **Reference Keynote (Straight)** tool on the Keynoting panel.

11. Select the *AecKeynotes (US).mdb* database from the drop-down list.

12.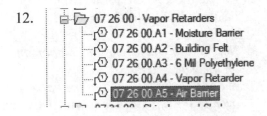

Locate the air barrier keynote.

Highlight and Click **OK**.

You can use the Filter tool at the bottom of the dialog to search for key words.

13.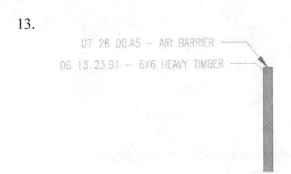

Left click to place the leader and text.

14.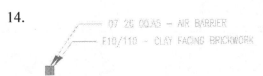

To turn off the shading on the keynotes, turn off the field display.

Type **FIELDDISPLAY** and then **0**.

15.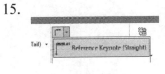

Select the **Reference Keynote (Straight)** tool on the Keynoting panel.

16.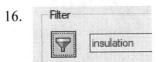

Select the top of the wall.

In the Filter text field, type **insulation**.
Left click on the Filter icon.

17.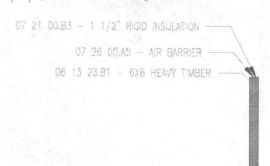

Locate the keynote for the rigid insulation and add to the view.

Note that if you select the component on the wall section, ACA will automatically take you to the correct area of the database to locate the proper keynote.

18.

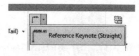

07 21 00.B3 - 1 1/2" RIGID INSULATION ————

07 26 00.A5 — AIR BARRIER ————

06 13 23.B1 — 6X6 HEAVY TIMBER ————

Left click to place the leader and the text.

19.

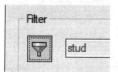

Select the **Reference Keynote (Straight)** tool on the Keynoting panel.

20.

Filter

stud

Select the top of the wall.

In the Filter text field, type s**tud**.
Left click on the Filter icon.

21.

Details\Details (US)\AecKeynotes (US).mdb

Select the *AecKeynotes (US).mdb*.

22.
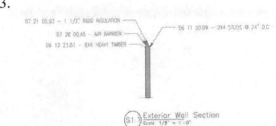

Division 06 - Wood, Plastics, and Composites
  06 11 00 - Wood Framing
    06 11 00.A3 - Existing Studs
    06 11 00.D7 - 2x4 Studs
    06 11 00.D8 - 2x4 Studs @ 16" O.C.
    06 11 00.D9 - 2x4 Studs @ 24" O.C.
    06 11 00.F6 - 2x6 Studs
    06 11 00.F7 - 2x6 Studs @ 16" O.C.
    06 11 00.F8 - 2x6 Studs @ 24" O.C.

Locate the keynote for the 2x4 Studs.

23.

Left click to place the leader and the text.

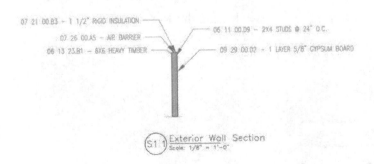

24.

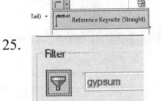

Select the **Reference Keynote (Straight)** tool on the Keynoting panel.

25.

Select the top of the wall.

In the Filter text field, type gypsum.
Left click on the Filter icon.

26.

09 29 00.C1 - 1/2" Gypsum Wallboard
09 29 00.C2 - 1/2" Type "X" Gypsum Wallboard
09 29 00.C3 - 1/2" M.R. Gypsum Board
09 29 00.D1 - 5/8" Gypsum Wallboard
09 29 00.D2 - 1 Layer 5/8" Gypsum Board
09 29 00.D3 - 1 Layer 5/8" Gypsum Board On Each Side
09 29 00.D4 - Add 1 Layer 5/8" Gypsum Board To Existing Wall

Locate the keynote for the **1 Layer 5/8" Gypsum Board.**

Click **OK.**

27.

Save the drawing.

EXTRA: If you created a section view of the interior wall, add keynotes to that section view.

Exercise 7-6:
Add Keynote Legend

Drawing Name: Exterior Wall Section.dwg
Estimated Time: 15 minutes

This exercise reinforces the following skills:

- ❑ Keynote Legend
- ❑ Keynote display
- ❑ View Manager

1. Open *the Exterior Wall Section.dwg*.

2.   **Model**   Activate the Model layout tab.

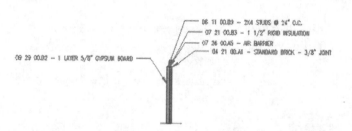

Notice the keynote numbers and text.

3. Switch to the Annotate ribbon.

Locate the **Reference Keynote Display** tool and select it.

┌ Reference Keynote (Straight) ▾
≡ Reference Keynote Legend ▾
┌ Text (Straight Leader) ▾

⌂ Keynote Editor
Select Database
Reference Keynote Display
Sheet Keynote Display
Keynoting

4.

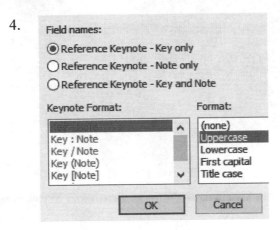

Enable **Reference Keynote – Key only**.

Enable **Uppercase**.

Click **OK**.

The keynotes update to display keys only.

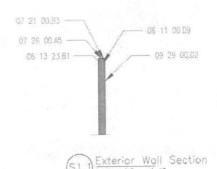

5.

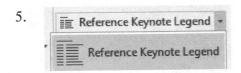

Select the **Reference Keynote Legend** tool from the Keynotes panel on the Annotate tab.

6.

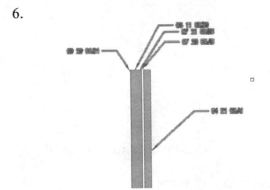

You will be prompted for the keynotes to be included in the legend.

Window around to select all the keynotes.

Click ENTER to complete the selection.

7.

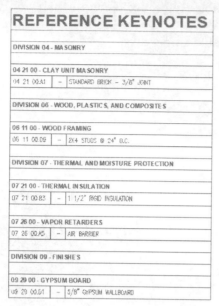

Click to place the keynote legend.

There will be a slight pause before the legend is placed while ACA compiles the data.

8.

Right
Front
Back
SW Isometric
SE Isometric
NE Isometric
View Manager...

Switch to the View tab.

Launch the **View Manager**.

9.

Set Current
New...
Update Layers
Edit Boundaries...
Delete

Click **New**.

10.

Type **Exterior Wall Keynotes Legend**.

Enable **Define Window**.

Click the window icon.

Window around the keynotes legend.

Click **ENTER**.

Click **OK**.

11.

The new named view is listed.

You can see a preview in the preview window.

12.

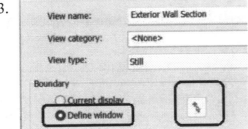

Click **New**.

13.

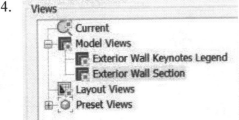

Type **Exterior Wall Section**.

Enable **Define Window**.

Click the window icon.

Window around the section view.

Click **ENTER**.

Click **OK**.

14.

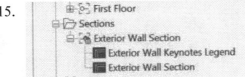

The new named view is listed.

You can see a preview in the preview window.

Click **OK**.

15.

Look in the Project Navigator.

The two named views are listed under the drawing.

16.

Save the drawing and close.

Section Views

Section views show a view of a structure or building as though it has been sliced in half or along an imaginary plane. This is useful because it gives a view through spaces and can reveal the relationships between different parts of the building which may not be apparent on the plan views. Plan views are also a type of section view, but they cut through the building on a horizontal plane instead of a vertical plane.

Exercise 7-7:

Create a Section View

Drawing Name: main floor plan – interior – Tahoe Cabin.dwg
Estimated Time: 20 minutes

This exercise reinforces the following skills:

- ❑   Sections
- ❑   Elevations
- ❑   WBLOCK
- ❑   View Manager

1.

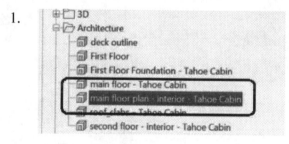

In the Project Navigator:
Open the Constructs tab.
Highlight *main floor plan – interior – Tahoe Cabin.dwg*.
Right click and select **Open**.

2.

Activate the Annotate ribbon.

Select the **Section (Sheet Tail)** tool on the Callouts panel.

3.

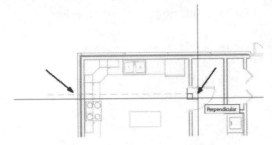

Draw a horizontal line through the kitchen as shown.

4.

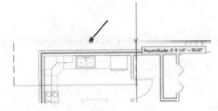

Select a point outside the building and above the kitchen area to define the view extents.

5.

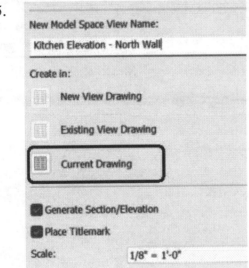

Name the View **Kitchen Elevation – North Wall**.
Enable **Generate Section/Elevation**.
Enable **Place Titlemark**.
Set the Scale to **1/2″ = 1′-0″**.

Click **Current Drawing**.

Place the view in the current drawing to the left of the building.

6.

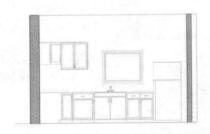

The view is created.

7.

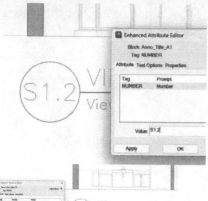

Double click on the View Number.

Change to **S1.2**.

Click **OK**.

8.

Double click on the View Title.
Change to **Kitchen- North Wall**.
Change the Scale to: **Scale: 1/8" = 1'-0"**.
Click **OK**.

9.

🗂 Right
🗂 Front
🗂 Back
◈ SW Isometric
◈ SE Isometric
◈ NE Isometric
View Manager...

Switch to the View tab.

Launch the **View Manager**.

10.

Current View: Current
Views
　◈ Current
　　🗂 Model Views
　　　🗂 Kitchen Elevation – North Wall
　　🗂 Layout Views
　　◈ Preset Views
　　　🗂 Top

The Kitchen Elevation view is listed.

Click **OK**.

11. Type**WBLOCK** to create a new view drawing.

12.

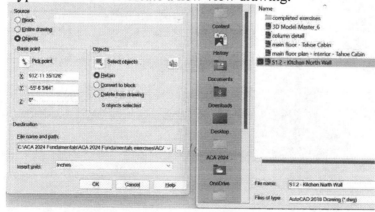

Select the lower left point of the view as the base point.
Window around the view to select it, include the view title.
Name the **file S1.2 – Kitchen North Wall**.
Click Save.
Click OK.

13. Save and Close the main floor plan-interior-Tahoe Cabin file.

14. Open the file *S1.2 – Kitchen North Wall.dwg*

15.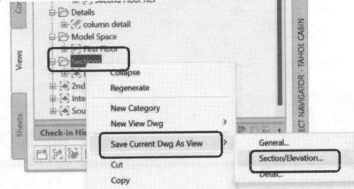

In the Project Navigator: Highlight the Sections category on the Views tab. Select **Save Current Dwg As View→Section/ Elevation**.

16.

Name	S1.2 - Kitchen North Wall
Description	S1.2 - Kitchen North Wall
Category	Views\Sections
File Name	S1.2 - Kitchen North Wall

Click **Next**.

17.

Level	Description	Division
Roof	Roof	☐
Second Floor	Second Floor	☐
First Floor	First Floor	☐
Ground Level	Ground Level	☐
Foundation	foundation	☐

Do not select any of the levels.

Click **Next**.

18.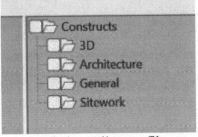

Click **Finish**.

19. Save and close all open files.

To create an elevation view with different visual styles, create a camera view.

Exercise 7-8:

Add a Door Elevation

Drawing Name: main floor – Tahoe Cabin
Estimated Time: 15 minutes

This exercise reinforces the following skills:

- Elevations
- View Titles
- View Manager
- Named Views

1.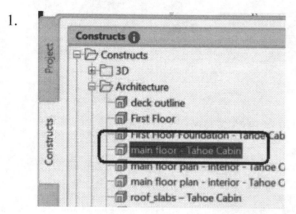

    Open the Project Navigator.

    Select the Constructs tab.

    Highlight main floor – Tahoe Cabin.

    Right click and select **Open**.

2.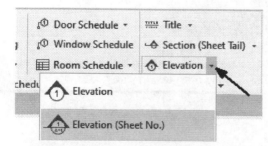

    Zoom into the laundry room area.

    We will create an elevation for the interior door frame.

3.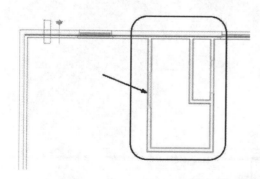

    Activate the Annotate ribbon.

    Select the **Elevation (Sheet No)** tool.

4.

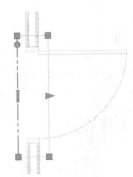

Left click to the left of the door to place the elevation marker.

Orient the marker so it is pointing toward the door.

5.

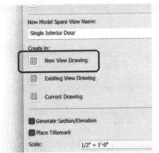

Name the view **Single Interior Door**.
Enable **Generate Section/Elevation**.
Enable **Place Titlemark**.
Set the Scale to **1/2″ = 1′-0″**.

Click **New View Drawing**.

6.

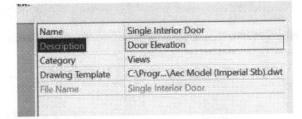

Type **Single Interior Door** for the Name.
Type **Door Elevation** for the Description.

Click **Next**.

7.

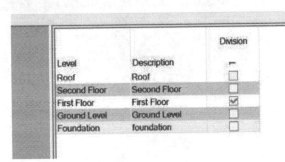

Enable **First Floor.**

Click **Next**.

8.

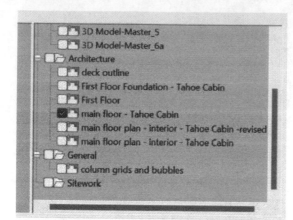

Enable main floor Tahoe Cabin.

Click **Finish.**

9.

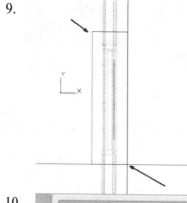

Draw a rectangle around the door to define the scope of the elevation.

Left click in the drawing to generate the elevation view.

10.

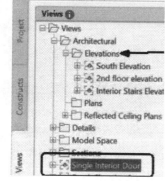

Open the Project Navigator.
Open the Views tab.
Locate the **Single Interior Door** Elevation.

Move the view into the Elevations category.

11.

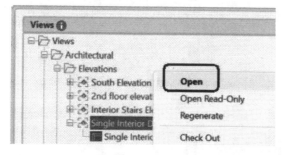

Highlight the **Single Interior Door** Elevation.
Right click and select **Open**.

12.

Open the View ribbon.
Select the **Single Interior Door** named view to zoom into the viewer.

13.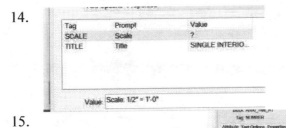

The view updates.

14.

Double click on the title attribute.

Change the Scale to **½" = 1'-0"**.

Click **OK.**

15.

Double click on the view number attribute.

Change the View Number to **D1.1**.

This is what the view should look like.

Save and close all the open drawings.

Callouts

Callouts are typically used to reference drawing details. For example, builders will often reference a door or window call-out which provides information on how to frame and install the door or window. Most architectural firms have a detail library they use to create these callouts as they are the same across building projects.

Exercise 7-9:
Add a Callout

Drawing Name:	Single Interior Door.dwg
Estimated Time:	30 minutes

This exercise reinforces the following skills:

- Callouts
- Viewports
- Detail Elements
- Layouts
- Hatching

1.
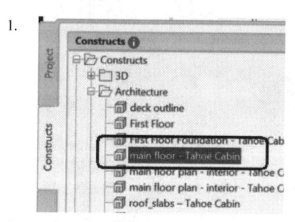

Open the Project Navigator.

Select the Constructs tab.

Highlight **Single Interior Door**.

Right click and select **Open**.

2.
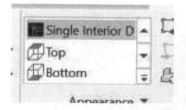

Open the View ribbon.
Select the **Single Interior Door** named view to zoom into the viewer.

3. Activate the Annotate ribbon.

Turn ORTHO **ON**.

Locate and select the **Detail Boundary (Rectangle)** tool on the Detail drop-down list on the Callouts panel.

4. Draw a small rectangle to designate the callout location around the top of the door.

Hint: Turn off OSNAPs to make it easier to place the rectangle.

Select the midpoint of the left side of the rectangle for the start point leader line.
Left click to the side of the callout to place the label.

Click **ENTER** or right click to accept.

5. Type **HD1** in the View Name field.

Enable **Generate Section/Elevation**.

Enable **Place Titlemark**.

Set the Scale to **1″ = 1′-0″**.

Click on the **Current Drawing** button.

6. Place the view to the right of the elevation.

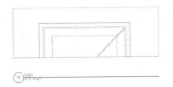

7. Select the view title for the door callout.

Type **1″ = 1′-0″** for the SCALE.

Type **Head Detail – Interior Door** for the TITLE.

8.  Select the View title bubble.

   Type **HD1** for the NUMBER.

   Click **OK.**

9.  Launch the **Design Tools** from the Home ribbon.

10. 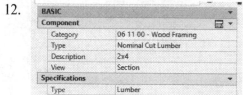 Right click on the Tool palette's bar.

    Enable the **Detailing** palette.

11. Select the **06- Woods, Plastics % Composites** detail tool on the Basic tab.

12. Select 2x4 from the Description list on the Properties palette.

13. 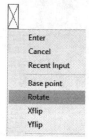 Right click and select Rotate to rotate the detail.

14. 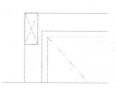 Place the stud in the upper door frame location.

15.

Geometry	
Position X	394'-9 25/32"
Position Y	20'-6 7/8"
Position Z	0"
Scale X	-1.30000
Scale Y	3.50000
Scale Z	1.00000

Select the stud detail.

Right click and select Properties.
Change the X scale to **1.3**.
Change the Y scale to **3.5**.

The stud size will adjust to fit the frame.

To get those values, I took the desired dimension 12" and divided it by the current dimension 3.5 to get 3.5. I took the desired dimension of 2.0 and divided it by the current dimension of 1.5 to get 1.3. To determine the dimensions, simply measure the current length and width and the desired length and width.

16.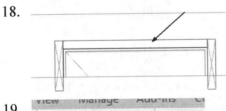

Copy and place the stud to the other side of the door.

17.

Activate the Home ribbon.
Create a layer called A-Detl-Wide.
Assign it the color BLUE and linetype CONTINUOUS.

Set the layer **A-Detl-Wide** current.
Draw a rectangle in the header area.

18.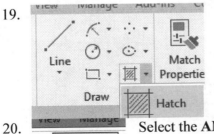

19.

Select the **Hatch** tool.

20.

Select the **ANSI31** hatch.

21.

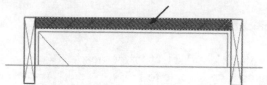

Place a hatch inside the rectangle.

22.

Select **Close Hatch Creation** on the ribbon.

23.

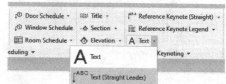

Activate the Annotate ribbon.

Select the **Text (Straight Leader)** tool.

24.

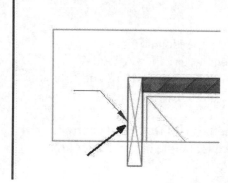

Select the left side of the stud as the start for the leader.

Click **ENTER** when prompted to Set the text width.

25.

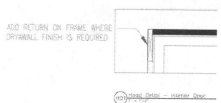

Text should read:

ADD RETURN ON FRAME WHERE DRYWALL FINISH IS REQUIRED

26.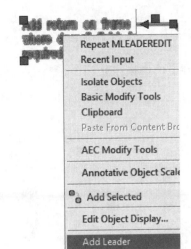

Select the text.
Right click and select **Add Leader**.

27.

Add a second leader to the other side of the door frame.

Click **ESC** to release the selection.

28.

Hover over the leader that was just placed.

Notice that it is on the **G-Anno-Note** layer. *This layer was created automatically when you placed the leader.*

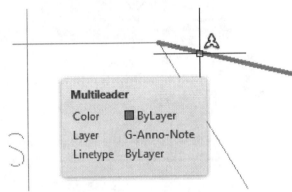

29. ⊢ Select the Linear Dimension tool on the Annotate ribbon.

30.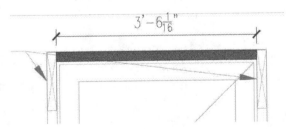

Place a dimension in the opening.

31.

Text	
Fill color	None
Fractional type	Horizontal
Text color	■ ByBlock
Text height	3/32"
Text offset	1/8"
Text outside align	Off

Select the dimension.
Right click and select **Properties**.

Change the Text height to **3/32"**.

32.

Text rotation	0.00
Text view direction	Left-to-Right
Measurement	2'-10 3/4"
Text override	THROAT OPENING = PARTITION WIDTH + 1/8" MIN.

Modify the text to read:

THROAT OPENING =
PARTITION WIDTH + 1/8" MIN.

33.

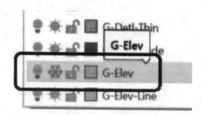

Freeze the **G-Elev** layer.

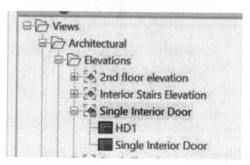

Go to the Project Navigator.
Notice that there are now two views
available under the Single Interior
Door.

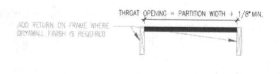

*This is what the view should look
like.*

Save and close all drawings.

Schedule Tags

Schedule tags are symbols that connect building objects with lines of data in a schedule table. Schedule tags display data about the objects to which they are attached. There are schedule tag tools with predefined properties provided with the software on the Scheduling tool palette in the Document tool palette set and in the Documentation Tool Catalog in the Content Browser. You can import schedule tags from the Content Browser to add them to your palette.

You can also create custom schedule tags to display specific property set data for objects in the building. After you create a schedule tag, you can drag it to any tool palette to create a schedule tag tool.

Exercise 7-10:
Adding Door Tags

Drawing Name: main floor – Tahoe Cabin.dwg
Estimated Time: 15 minutes

This exercise reinforces the following skills:

☐   Tags

1.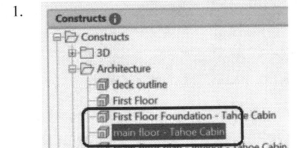

In the Project Navigator:
Open the Constructs tab.
Highlight *main floor– Tahoe Cabin.dwg*.
Right click and select **Open**.

2.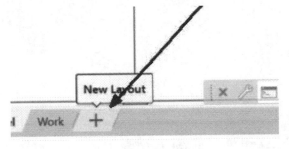

Add a new **Layout**.

3.

Rename **the Layout Main Floor – Plan 1**.

4.

Create a new layer called Viewport. Set the layer to use the color **Cyan**. Set it to **NO PLOT**.

5.

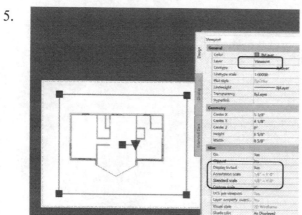

Place the viewport on the viewport layer.
Set the Scale to 1/8" = 1'-0".
Set the Annotation Scale to 1/8" = 1'-0".

Adjust the view inside the viewport so that you see the entire floor plan. Lock the display so that the floor plan doesn't move around.

Release the viewport.

6. Double click inside the viewport to activate model space.
7. Activate the Annotate ribbon.
Select the **Door Tag** tool.

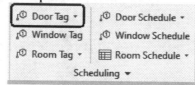

8.
9.

Select a door.

Click ENTER to center the tag on the door.

10. Click **OK**.

Edit the property set data for the object:

PROPERTY SETS	−
DoorObjects	−
Data source	C:\Users\Elise\SkyDr...
Number	01
NumberProject...	"Space not found"A
RoomNumber	"Space not found"
NumberSuffix	A
Style	Hinged - Single

Some students may be concerned because they see the Error getting value next to the Room Number field.

If you tag the spaces with a room tag assigning a number, you will not see that error.

11. Right click and select **Multiple**.
    Window around the entire building.

    All the doors will be selected.

    Click **ENTER**.

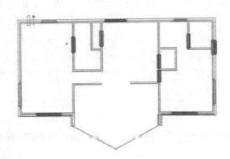

12.

1 object was already tagged with the same tag. Do you want to tag it again?

Click **No**.

13.

Click **OK**.

14.

All the doors are tagged.

Exit to paper space.

If the door tags don't show the correct numbers, try closing the project and then re-opening it.

15. Save and close all the drawings.

Extra: *Tag all the windows in the building.*

Schedule Styles

A schedule table style specifies the properties that can be included in a table for a specific element type. The style controls the table formatting, such as text styles, columns, and headers. To use a schedule table style, it must be defined inside the building drawing file. If you copy a schedule table style from one drawing to another using the Style Manager, any data format and property set definitions are also copied.

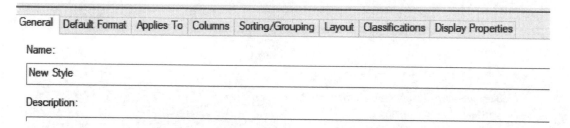

When you create a new Schedule Table Style, there are eight tabs that control the table definition.

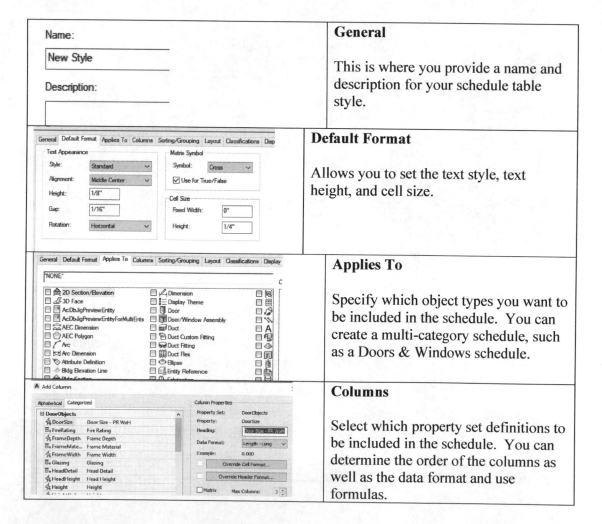

	General This is where you provide a name and description for your schedule table style.
	Default Format Allows you to set the text style, text height, and cell size.
	Applies To Specify which object types you want to be included in the schedule. You can create a multi-category schedule, such as a Doors & Windows schedule.
	Columns Select which property set definitions to be included in the schedule. You can determine the order of the columns as well as the data format and use formulas.

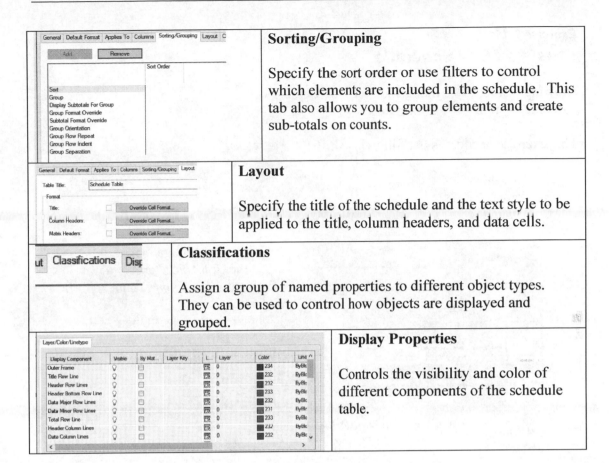

Sorting/Grouping

Specify the sort order or use filters to control which elements are included in the schedule. This tab also allows you to group elements and create sub-totals on counts.

Layout

Specify the title of the schedule and the text style to be applied to the title, column headers, and data cells.

Classifications

Assign a group of named properties to different object types. They can be used to control how objects are displayed and grouped.

Display Properties

Controls the visibility and color of different components of the schedule table.

Exercise 7-11:

Create a Door Schedule

Drawing Name: main floor – Tahoe Cabin.dwg
Estimated Time: 15 minutes

This exercise reinforces the following skills:

- ❑ Schedules
- ❑ Named Views
- ❑ Layouts

1.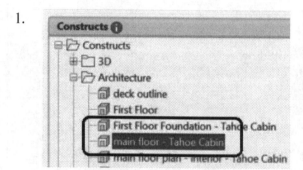
In the Project Navigator:
Open the Constructs tab.
Highlight *main floor – Tahoe Cabin.dwg.*
Right click and select **Open.**

2. **Model**   Activate the **Model** layout tab.

3.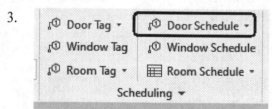
Activate the Annotate ribbon.

Select the **Door Schedule** tool.

4. Window around the building model.

ACA will automatically filter out any non-door entities and collect only doors.

Click **ENTER.**

5. ![Door and Frame Schedule]
Left click to place the schedule in the drawing window.

Click **ENTER.**

6. *A ? will appear for any doors that are missing tags. So, if you see a ? in the list go back to the model and see which doors are missing tags.*

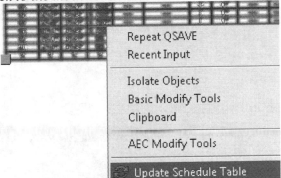

Once you add door tags, you need to update the schedule. Select the schedule, right click and select Update Schedule Table.

7.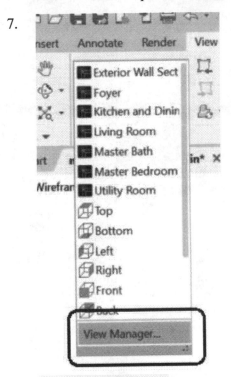

Go to the View ribbon.

Launch the **View Manager**.

8.

Select **New**.

9.

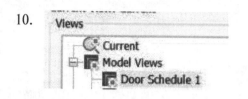

Name the view **Door Schedule 1**. Click Define Window and window around the door schedule.

Click **OK**.

10.

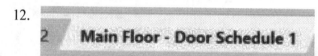

The door schedule now is included in the list of named views.

Click **OK**.

11.

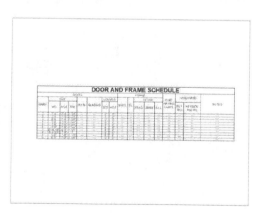

Add a new **Layout**.

12.

Main Floor - Door Schedule 1

Rename the Layout **Main Floor – Door Schedule 1**.

13.

Move the viewport to the Viewport layer.

14.

Double click inside the viewport. Restore the door schedule named view.

15.

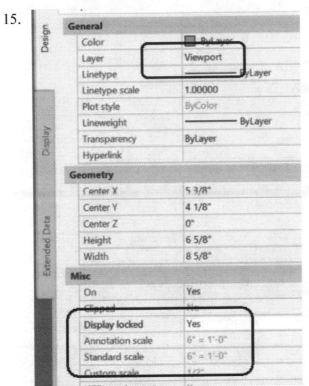

Select the viewport.
Right click and select **Properties**.

Change the Annotation scale to **6" = 1'-0"**.
Change the Standard scale to **6" = 1'-0"**.
Set the Display locked to **Yes.**

Release the selection.

16. Save and close all open drawings.

Exercise 7-12:

Create a Door Schedule Style

Drawing Name: main floor – Tahoe Cabin.dwg
Estimated Time: 30 minutes

This exercise reinforces the following skills:

- Schedules
- Style Manager
- Text Styles
- Update Schedule

1.

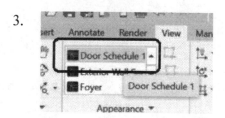

In the Project Navigator:
Open the Constructs tab.
Highlight *main floor – Tahoe Cabin.dwg*.
Right click and select **Open**.

2.    Activate the **Model** layout tab.

3.

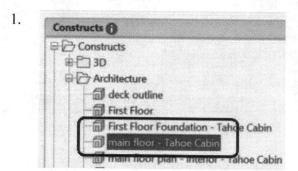

Activate the View ribbon.

Restore the **Door Schedule 1** view.

4.

Select the door schedule.

Right click and select **Copy Schedule Table Style and Assign…**.

This will create a new style that you can modify.

5. 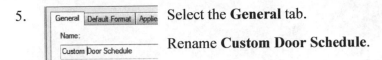 Select the **General** tab.

Rename **Custom Door Schedule**.

6. Select the **Columns** tab.

7.  Highlight Louver, WD, and HGT.

8. Click **Delete**.

9. 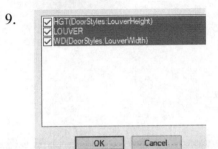 You will be asked to confirm the deletion.

Click **OK**.

10. 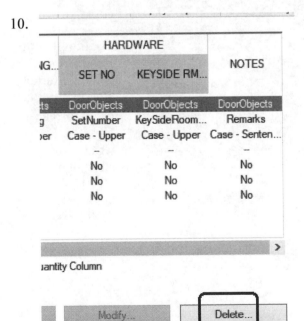 Highlight the columns under Hardware.

Click **Delete**.

11.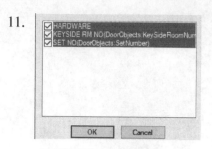

Click **OK**.

12.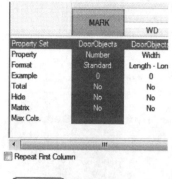

Highlight **MARK**.

Click the **Add Column** button.

13.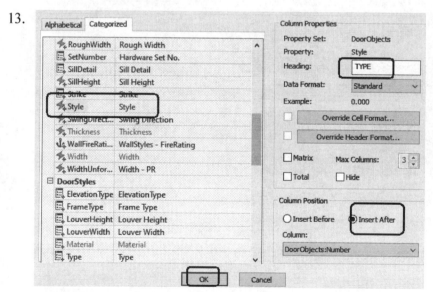

Scroll down until you see Style.

Highlight **Style**.

Confirm that the Column Position will be inserted after the DoorObjects: Number.
Change the Heading to **TYPE**.

Click **OK**.

14.

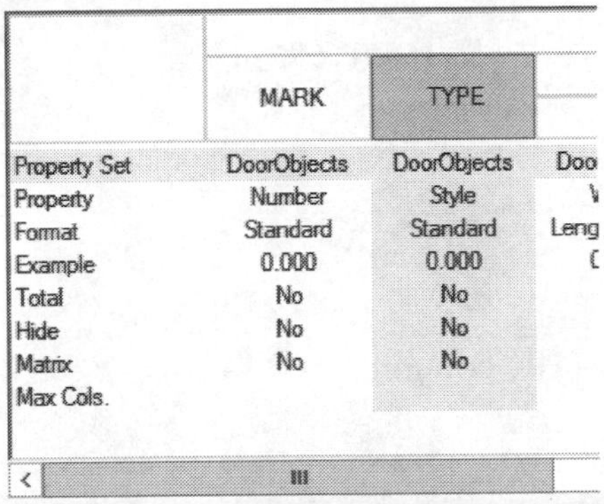

The new column is now inserted.

15.

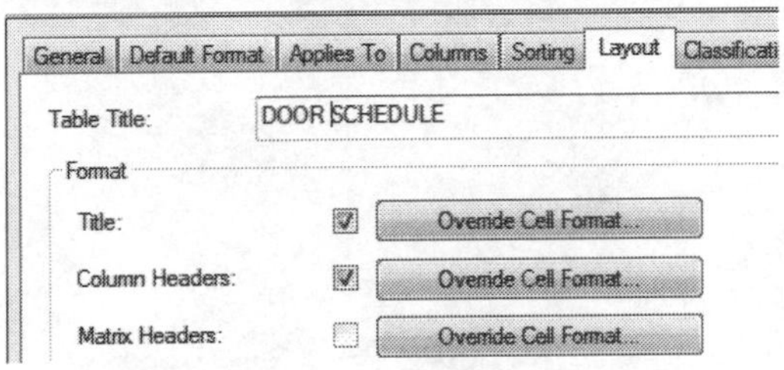

Select the Layout tab.

Change the Table Title to **DOOR SCHEDULE**.

16.

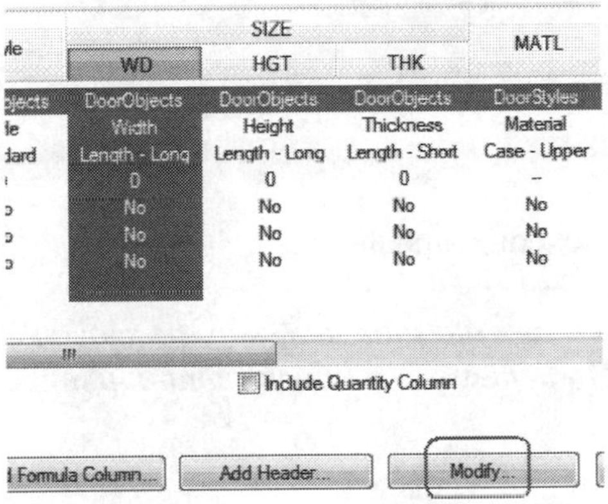

Activate the Columns tab.

Highlight the **WD** column.

Select **Modify**.

17.

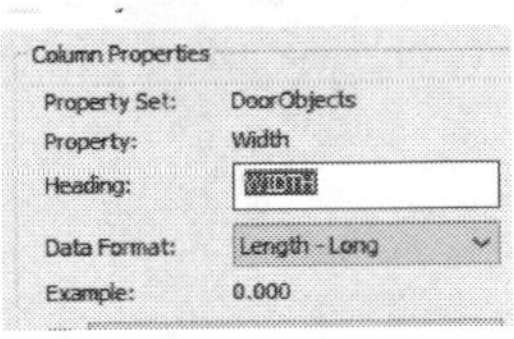

Change the Heading to **WIDTH**.

Click **OK**.

18.

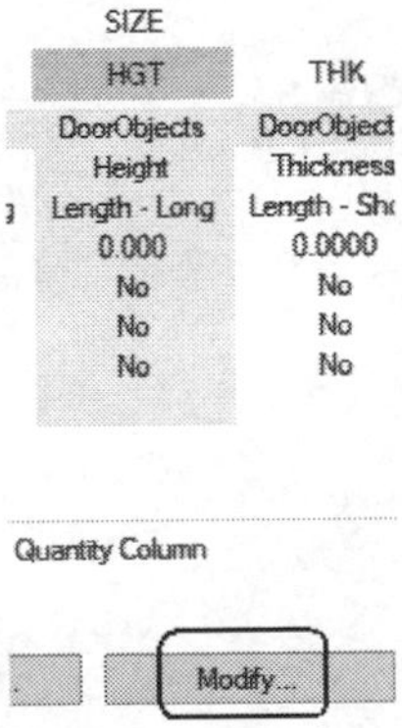

Highlight the **HGT** column.

Select **Modify**.

19.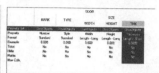

Change the Heading to **HEIGHT**.

Click **OK**.

20.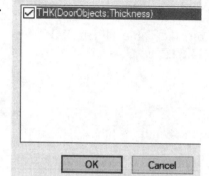

Delete the THK column.

21.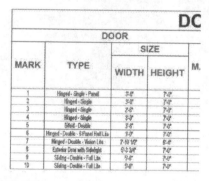

Click **OK**.

22. Click **OK** to close the dialog.

23.

The schedule updates.

If there are question marks in the schedule, that indicates doors which remain untagged.

24.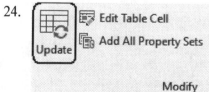

Select the schedule.
Select **Update** from the ribbon.
Click **ESC** to release the selection.

25. 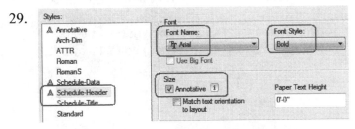 Activate the Home ribbon.

   Select the **Text Style** tool on the Annotation panel.

26. RomanS
   Schedule-Data
   Schedule-Header
   Schedule-Title

   Notice there are text styles set up for schedules.
   Highlight each one to see their settings.

27. Highlight **Schedule-Data.**
   Set the Font Name to **Arial Narrow**.
   Enable **Annotative**.

   Click **Apply**.

28. If this dialog appears, Click **Yes**.

   AutoCAD
   The current style has been modified.
   Do you want to save your changes?
   Yes   No

29. Highlight Schedule-Header.
   Set the Font Name to **Arial**.
   Set the Font Style to **Bold**.
   Enable **Annotative**.

   Click **Apply**.

30. Highlight Schedule-Title.
   Set the Font Name to **Arial**.
   Set the Font Style to **Bold**.
   Enable **Annotative**.

   Click **Apply** and **Close**.

31. Note that the schedule automatically updates.

DOOR				
		SIZE		**DC**
MARK	**TYPE**	**WIDTH**	**HEIGHT**	**M.**
1	Hinged - Single - Panel	3'-6"	7'-0"	
2	Hinged - Single	3'-0"	7'-0"	
3	Hinged - Single	2'-8"	7'-0"	
4	Hinged - Single	3'-0"	7'-0"	
5	Bifold - Double	3'-6"	7'-0"	
6	Hinged - Double - 6 Panel Half Lite	3'-0"	7'-0"	
7	Hinged - Double - Vision Lite	2'-10 1/2"	6'-4"	
8	Exterior Door with Sidelight	5'-3 3/4"	7'-0"	
9	Sliding - Double - Full Lite	5'-8"	7'-0"	
10	Sliding - Double - Full Lite	5'-8"	7'-0"	

32. Save and close all open drawings.

Legends

Component legends are used to display symbols that represent model components, such as doors and windows. Some architectural firms may already have tables with these symbols pre-defined. In which case, you can simply add them to your drawing as an external reference or as a block. However, if you want to generate your own legends using the components in your ACA model, you do have the tools to do that as well.

Exercise 7-13:

Create Door Symbols for a Legend

Drawing Name: main floor – Tahoe Cabin.dwg
Estimated Time: 20 minutes

This exercise reinforces the following skills:

- ❑ Doors
- ❑ Images
- ❑ Object Viewers

1.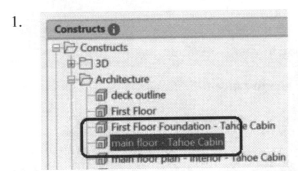
In the Project Navigator:
Open the Constructs tab.
Highlight *main floor – Tahoe Cabin.dwg.*
Right click and select **Open**.

2.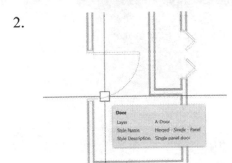

Set the Display Configuration to **Medium Detail**.

Hover over the interior door in the laundry room.

Note that it is a Single hinged door.

3.

Select the interior door in the laundry room.

Right click and select **Object Viewer**.

4.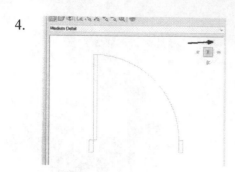
Use the arrows above the view cube to rotate the view so the door appears to be horizontal.

5.
Select the **Save** button in the upper left.

6.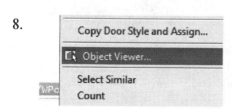
Browse to your work folder.

Type **single-hinged** for the file name.

Click **Save**.
Close the Object Viewer.

7.
Hover over the closet door.

Note that it is a Bifold Double door.

8.

Copy Door Style and Assign...

Object Viewer...

Select Similar
Count

Select the closet door.

Right click and select **Object Viewer**.

9.
Use the arrows above the view cube to rotate the view so the door appears to be horizontal.

10. Object Viewer
Hidden

Select the **Save** button in the upper left.

11.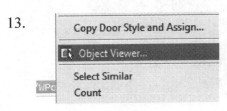

Browse to your work folder.

Type **Bifold-Double** for the file name.

Click **Save**.
Close the Object Viewer.

12.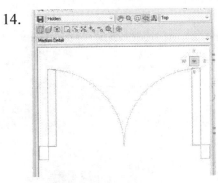

Hover over the Hinged Double-Vision door.

13.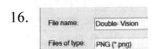

Select the door.

Right click and select **Object Viewer**.

14.

The door should appear correctly.

If not, use the arrows over the Viewcube to orient it correctly.

15.

Select the **Save** button in the upper left.

16.

Browse to your work folder.

Type **Double-Vision** for the file name.

Click **Save**.
Close the Object Viewer.

17. 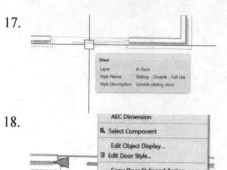 Hover over the sliding doors on the south wall.

18.  Select the door.

Right click and select **Object Viewer**.

19. Select the **Save** button in the upper left.

20. Browse to your work folder.

Type **Sliding-Double** for the file name.

Click **Save**.
Close the Object Viewer.

21. Save and close all drawings.

Exercise 7-14:
Convert Images to Blocks

Drawing Name: main floor – Tahoe Cabin.dwg
Estimated Time: 20 minutes

This exercise reinforces the following skills:

- ❑ Images
- ❑ Blocks
- ❑ Attach

1. 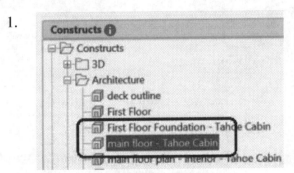 In the Project Navigator:
Open the Constructs tab.
Highlight *main floor – Tahoe Cabin.dwg*.
Right click and select **Open**.

2.  Activate the Insert ribbon.

   Select **Attach**.

3. Locate the **Bifold-Double** image file.
Click **Open**.

4. Enable **Specify Insertion point**.
Disable **Scale**.
Disable **Rotation**.

   Click **OK**.

   Left click to place in the display window.

5.  Select **Create Block** from the Insert ribbon.

6.

Type **bifold-double** for the Name.

Select a corner of the image as the base point.
Select the image.
Enable **Delete.**
Enable **Scale Uniformly.**
Disable **Allow Exploding.**
Click **OK.**

7.

Select **Attach.**

8.

| File name: | Double- Vision |
| Files of type: | All image files |

Locate the **Double-Vision** image file.
Click **Open.**

9.

Enable **Specify Insertion point.**
Disable **Scale.**
Disable **Rotation.**

Click **OK.**

Left click to place in the display window.

10.

Select **Create Block** from the Insert ribbon.

11.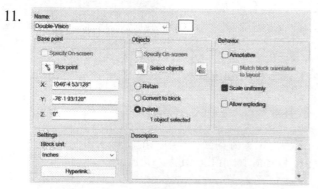

Type **Double-Vision** for the Name.

Select a corner of the image as the base point.
Select the image.
Enable **Delete.**
Enable **Scale Uniformly.**
Disable **Allow Exploding.**

Click **OK.**

12.

Select **Attach.**

13.

Locate the **single-hinged** image file.
Click **Open**.

14.

Enable **Specify Insertion point**.
Disable **Scale**.
Disable **Rotation**.

Click **OK**.

Left click to place in the display
window.

15.
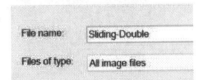

Select **Create Block** from the Insert ribbon.

16.

Type **Single-Hinged** for the
Name.

Select a corner of the image as the
base point.
Select the image.
Enable **Delete.**
Enable **Scale Uniformly**.
Disable **Allow Exploding**.

Click **OK**.

17.

Select **Attach**.

18.

Locate the **Sliding-Double** image file.
Click **Open**.

File name: Sliding-Double

Files of type: All image files

19.

Enable **Specify Insertion point**.
Disable **Scale**.
Disable **Rotation**.

Click **OK**.

Left click to place in the display
window.

20. 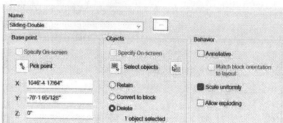 Select **Create Block** from the Insert ribbon.

21. 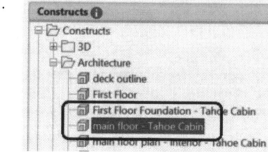 Type **Sliding-Double** for the Name.

Select a corner of the image as the base point.
Select the image.
Enable **Delete.**
Enable **Scale Uniformly**.
Disable **Allow Exploding**.

Click **OK**.

22. Save and close all drawings.

Exercise 7-15:

Create a Door Legend

Drawing Name: main floor – Tahoe Cabin.dwg
Estimated Time: 20 minutes

This exercise reinforces the following skills:

- ❑ Legends
- ❑ Tables
- ❑ Blocks
- ❑ Text Styles
- ❑ Viewports
- ❑ Layouts

1.

In the Project Navigator:
Open the Constructs tab.
Highlight *main floor – Tahoe Cabin.dwg.*
Right click and select **Open**.

2.

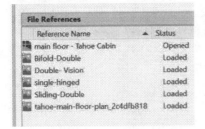

In order for the blocks to appear properly, you need to re-path all the image files.

Bring up the XREF Manager and verify that all the images are loaded into the drawing properly.
If not, assign the images to the correct file location.

3.

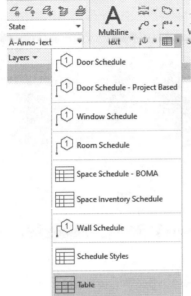

Activate the Home ribbon.

Select the **Table** tool located in the drop-down on the Annotation panel.

4.

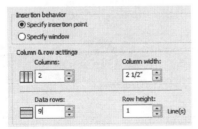

Set the Columns to **2**.
Set the Data rows to **9.**

Click **OK.**

5.

Left click to place the table next to the door schedule.

Type **DOOR LEGEND** as the table header.

6.

Highlight the second row, first cell.
Right click and select **Insert Block**.

7.

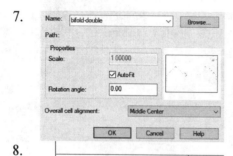

Select **bifold-double** from the drop-down list.
Enable **Auto-fit**.
Set the Overall cell alignment to **Middle Center**.
Click **OK**.

8.

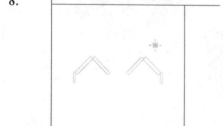

The image appears in the cell.

To turn off the IMAGEFRAME, type IMAGEFRAME, 0.

9.

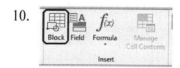

Type **Bifold Double** in the second cell, second row.

Change the text height to **1/16"**.

Adjust the width of the columns.

10.

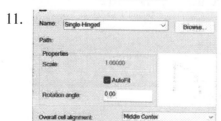

Highlight the third row, first cell.
Select **Insert Block** from the ribbon.

11.

Select **Single-Hinged** from the drop-down list.
Enable **Auto-fit**.
Set the Overall cell alignment to **Middle Center**.
Click **OK**.

12.

The image appears in the cell.

13.

Type **Single Hinged** in the second cell, third row.

Change the text height to **1/16"**.

14.

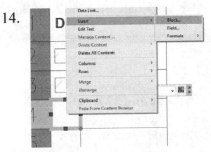

Highlight the fourth row, first cell.
Right click and select **Insert Block**.

15.

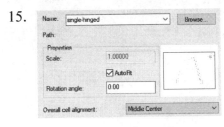

Select **single-hinged** from the drop-down list.
Enable **Auto-fit**.
Set the Overall cell alignment to **Middle Center**.
Click **OK**.

16.

The image appears in the cell.

17.

Type **Single Hinged** in the second cell, fourth row.

Change the text height to **1/16"**.

Bifold Double

Single Hinged

18.

DOOR LEGEND

Bifold Double

Single Hinged

Sliding Double

Continue inserting the blocks and text until the door legend is completed as shown.

Unsaved Layer State

A-Anno-Schd-Bdr

Place the table on the **A-Anno-Schd-Bdr** layer.

Type IMAGEFRAME and then 0 to turn off the border around the images.

19.

Top
Bottom
Left
Right
Front
Back
SW Isometric
SE Isometric

View Manager...

Go to the View ribbon.

Launch the **View Manager**.

20.

New...

Select **New**.

21.

View name:	Door Legend
View category:	<None>
View type:	Still

Boundary
○ Current display
◉ Define window

Type **Door Legend** in the View name field.
Enable **Define window**.

Window around the legend table.
Click **ENTER**.
Click **OK**.
Close the View Manager.
Activate the View ribbon.

22.

Insert Annotate Render Solids Parametric View

Door Legend
Door Schedule
Exterior Wall Sec

World

Appearance

Coordinates

ex8-33*

Double click on the **Door Legend** view.

The display in the viewport updates with the door legend.

23.

Create a layout called Door Legend.

Place the door legend in a view on the layout tab.

24.

Save and close all drawings.

Exercise 7-16:

Creating a Custom Imperial Multi-Leader

Drawing Name: Leader Symbols (Imperial).dwg
Estimated Time: 40 minutes

This exercise reinforces the following skills:

- Tool palette
- Multi-Leader Styles
- Content Browser
- Tool Properties
- Text Styles
- Attributes
- Block Editor

1.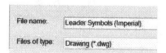

   Browse to the folder where the styles drawing is located.

   ProgramData is a hidden folder by default. You may need to use File Explorer to make the ProgramData file visible. If you are in a classroom environment, your IT group may make this folder uneditable. If this is the case, make a copy of the file and locate it in your work folder. You will then need to set the Properties to point to the correct folder location.

2.
File name:	Leader Symbols (Imperial)
Files of type:	Drawing (*.dwg)

   Open the *Leader Symbols (Imperial).dwg.*

3.

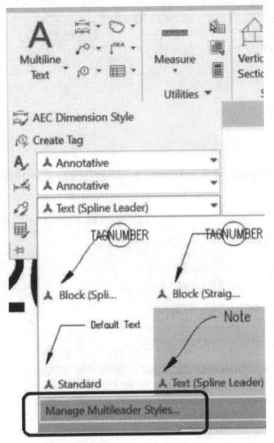

On the Home ribbon:

Locate the Annotation panel.

In the drop-down, select **Manage Multileader Styles**.

4.

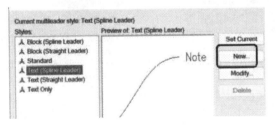

Select **New**.

5.

Type **Detail Callout** in the New style name field.

Select **Standard** under Start with.

Click **Continue**.

6.

On the Leader Format tab:
Set the Type to Straight.
Set the Arrowhead Symbol to **Open**.
Set the Size to **3/8"**.

7.

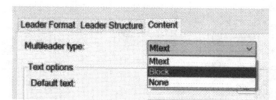

Select the Content tab.

Under Multileader type:

Select **Block**.

8.

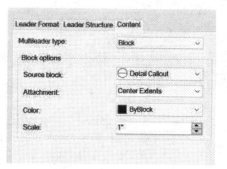

Set the Source block to **Detail Callout**.

Set the Scale to **1"**.

Click **OK**.

9.

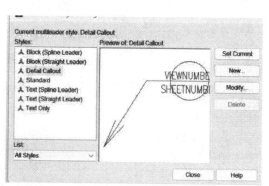

The new multi-leader style is now listed.

Click **Close**.

10. Save your drawing.

This saves the style you just created.

11.

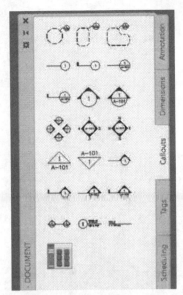

Verify that the Annotation tool palette is open.

Select the **Callouts** tab.

12.

Go to the Home ribbon.

Open the **Content Browser**.

13.

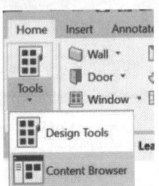

Locate and left click to open the **Documentation Tool Catalog – Imperial**.

14.

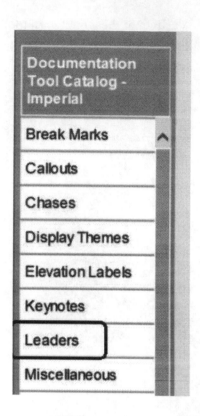

Click on the **Leaders** category.

15.

Locate the **Circle (Straight Leader)** tool.

Highlight and right click **to Add to Tool Palette**.

16.

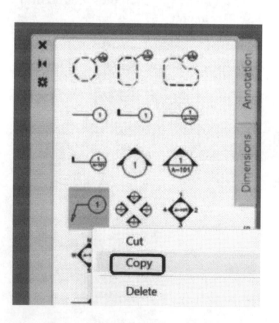

The tool should be added to the **Document/Callouts** tool palette.

Move it to a good location on the palette.

Highlight the **Circle (Straight Leader)** tool.

Right click and select **Copy.**

17.

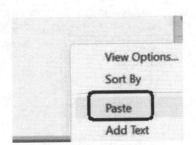

Move the cursor to a blank area on the palette.

Right click and select **Paste.**

18.

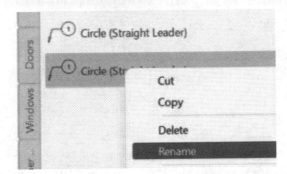

Rename the copied tool **Detail Callout**.

19.

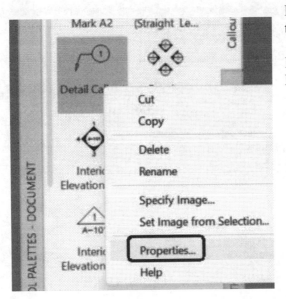

Highlight the **Detail Callout** tool.

Right click and select **Properties**.

20.

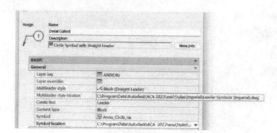

Notice that the tool uses blocks located in the Leader Symbols (Imperial).dwg that you modified earlier.

21.

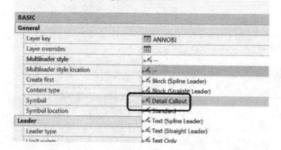

Select the **Detail Callout** style from the drop down list.

This style will only appear if you saved the file earlier.

22.

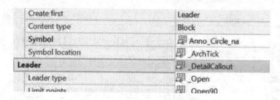

Set the Symbol to **_DetailCallout**.

23.

Set the **Attribute Style to As defined by tag**.

24.

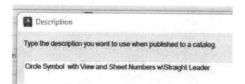

Change the Description to indicate the multileader style correctly.

Click **OK**.

Note that the preview icon is not correct.

25.

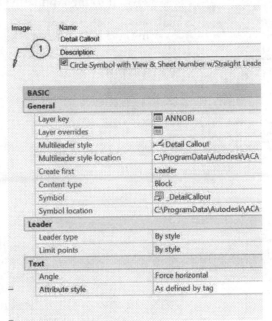

Review the Properties panel to ensure that everything looks correct.

Click **OK** to close the Properties dialog.

26.

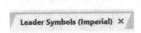

Return to the **Leader Symbols (Imperial).dwg** file.

Verify that you arc in thc **Model** spacc tab.

27.

Click on the Detail Callout tool to place a detail callout in the drawing.

Click to select the start and end point for the leader.

Enter in a View Number and Sheet Number when the attribute dialog appears.

Click **OK**.
The multileader is placed.

Select the multileader.
Right click and select **Properties**.
Set the Scale for the block to **1.0**.
Click ESC to release the selection.

28.

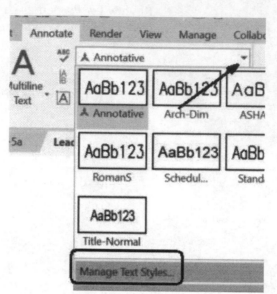

Switch to the Annotate ribbon.

In the Text Style box, select the drop down and click on **Manage Text Styles**.

29.

Click **New**.

30.

Type **Comic Sans**.

Click **OK**.

If you prefer a different font, choose a different font.

31.

Set the Font Name to **Comic Sans MS.**

Click **Close**.

32.

Click **Yes** to save your changes.

33.

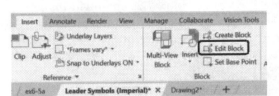

Switch to the Insert ribbon.

Click **Edit Block**.

34.

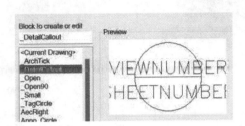

Highlight the **_DetailCallout** block.

Click **OK**.

35.

Set the view scale to **1'-0":1'-0"**.

36.

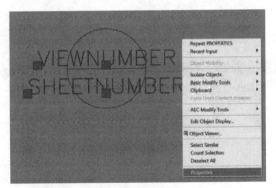

Select the two attributes. *This is the text.*

Right click and select **Properties**.

37.

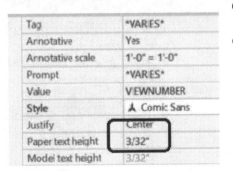

Change the Style to **Comic Sans**.

Change the Paper text height to **3/32"**.

38.

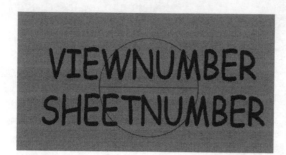

The attributes update.
Adjust the position of the attributes.

Click **Save Block** on the ribbon.

39.

Click **Close Block Editor** on the ribbon.

The multileader updates.

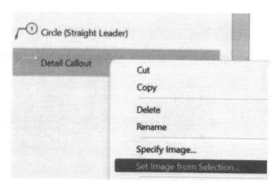

40.

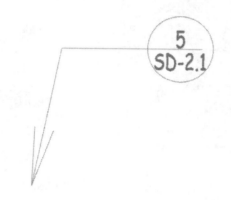

Highlight the Detail Callout on the tool palette.

Right click and select **Set Image from Selection**.

41.

Select the multileader and press ENTER.

The icon on the tool palette updates.

42. Save *Leader Symbols (Metric).dwg*.

Create a Framing Plan View

Drawing Name: new
Estimated Time: 30 minutes

This exercise reinforces the following skills:

- ❑ Project Browser
- ❑ Project Navigator
- ❑ Drawing Utilities
- ❑ Insert Image
- ❑ Insert Hyperlink

1. 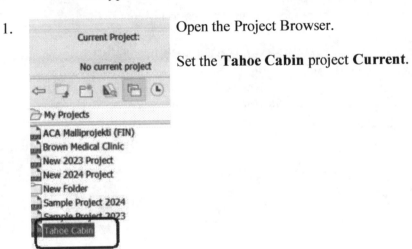 Open the Project Browser.

   Set the **Tahoe Cabin** project **Current**.

2. Start a new drawing.
   Name it *framing plan – Tahoe Cabin.dwg*.

3.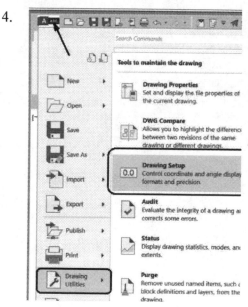

Go to the Application menu.

Select **Drawing Utilities→Drawing Setup**.

4.

Go to the Application Menu.

Select **Drawing Utilities→Drawing Setup**.

*You can also type **aecdwgsetup** to launch the dialog.*

5.

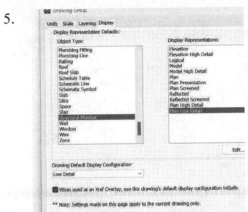

Highlight **Structural Member** in the Object Type list.
Highlight **Plan Low Detail**.
Set the Drawing Default Display Configuration to **Plan Low Detail**.
Enable **When used as an Xref Overlay, use this drawing's default display** configuration initially.

Click **OK**.

6.

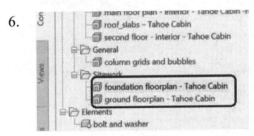

Highlight the **foundation floorplan – Tahoe Cabin**.
Right click and select **Xref Overlay**.

7.

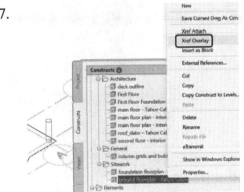

Highlight the **ground floorplan -Tahoe Cabin**.
Right click and select **Xref Overlay**.

8.

Highlight the **first floor foundation - Tahoe Cabin**.
Right click and select **Xref Overlay**.

9.

Select the **ground floorplan -Tahoe Cabin overlay**.
Right click and select **Properties**.
Set the Insertion point Z to **4'-0"**.
Click **ESC** to release the selection.

10.

Select the **First Floor Foundation -Tahoe Cabin overlay**.
Right click and select **Properties**.
Set the Insertion point Z to **6'-5 3/4"**.
Click **ESC** to release the selection.

11.

Freeze the following layers:

- Deck outline
- Gravel outline
- Grids
- S-Grid
- S-Grid-Iden
- G-Elev
- G-Elev-Iden
- G-Elev-Line
- A-Slab

12. On the Home ribbon:

Select **New Layer State**.

13. Type **framing plan**.

Click **OK**.

14. [−][Top][2D Wireframe]

Switch to a **Top|2D Wireframe** view.

15. Set the Display Configuration to **Medium Detail**.

Medium Detail ▾

16.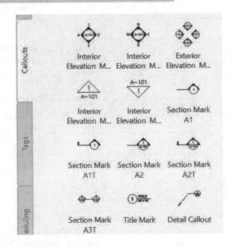

Open the Document tool palette.

Select the **Callouts** tab.

Select the **Detail Callout** tool.

17.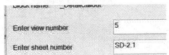

Place the callout.
Start the leader at the column located at the upper right of the layout.
Change the attributes as shown.
The View Number should be **5**.
The sheet number should be **SD-2.1**.

18.

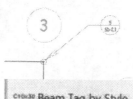

Change the scale of the block to **1**.

19.

Switch to the Tags tab on the Document tool palette.
Select **Beam Tag by Style**.

20.

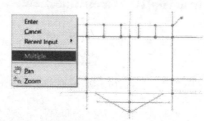

Select a beam.
Click ENTER.
Click ENTER to accept the default.
Click **OK** when the Property Set Data dialog appears.

21.

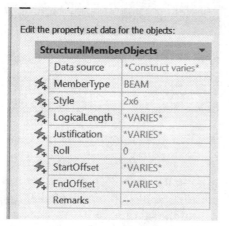

Right click and select **Multiple**.

Window around the drawing.
Click ENTER to complete the selection.

22.

1 object was already tagged with the same tag.
Do you want to tag it again?

Yes   No

Click **No**.

23.

Edit the property set data for the objects:

StructuralMemberObjects	
Data source	*Construct varies*
MemberType	BEAM
Style	2x6
LogicalLength	*VARIES*
Justification	*VARIES*
Roll	0
StartOffset	*VARIES*
EndOffset	*VARIES*
Remarks	--

Click **OK**.

Click ENTER to exit the command.

24.

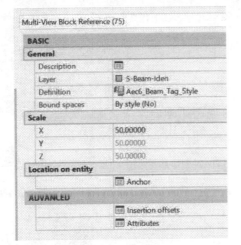

Select one of the beam tags.
Right click and select **Properties**.
Set the Scale to **50**.
Click ESC to release the selection.

25.

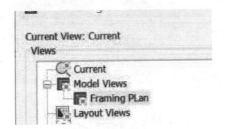

Create a Named View called **Framing Plan**.

26.

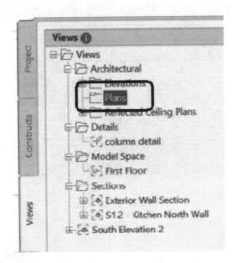

Go to the View tab of the Project Navigator.
Highlight the Plans category.

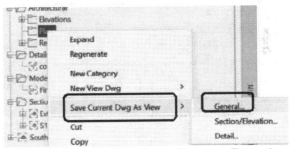

Right click and select **Save Current Dwg As View→General**.

27.

Name	framing plan _ Tahoe Cabin
Description	framing plan _ Tahoe Cabin
Category	Views\Architectural\Plans
File Name	framing plan _ Tahoe Cabin

Click **Next**.

28.

Level	Description	DIVISION
Roof	Roof	☐
Second Floor	Second Floor	☐
First Floor	First Floor	☐
Ground Level	Ground Level	☑
Foundation	foundation	☑

Enable **Ground Level** and **Foundation**.
Click **Next**.

29.

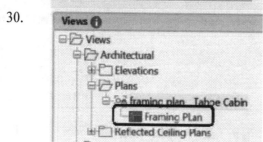

Disable the **3D** category.

Click **Finish**.

30.

The Named View is listed under the drawing.

31.

Save and close all drawings.

Extra:

1) Create additional elevation and section views for the kitchen, the second bathroom, the exterior elevations, and windows.

2) Add the new views to layouts.

3) Add window tags to all the windows.

4) Create a window schedule.

5) Create a window legend.

6) Create a detail view for window framing.

7) Create a callout to the window framing detail view.

8) Create a section elevation view of an interior wall.

9) Add keynotes to the interior wall section elevation view.

Notes:

QUIZ 7

True or False

1. You can create a view drawing by selecting constructs from the Project Navigator palette and using them as external references in a View drawing.

2. A construct drawing can be used as a view drawing.

3. View drawings are usually created after you have defined the project's levels and basic construct drawings.

4. In order to create a schedule, you need to add schedule tags to the objects to be included in the schedule first.

5. Property set definitions must be attached to an object or style before they can be added to a schedule.

Multiple Choice

6. The three types of views that can be created in a project are:

   A. General
   B. Architectural
   C. Detail
   D. Section/Elevation
   E. Floor Plan
   F. Reflected Ceiling Plan

7. To display the external references for a view drawing:

   A. Highlight the view drawing in the Project Navigator, right click and click External References
   B. Open the view drawing, bring up the XREF Manager
   C. Launch the Layers Properties Manager, highlight XREFs
   D. Go to the View ribbon, click on XREFs

8. Space Styles: (select all that apply)

   A. Control the hatching displayed
   B. Control the name displayed when spaces are tagged
   C. Cannot be modified
   D. Have materials assigned

9. When placing a wall tag:

> A. Tags are positioned relative to the midpoint of a wall
> B. Tags are positioned relative to the end of a wall
> C. Tags are positioned relative to the center of a wall
> D. Tags are positioned based on user input.

10. When placing a space tag:

> A. Tags are positioned based on user input
> B. Tags are hidden behind the space hatch and need to have the display order changed
> C. Tags are positioned relative to the center of the geometric extents of the space
> D. Tags are positioned above the space hatch

ANSWERS:

1) T; 2) F; 3) T; 4) F; 5) T; 6) B, C, and D; 7) A; 8) A, B; C, and D 9) A; 10) C;

Lesson 8
Spaces & Sheets

Layouts

Layouts act as the sheets for your documentation set. You name the layouts, add views to the layouts, add text and dimensions, and any other relevant annotation.

You place one or more viewports on a layout. The area outside the viewport is called *paper space*. Normally, paper space is used to add a title block and possibly notes. Views are created in model space for any schedules, elevations, plans, etc. These are then added to the layout. The viewport controls the scale and display of the model elements inside the viewport.

You can access one or more layouts from the tabs located at the bottom-left corner of the drawing area to the right of the Model tab. You can use multiple layout tabs to display details of the various components of your model at several scales and on different sheet sizes.

If the view does not generate properly for you, use the Current Drawing option, then use WBLOCK to create a new drawing using the view.

Exercise 8-1:
Creating a View Drawing

Drawing Name: new
Estimated Time: 10 minutes

This exercise reinforces the following skills:

- ❏ Project Browser
- ❏ Project Navigator
- ❏ Layers

1.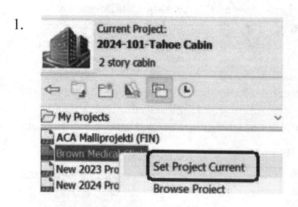

   Launch the Project Browser.

   Set the **Brown Medical Clinic** as the current project.

2. Start a new drawing.

3.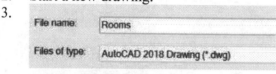

   Save as *Rooms.dwg*.

4.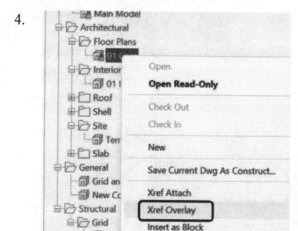

   Right click on the **01 Core** construct. Select **Xref Overlay**.

   Double click the mouse wheel to **Zoom Extents**.

5.

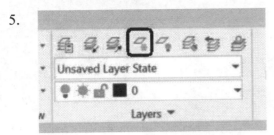

Select the **Freeze Layer** tool on the Home ribbon.

6.

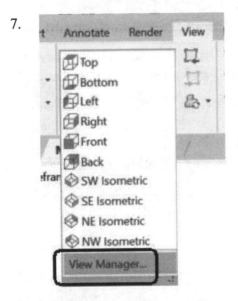

Select the grid lines indicated located at the lower left of the drawing.

7.

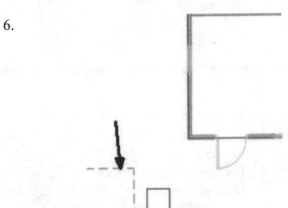

Go to the View ribbon.

Launch the **View Manager**.

8.

View name:	Top Model Plan
View category:	<None>
View type:	Still

Boundary
○ Current display
● Define window

Create a Named View of the top plan view of the building.

Click **OK**.

9. Click **OK**.

Current View: Current
Views
- Current
- Model Views
  - Top Model Plan
- Layout Views
- Preset Views

General	
Name	Top Model Plan
Category	<None>
UCS	World
Layer snapshot	Yes
Annotation s...	1:100
Visual Style	2D Wireframe
Background...	<None>
Live Section	<None>

Set Current
New...
Update Layers
Edit Boundaries...
Delete

Animation	
View type	Still

10. Set the Display Configuration to **Medium Detail**.

Medium Detail

11. Launch the Layer Properties Manager.
Do a search for layers with the word **Door**.
Change the layer color to **Blue**.

12. Do a search for layers with the word **Glaz**.
Change the layer color to **Green**.

This changes the layer colors in the host drawing only – not in the external reference files.

Close the Layer Properties Manager.

The view should look similar to this.

Can you create and save a new layer state called Rooms?

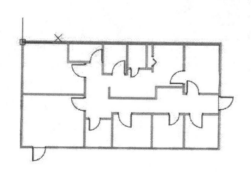

13. Save the file as *Rooms_2.dwg*.
Close all open drawings.

Space Styles

Depending on the scope of the drawing, you may want to create different space styles to represent different types of spaces, such as different room types in an office building. Many projects require a minimum square footage for specific spaces – such as the size of bedrooms, closets, or common areas – based on building code. Additionally, by placing spaces in a building, you can look at the traffic flow from one space to another and determine if the building has a good flow pattern.

You can use styles for controlling the following aspects of spaces:

• Boundary offsets: You can specify the distance that a space's net, usable, and gross boundaries will be offset from its base boundary. Each boundary has its own display components that you can set according to your needs.

• Name lists: You can select a list of allowed names for spaces of a particular style. This helps you to maintain consistent naming schemes across a building project.

• Target dimensions: You can define a target area, length, and width for spaces inserted with a specific style. This is helpful when you have upper and lower space limits for a type of room that you want to insert.

• Displaying different space types: You can draw construction spaces, demolition spaces, and traffic spaces with different display properties. For example, you might draw all construction areas in green and hatched, and the traffic areas in blue with a solid fill.

• Displaying different decomposition methods: You can specify how spaces are decomposed (trapezoid or triangular). If you are not working with space decomposition extensively, you will probably set it up in the drawing default.

Space Styles are created, modified, or deleted inside the Style Manager.

Exercise 8-2:
Import Space Styles

Drawing Name: Rooms_2.dwg
Estimated Time: 10 minutes

This exercise reinforces the following skills:

- ❑ Project Navigator
- ❑ Styles Browser

1.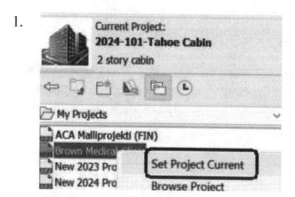

   Launch the Project Browser.

   Set the **Brown Medical Clinic** as the current project.

   Open *Rooms_2.dwg (this is a downloaded exercise file or saved from the previous exercise. If you download it, you may need to re-path the external references).*

2.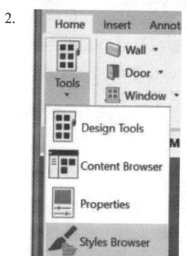

   Go to the Home ribbon.

   Launch the **Styles Browser**.

3.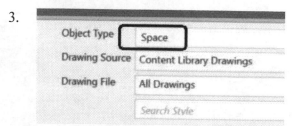

   Change the Object Type to **Space**.

4.

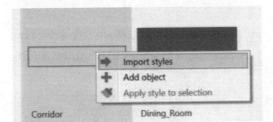

Highlight Corridor.

Right click and select **Import Styles**.

A green check appears next to the Style Name to indicate it exists in the drawing.

5.

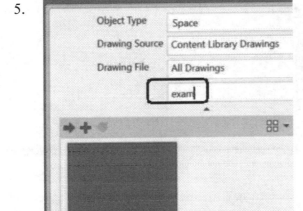

Type **exam** in the search field.

A Space Style called **General Exam** is displayed.

Right click and select **Import Styles**.

6.

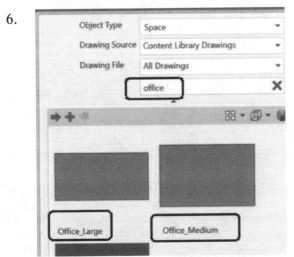

Type **office** in the search field.

Space Styles with Office in their name are displayed.

7.

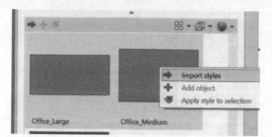

Hold down the CTL key to select the **Office_Large** and **Office_Medium** space styles.

Right click and select **Import Styles**.

8.

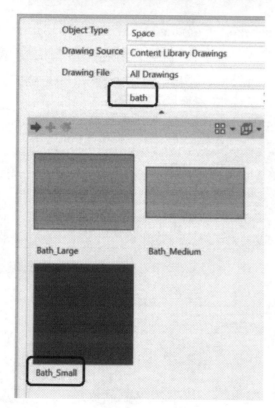

Type **bath** in the search field.

Highlight **Bath_Small**.
Right click and select **Import Styles**.

9.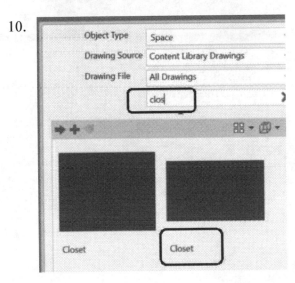

Type **kitch** in the search field.

Highlight **Kitchen_Small**.
Right click and select **Import Styles**.

10.

Type **clos** in the search field.

Highlight **Closet**.
Right click and select **Import Styles**.

11.

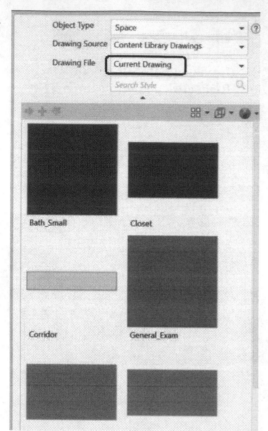

Change the Drawing File to **Current Drawing**.

You will see all the space styles that exist in the current drawing.

Close the Styles Browser.

12.

Save as *Rooms_3.dwg*.
Close the drawing.

Exercise 8-3:
Create Space Styles

Drawing Name: Rooms_3.dwg
Estimated Time: 30 minutes

This exercise reinforces the following skills:

- ❑ Project Navigator
- ❑ Style Manager

1.   Open *Rooms_3.dwg (this is a downloaded exercise file or saved from the previous exercise. If you download it, you may need to re-path the external references).*

2.      Open the Manage ribbon.

Launch the **Style Manager**.

3.   Locate the Space Styles category.

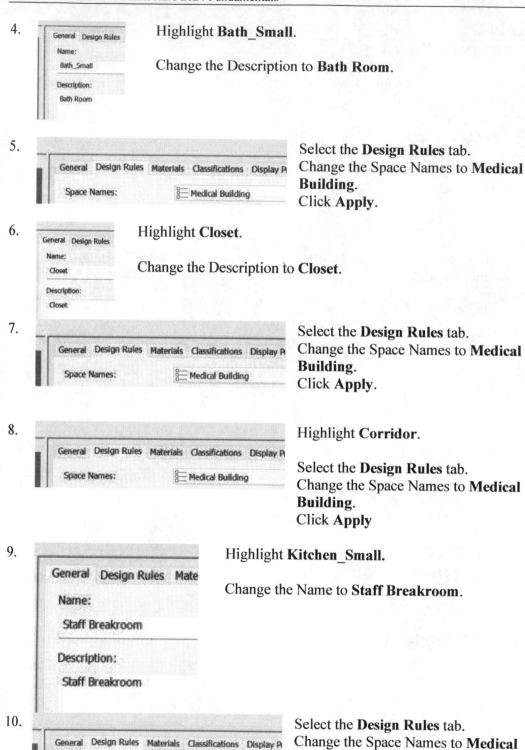

4. Highlight **Bath_Small**.

Change the Description to **Bath Room**.

5. Select the **Design Rules** tab.
Change the Space Names to **Medical Building**.
Click **Apply**.

6. Highlight **Closet**.

Change the Description to **Closet**.

7. Select the **Design Rules** tab.
Change the Space Names to **Medical Building**.
Click **Apply**.

8. Highlight **Corridor**.

Select the **Design Rules** tab.
Change the Space Names to **Medical Building**.
Click **Apply**

9. Highlight **Kitchen_Small**.

Change the Name to **Staff Breakroom**.

10. Select the **Design Rules** tab.
Change the Space Names to **Medical Building**.
Click **Apply.**

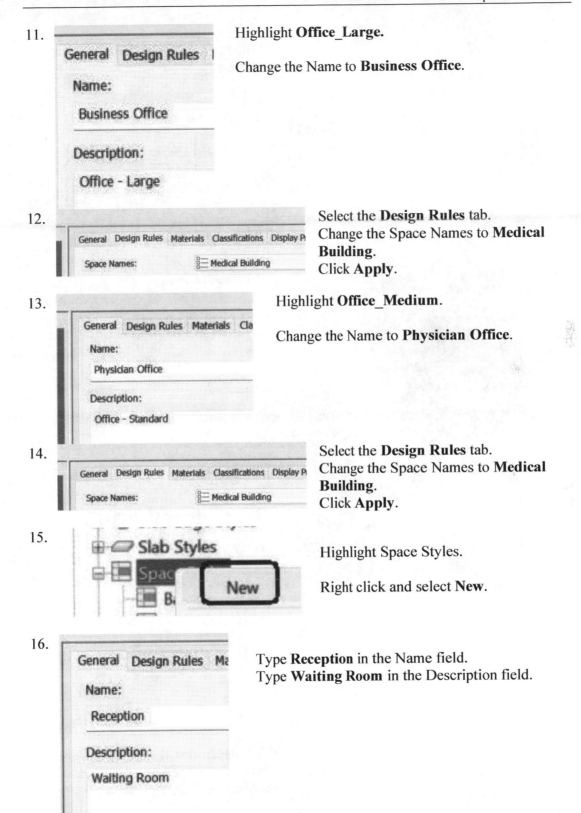

11. Highlight **Office_Large.**

Change the Name to **Business Office**.

12. Select the **Design Rules** tab.
Change the Space Names to **Medical Building**.
Click **Apply**.

13. Highlight **Office_Medium**.

Change the Name to **Physician Office**.

14. Select the **Design Rules** tab.
Change the Space Names to **Medical Building**.
Click **Apply**.

15. Highlight Space Styles.

Right click and select **New**.

16. Type **Reception** in the Name field.
Type **Waiting Room** in the Description field.

17.

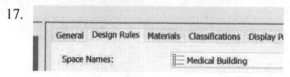

Select the **Design Rules** tab.
Change the Space Names to **Medical Building**.

18.

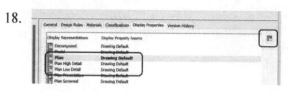

Select the Display Properties tab.

Click on **Edit Display Properties**.

19.

Highlight the **Base Boundary** and **Base Hatch** Display Components.
Set the Color to **100/Green**.

20.

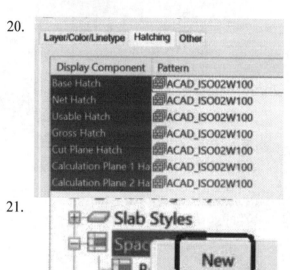

Select the **Hatching** tab.

Highlight all the Display Components.
Assign the **ACAD_ISO02W100** hatch pattern.

Click **OK**.

Click **Apply**.

21.

Highlight Space Styles.

Right click and select **New**.

22.

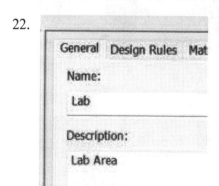

Type **Lab** in the Name field.
Type **Lab Area** in the Description field.

23.

Select the **Design Rules** tab.
Change the Space Names to **Medical Building**.

24.

Select the Display Properties tab.

Click on **Edit Display Properties**.

25.

Highlight the **Base Boundary** and **Base Hatch** Display Components.
Set the Color to **200/Magenta**.

26.

Select the **Hatching** tab.

Highlight all the Display Components.
Assign the **DOTS** hatch pattern.

Click **OK**.

Click **Apply**.

27.

Highlight Space Styles.

Right click and select **New**.

28.

Type **Nurse Station** in the Name field.
Type **Nurse Station** in the Description field.

29.

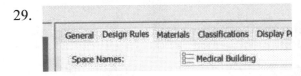

Select the **Design Rules** tab.
Change the Space Names to **Medical Building**.

30.

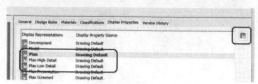

Select the Display Properties tab.

Click on **Edit Display Properties**.

31.

Highlight the **Base Boundary** and **Base Hatch** Display Components.
Set the Color to **240/Red**.

32.

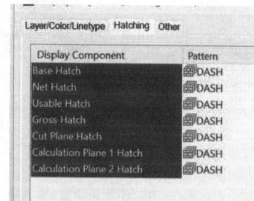

Select the **Hatching** tab.

Highlight all the Display Components.
Assign the **DASH** hatch pattern.

Click **OK**.

Click **Apply**.

33.

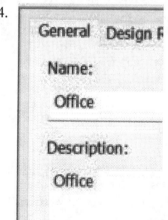

Highlight Space Styles.

Right click and select **New**.

34.

Type **Office** in the Name field.
Type **Office** in the Description field.

35.

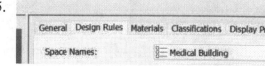

Select the **Design Rules** tab.
Change the Space Names to **Medical Building**.

36.

Select the Display Properties tab.

Click on **Edit Display Properties**.

37.

Highlight the **Base Boundary** and **Base Hatch** Display Components.
Set the Color to **46/Brown**.

38.

Select the Hatching tab.

Highlight all the Display Components.
Assign the **AR-CONC** hatch pattern.

Click **OK**.

Click **Apply**.

39.

The list of space styles should be as shown.

Click OK to close the Styles.

40.

Save as *Rooms_4.dwg*.
Close all open drawings.

Exercise 8-4:
Adding Spaces

Drawing Name: Rooms_4.dwg
Estimated Time: 20 minutes

This exercise reinforces the following skills:

- ❏ Layouts
- ❏ Spaces

1. 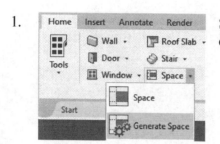 Select the **Generate Space** tool from the Home ribbon on the Build panel.

2. On the Properties palette, select the **Reception** space style. Change the Name to **Lobby**.

   Left click inside the lower left room to place the space.

3. If you hover over the space, you should see a tool tip with the space style name.

4. 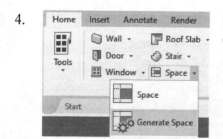 Select the **Generate Space** tool from the Home ribbon on the Build panel.

5.  On the Properties palette, select the **Business Office** space style.
   Change the Name to **Office**.

   Left click inside the upper left room to place the space.

6. 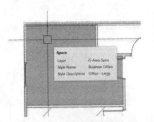 If you hover over the space, you should see a tool tip with the space style name.

7. 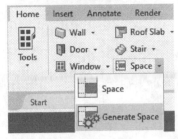 Select the **Generate Space** tool from the Home ribbon on the Build panel.

8. 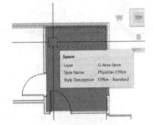 On the Properties palette, select the **Physician Office** space style.
Change the Name to **Office**.

Left click inside the upper right room to place the space.

9. 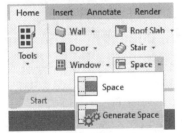 If you hover over the space, you should see a tool tip with the space style name.

10. 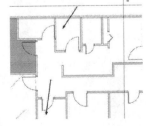 Select the **Generate Space** tool from the Home ribbon on the Build panel.

11. On the Properties palette, select the **Bath Small** space style.

Change the Name to **Bath**.

12. Left click inside the two rooms indicated to place the space.

13.

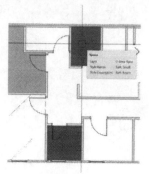

If you hover over the space, you should see a tool tip with the space style name.

14.

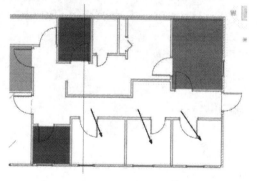

Select the **Generate Space** tool from the Home ribbon on the Build panel.

15.

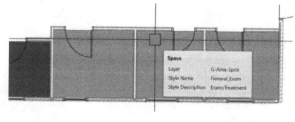

On the Properties palette, select the **General_Exam** space style.

Change the Name to **General Exam**.

16.

Left click inside the three rooms indicated to place the spaces.

17.

If you hover over the space, you should see a tool tip with the space style name.

18. 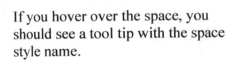 Select the **Generate Space** tool from the Home ribbon on the Build panel.

19. 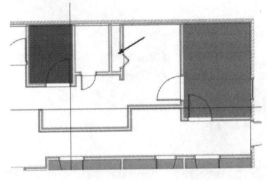 On the Properties palette, select the **Closet** space style.

Change the Name to **Closet**.

20. Left click inside the room indicated to place the space.

21. If you hover over the space, you should see a tool tip with the space style name.

You may need to zoom in because the closet space is small.

22.  Select the **Generate Space** tool from the Home ribbon on the Build panel.

23.  On the Properties palette, select the **Corridor** space style.

Change the Name to **Corridor**.

24. 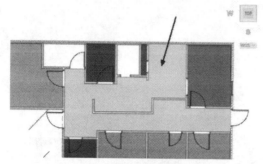 Left click inside the room indicated to place the space.

The corridor space fills in areas that are designated for other uses. This will be corrected in the next exercise.

25. 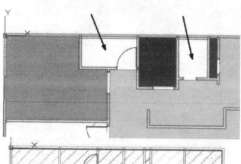 Assign the remaining two enclosed rooms as office spaces.

26. 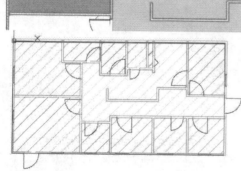 Change the Display Configuration to **Low Detail**.

Save as *Rooms_5.dwg*.
Close all drawings.

Exercise 8-5:
Modifying Spaces

Drawing Name: Rooms_5.dwg
Estimated Time: 10 minutes

This exercise reinforces the following skills:

- Spaces
- Divide Space

1.

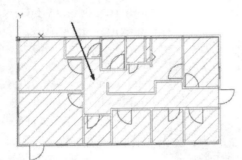

Select the **Corridor** space.

2.

On the Space contextual ribbon,
select **Divide Space**.

3.

Set the **G-Area-Spce** layer current.

4.

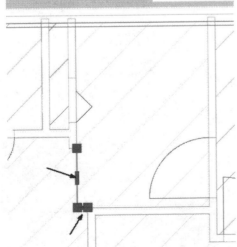

Turn ORTHO ON.
Draw two lines to designate the staff break room area.

5.

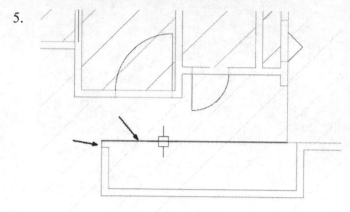

Draw two lines to designate the lab area.

6.

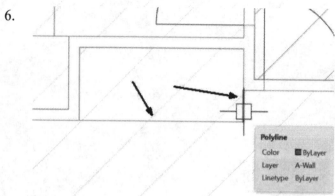

Draw two lines to designate the Nurse Station area.

Notice that the polylines are placed on the A-Wall layer, not the current layer.
You can use Properties to move them to the G-Area-Spc layer.

7.

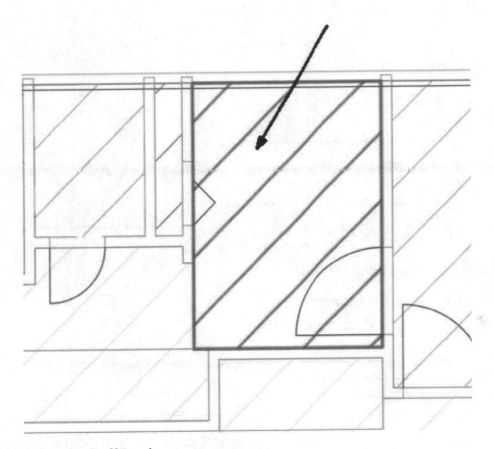

Select the Staff Breakroom area space.

Right click and select **Properties**.

Change the Style to **Staff Breakroom**.
Change the Name to **Staff Breakroom.**

Click **ESC** to release the selection.

8.

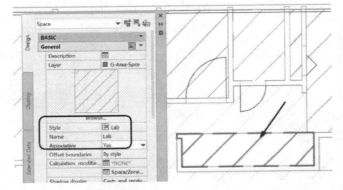

Select the lab area space.

Right click and select **Properties**.

Change the Style to **Lab**.
Change the Name to **Lab**.

Click **ESC** to release the selection.

9.

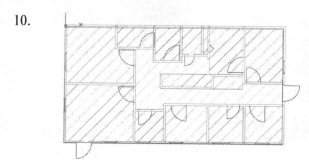

Select the Nurse Station area space.

Right click and select **Properties**.

Change the Style to **Nurse Station**.
Change the Name to **Nurse**.

Click **ESC** to release the selection.

10.

Inspect the floor plan to verify that all spaces have been placed.

Note that different space styles use different hatch patterns and colors.

If your spaces don't display with different hatch patterns and colors, verify that the display override is checked for that display style.

11. Save as *Rooms_6.dwg*.
Close all open drawings.

Property Set Definitions

Property set definitions are used to assign properties to AEC objects. These properties can then be used in tags and schedules.

Exercise 8-6:

Create a Room Tag

Drawing Name: Rooms_6.dwg
Estimated Time: 60 minutes

This exercise reinforces the following skills:

- ❑ Spaces
- ❑ Property Set Definitions
- ❑ Style Manager
- ❑ Property Set Formats
- ❑ Attributes
- ❑ Add/Delete Scales
- ❑ Tags
- ❑ Palettes
- ❑ Blocks

1.

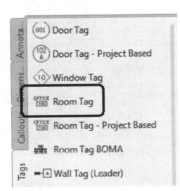

Launch the Document tools palette.

Click on the Tags tab.

Locate the **Room Tag**.

Place on the top space in the drawing.

Click **ENTER** to accept placing in the center of the space.

2.

The available properties will display.

Click **OK**.

Right click to select **Multiple**.

3.

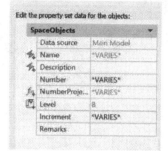

Window around the building.
All spaces are automatically selected.
Click **ENTER** to complete the selection.

Click **No**.

4.

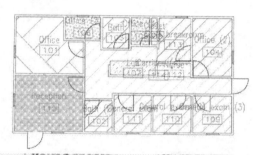

Click **OK**.

Click ENTER to exit the command.

5.

The tags are placed, but they are
a bit large.

Select one of the tags.
Right click and **Select Similar**.
This will select all placed tags.

6.

Right click and select
Properties.

Select **Multi-View Block
Reference** from the drop-down
list.
Change the Scale to **0.50**.

Click **ESC** to release the
selection.

7.

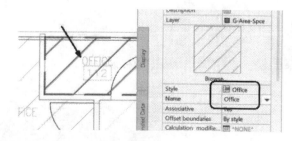

You can modify the displayed
names by selecting the space.
Right click and select
Properties.
Then modify the Name.

8.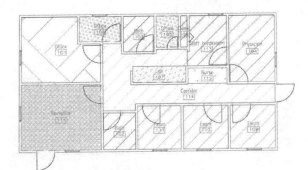

Modify the names the way
you would like them to
appear.

9.

Switch to the Manage ribbon.

Click on **Style Manager**.

Style Manager   Display Manager   Renovation Mode

Style & Display ▾

10.

Documentation Objects
  2D Section/Elevation Styles
  AEC Dimension Styles
  Calculation Modifier Styles
  Display Theme Styles
  Property Data Formats
  Property Set Definitions
    DoorStyles
    FrameStyles
    GeoObjects
    ManufacturerStyles
    RoomFinishObjects
    SpaceEngineeringObjects
    SpaceObjects
    SpaceStyles
    ThermalProperties
    WallStyles
    WindowStyles
    ZoneEngineeringObjects
  Schedule Table Styles
  Zone Styles
  Zone Templates

Expand the **Documentation Objects** folder.

Expand **Property Set Definitions**.

Click on **SpaceObjects**.

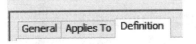

Select the **Definition** tab.

11. Click on **Add Automatic Property Set Definitions**.

12. Alphabetical | Categorized

Base Area ✓
Base Area Minus Int... ☐
Base Ceiling Area ☐

Click on the Alphabetical tab.

Enable **Base Area**.

13.

Enable **Length**.

14.

Enable **Name** and **Style**.

15.

Enable **Width**.

16.

Click **OK**.
You can see a list of properties which will be applied to space objects.

Click **Apply** to save your changes.

17.

Set the Format for the Property Set Definitions.

Highlight BaseArea.

In the Format column: use the drop-down list to select **Area**.

18.

Set BaseArea to **Area**.
Set Length to **Length – Short**.
Set Name to **Standard**.
Set Style to **Standard.**
Set Width to **Length – Short**.
Click **Apply**.

19.

Expand the **Property Data Formats** category.

20.

Highlight **Area** in the list.

Click on the **Formatting** tab.

Set the Precision to **0.0**.

Verify that the other fields are set correctly.

21.

Verify that the Suffix is set to **M2**.

Click **Apply**.

22.

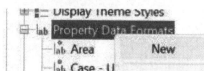

Highlight Length – Short.

Set the Precision to **0**.

Click **Apply**.

23.

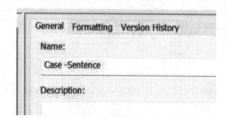

Highlight Property Data Formats.
Right click and select **New**.

24.

On the General tab:
Type **Case – Sentence** in the Name field.

25.

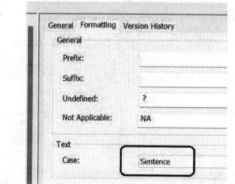

On the Formatting tab:

Set the Case to **Sentence**.

Click **Apply**.

26.

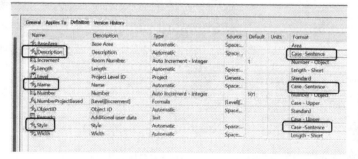

Highlight **SpaceObjects** under Property Set Definitions.

Select the Definition tab. Change the Format for **Description, Name**, and **Style** to use **Case-Sentence**.

27.

Click **OK** to close the Style Manager.

28.

Open the Home ribbon.

Set the **G-Area-Spce-Iden** layer current.

29.

Switch to the Insert ribbon.

Click **Define Attributes**.

30.

In the Tag field:
Type **SPACEOBJECTS:STYLE**.
In the Prompt field:
Type **Style**.
In the Default field:
Type **Lobby**.
Set the Justification to **Center**.
Set the Text style to **RomanS**.
Disable **Annotative**.
Set the Text height to **300.**
Enable **Specify on-screen** for the insertion point.
Click **OK**.

31.

Click to place the attribute in the drawing.

Zoom into the attribute.

32.

Click **Define Attributes**.

You can also right click and select **ATTDEF** under Recent.

33.

In the Tag field:
Type **SPACEOBJECTS:BASEAREA**.
In the Prompt field:
Type **Area**.
In the Default field:
Type **0000 M2**.
Set the Justification to **Center**.
Set the Text style to **RomanS**.
Disable **Annotative**.
Set the Text height to **300.**

34. 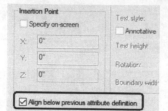 Enable **Align below previous attribute definition** for the insertion point.
Click **OK**.

The attribute is placed.

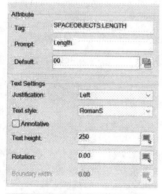

35. Place some text below the existing attributes to help you locate the length and width attributes.

You can then modify the text to only show the X.

36.  Switch to the Insert ribbon.

Click **Define Attributes**.

37. In the Tag field:
Type **SPACEOBJECTS:LENGTH**.
In the Prompt field:
Type **Length**.
In the Default field:
Type **0000**.
Set the Justification to **Left**.
Set the Text style to **RomanS**.
Disable **Annotative**.
Set the Text height to **250**

38. 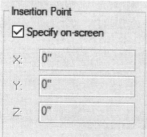 Enable **Specify on-screen** for the insertion point.
Click **OK**.

39.

SPACEOBJECTS:STYLE
SPACEOBJECTS:BASEAREA
SPACEOBJECTS:LENGTH

Place below and slightly left of the previous attributes.

Use the text to help you locate the attribute.

40.

SPACEOBJECTS:STYLE
SPACEOBJECTS:BASEAREA
SPACEOBJECTS:LENGTH X SPACEOBJECTS:WIDTH

Copy the length attribute and place to the right.

41.

Double click on the copied attribute.
Change the tag to **SPACEOBJECTS:WIDTH**.
Change the prompt to **Width**.
Click **OK**.
Click **ENTER** to exit the command.

42.

SPACEOBJECTS:STYLE
SPACEOBJECTS:BASEAREA
SPACEOBJECTS:LENGTH X SPACEOBJECTS:WIDTH

Double click on the text and edit to leave only the X.

43.

Activate the Annotate ribbon.

Select the **Create Tag** tool from the Scheduling panel in the drop-down area.

44.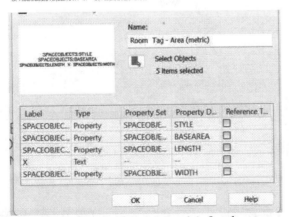

Select the attributes and the text.

45. Type **Room Tag – Area (metric)** for the name.
Verify that the property sets and definitions have been assigned correctly.

If the property definitions are not correct, you may need to double check your attribute definitions.

46.   Click **OK**.
        Select the center of the attribute list to set the insertion point for the tag.

47.   Save and close all drawings.

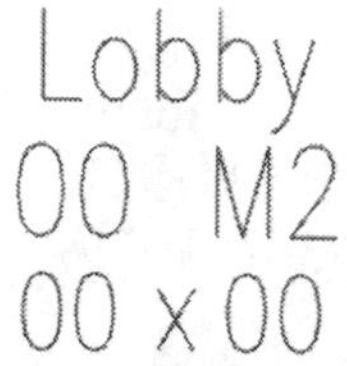

Lobby
00 M2
00 x 00

Exercise 8-7:

Create a Room Tag Tool

Drawing Name: Rooms_7.dwg, schedule tags_metric.dwg
Estimated Time: 5 minutes

This exercise reinforces the following skills:

- ❑ Spaces
- ❑ Property Set Definitions
- ❑ Style Manager
- ❑ Property Set Formats
- ❑ Attributes
- ❑ Add/Delete Scales
- ❑ Tags
- ❑ Palettes
- ❑ Blocks

1. 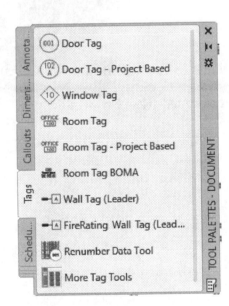 Launch the **Document** tool palette.

2. 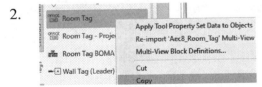 Locate the **Room Tag** tool on the Document tool palette.

   Right click and select **Copy**.

3. Move the cursor to a blank space on the palette.

Right click and select **Paste**.

4. Highlight the copied Room Tag tool.

Right click and select **Rename**.

5. Rename it **Room Tag (w/Dimensions)** and move it below the original wall tag.

6. Highlight the **Room Tag (w/Dimensions) -**

Right click and select **Properties**.

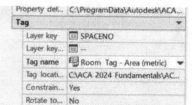

7. Set the Tag Location as *schedule_tag_metric.dwg*.
   Set the Tag Name to **Room Tag-Area (metric)**.

You may want to create a drawing to store all your custom schedule tags and then use that drawing when setting the tag location.

Click **OK**.

Exercise 8-8:

Adding Room Tags

Drawing Name: Rooms_8.dwg
Estimated Time: 5 minutes

This exercise reinforces the following skills:

- ❑ Tags
- ❑ Tool Palette
- ❑ Content Browser

1.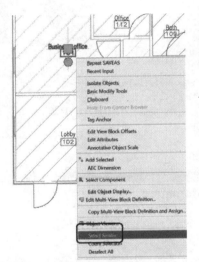

   Select an existing room tag.
   Right click and **Select Similar**.

   All the room tags are selected.

   Right click and select **Properties**.

2. Change the tag definition to **Room Tag – Area (metric).**

   Change the scale to **.001.**

   Click **ESC** to release the selection.

3. The tags all update.

4. Save as Rooms_9.dwg
   Close all drawings.

Extra: Create room plans with tags for the first and second floors in the Tahoe Cabin project.

Exercise 8-9:

Create a Room Schedule

Drawing Name: Rooms_9.dwg
Estimated Time: 50 minutes

This exercise reinforces the following skills:

- ❑ Schedules
- ❑ Tool Palettes
- ❑ Creating Tools
- ❑ Schedule Styles
- ❑ Property Set Definitions
- ❑ Add viewport
- ❑ Named Views
- ❑ Project Navigator
- ❑ Project Views
- ❑ Room Tags

1.    Activate the Manage ribbon.

   Select **Style Manager**.

2.   Locate the Schedule Table Styles category under Documentation.

3.   Highlight the Schedule Table Styles category.

   Right click and select **New**.

4.   On the General tab:

   Type **Room Schedule** for the Name.

   Type **Uses Space Information** for the Description.

5. Select the Applies To tab.

Enable **Space**.

Note that Spaces have Classifications available which can be used for sorting.
Select the Columns tab.

6.

Select **Add Column**.

7. Select **Style**.

Change the Heading to **Room Name**.

Set the Data Format to **Case – Sentence**.

Click **OK**.

If your property set definitions have not been modified, you will not be able to access these properties.

8.

Select **Add Column**.

9. Select the Alphabetical tab to make it easier to locate the property definition.

Locate and select **Length**.

Click **OK**.

10.

Select **Add Column**.

11. Select the Alphabetical tab to make it easier to locate the property definition.

Locate and select **Width**.

Click **OK**.

12.

Select Add Column.

13. Select the Alphabetical tab to make it easier to locate the property definition.

Select **BaseArea**.

Change the Heading to **Area**.

Click **OK**.

14.

	Room Name	Length	Width
Property Set	SpaceObjects	SpaceObjects	SpaceObjects
Property	Style	Length	Width
Format	Case - Senten...	Length - Short	Length - Short
Example			
Total	No	No	No
Hide	No	No	No
Matrix	No	No	No
Max Cols.			

The columns should appear as shown.

15.

Table Title:	Room Schedule

Format

Title: ☐ Override Cell Format...

Column Headers: ☐ Override Cell Format...

Matrix Headers: ☐ Override Cell Format...

Select the Layout tab.

Change the Table Title to **Room Schedule.**

16.

Add...	Remove

Sort Order

Sort
Group
Display Subtotals For Group

Select the **Sorting/Grouping** tab.
Highlight **Sort**.
Select the **Add** button.

17.

SpaceObjects:BaseArea
SpaceObjects:Length
SpaceObjects:Style
SpaceObjects:Width

OK

Select **SpaceObjects: Style** to sort by Style name.
Click **OK**.

	Sort Order
	SpaceObjects:Style
Sort	Ascending
Group	☐
Display Subtotals For Group	☐
Group Format Override	"None"
Subtotal Format Override	"None"
Group Orientation	Column
Group Row Repeat	☐
Group Row Indent	
Group Separation	None

The panel should fill in with the sort information.

18.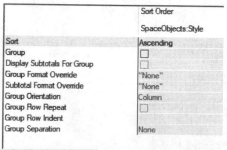

Place a check next to **Group**.
Place a check next to **Display Subtotals for Group.**

Note: To enable subtotals modify at least one column on the columns tab to include a total.

A small red note will appear at the bottom of the dialog.

19. Select the Columns tab.

Area

s SpaceObjects
BaseArea
t Area
SF
No
No
No

Modify...

Highlight **Area** and select **Modify**.

20.

Column Properties

Property Set: SpaceObjects
Property: BaseArea
Heading: Area

Data Format: Area
Example: M2

Override Cell Format...

Override Header Format...

Matrix   Max Columns: 3

Total   Hide

Enable **Total.**

Click **OK** to close the dialog.

Click **OK** to close the Style Manager dialog.

Save the current drawing.

21.

Annotation
Tools

Tools

Launch the Annotation Tools palette from the Annotate ribbon.

22.

Edit Property Set Data
Browse Property Data
Door Schedule
Door Schedule Project Based
Window Schedule
Wall Schedule
Space List - BOMA
Space Inventory Schedule
Room Finish Schedule
Space Evaluation

Select the Scheduling tab.

23.

Highlight the Space List – BOMA.

Right click and select Copy.

24.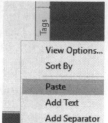

Move the mouse off the tool.

Right click and select **Paste**.

25.

Select the copied tool.

Right click and select **Rename**.

26.

Type **Room Schedule w/Area** for the name.

27.

Highlight the new tool.

Right click and select **Properties**.

28.

Layer overrid...	▦ --
Style	▦ Room Schedule
Style location	C:\ACA 2024 Fund
Title	--

In the Style field, select the **Room Schedule** style which you just created.

In order to use the Room Schedule style you created, you need to reference the drawing you saved the style in.
By default, ACA will reference a different drawing.
If you want to be able to use this style in other projects, use the Style Manager to copy and paste the style to the default schedule style drawing.

Click **OK**.

29.

💡 ☀ 🔓 ⬛ G-Anno-

💡 ☀ 🔓 ⬛ G-Area-

💡 ☀ 🔓 ◻ G-Area-

Thaw the **G-Area-Spce** layer if it is not thawed.

30.

Room Schedule			
Room Name	Length	Width	Area
	2086	2167	4.5 N
Bathroom_small	1814	2370	4.3 N
			8.8 N
Business office	4939	4024	17.7
			17.7
Closet	344	1911	0.7 N
			0.7 N
Corridor	10673	4316	22.0
			22.0
General_exam	2463	2595	6.4 N
	2700	2595	7.0 N
	3008	2595	7.8 N
			21.2
Lab	3755	909	3.3 N
			3.3 N
Nurse station	2081	909	1.9 N
			1.9 N
Office	1309	1911	2.7 N
	2650	1441	3.8 N
			6.5 N
Physician office	2779	3858	10.7
			10.7
Reception	4956	4036	20.0
			20.0
Staff breakroom	2404	3343	8.0 N
			8.0 N
			120.7

Select the Room Schedule with Area tool.

Window around the building.

Click ENTER.

Pan over to an empty space in the drawing and left pick to place the schedule.

Click ENTER.

If there are any ? symbols in the table, you have some spaces which are unassigned.

You may have placed duplicate spaces or empty spaces.

To update: select the schedule. Select Edit Table Cell. Click on the cell with the question mark. The table should update with the correct information.

It may be easier to select all the spaces using QSELECT and delete them and then replace all the spaces.

31.

Go to the View ribbon.
Launch the **View Manager**.

32.

Create a Named View for the **Room Schedule**.

Click **OK**.

33.

Save the drawing as **Rooms with Schedule**.dwg.

34.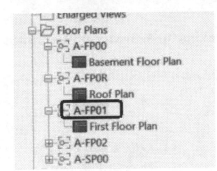

Go to the Project Navigator.

Open the Views tab.

Right click on **A-FP01**.

35.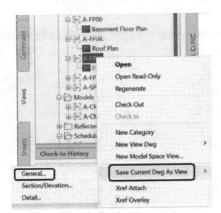

Select **Save Current Dwg As View→General**.

36.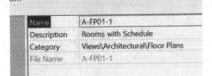

Change the Name to **A-FP01-1**.

Click **Next**.

37.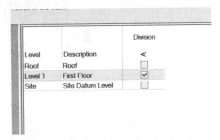

Enable **First Floor**.

Click **Next**.

38.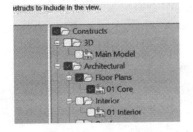

Enable the 01 Core construct only.

Click **Finish**.

39.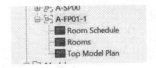

The named views that are in the drawing are listed.

40. 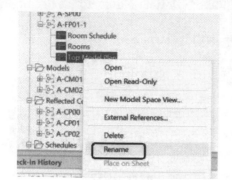 Highlight the **Top Model Plan** named view.
Right click and select **Rename**.

41. 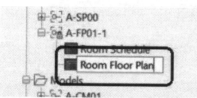 Rename the view **Room Floor Plan**.

42. 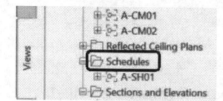 Right click on **Schedules**.

43. 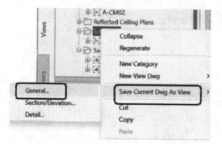 Select **Save Current Dwg As View→General**.

44. 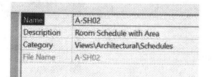 Change the Name to **A-SH02**.
Change the Description to **Room Schedule with Area**.
Click **Next**.

45. 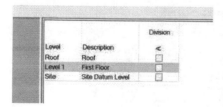 Do not select any of the levels.

Click **Next**.

Click **Finish**.

46.

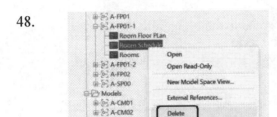

Right click on the **Room Floor Plan** named view under **A-SH02**.
Select **Delete**.

47.

Delete model space view
Views\Architectural\Schedules\A...\Room
Floor Plan ?

| Yes | No |

Click **Yes**.

48.

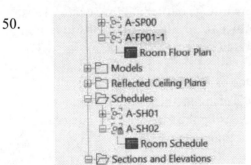

Right click on the **Room Schedule** named view under **A-FP01-1**.
Select **Delete**.

49.

Delete model space view
Views\Architectural\Floor
Plans\A-FP01-1\Room Schedule ?

| Yes | No |

Click **Yes**.

50.

⊞ A-SP00
⊟ A-FP01-1
  Room Floor Plan
⊞ Models
⊞ Reflected Ceiling Plans
⊟ Schedules
  ⊞ A-SH01
  ⊟ A-SH02
    Room Schedule
⊟ Sections and Elevations

Each named view is now saved in a separate drawing file.

51.

Save and close any open drawings.

Extra: Create a named view for the room schedule.

Create a different room schedule style.

Create a new layout with the floor plan and the new room schedule.

You can insert blocks into model or paper space. Note that you did not need to activate the viewport in order to place the Scale block.

➢ By locking the Display you ensure your model view will not accidentally shift if you activate the viewport.

➢ The Annotation Plot Size value can be restricted by the Linear Precision setting on the Units tab. If the Annotation Plot Size value is more precise than the Linear Precision value, then the Annotation Plot Size value is not accepted.

Most projects require a floorplan indicating the fire ratings of the building walls. These are submitted to the planning department for approval to ensure the project is to building code. In the next exercises, we will complete the following tasks:

- Load the custom linetypes to be used for the exterior and interior walls
- Apply the 2-hr linetype to the exterior walls using the Fire Rating Line tool
- Apply the 1-hr linetype to the interior walls using the Fire Rating Line tool
- Add wall tags which display fire ratings for the walls
- Create a layer group filter to save the layer settings to be applied to a viewport
- Apply the layer group filter to a viewport
- Create the layout of the floorplan with the fire rating designations

Exercise 8-10:
Loading a Linetype

Drawing Name: new
Estimated Time: 25 minutes

This exercise reinforces the following skills:

- ❑ Loading Linetypes
- ❑ Layer States
- ❑ Views
- ❑ Project Navigator

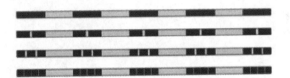

These are standard line patterns used to designate different fire ratings.

I have created the linetypes and saved them into the firerating.lin file included in the exercise files. You may recall we used the Express Tools→ Make Linetype tool to create custom linetypes.

1.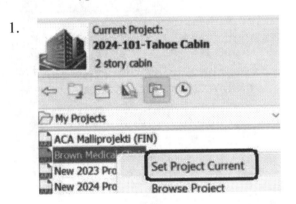

Launch the Project Browser.

Set the **Brown Medical Clinic** as the current project.

2. Start a new drawing using a Metric template.

3.   Save as *Level 1 Fire Rating.dwg.*

4.

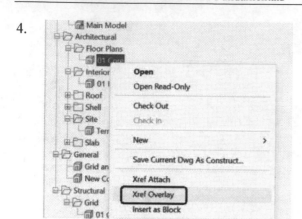

Right click on the **01 Core** construct. Select **Xref Overlay**.

Double click the mouse wheel to **Zoom Extents**.

5.

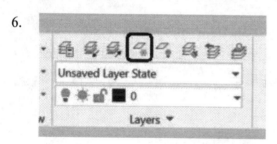

Freeze the **PDF Images** layer.

6.

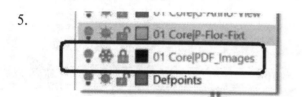

Select the **Freeze Layer** tool on the Home ribbon.

7.

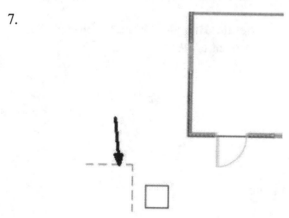

Select the grid lines indicated located at the lower left of the drawing.

8.

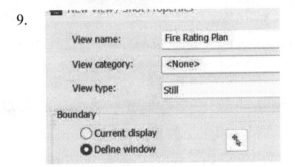

Go to the View ribbon.

Launch the **View Manager**.

9.

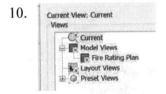

Create a Named View of the top plan view of the building.

Click **OK**.

10.

Click **OK**.

11.

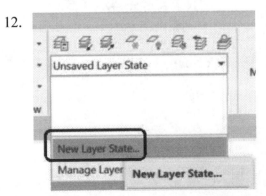

Set the Display Configuration to **Medium Detail**.

12.

On the Home ribbon/Layer panel:

Select **New Layer State**.

13.

New layer state name:

Fire Rating

Description

fire rating for walls

Type **Fire Rating** for the layer state name.

Type **fire rating for walls** for the description.

Click **OK**.

14. Type **LINETYPE** to load the linetypes.

15. Select **Load**.

16.

File...   acad.lin

Select the **File** button to load the linetype file.

17.

File name: fire rating.lin

Files of type: Linetype (*.lin)

Locate the *fire rating.lin* file in the exercise files.

Click **Open**.

18.

Available Linetypes

This is a txt file you can open and edit with Notepad.
Hold down the control key to select both linetypes
available.

19.

1-HR
2-HR
CENTER2
Continuous
DASHED

Click **OK** to load.
The linetypes are loaded.

Click **OK** to close the dialog box.

20.

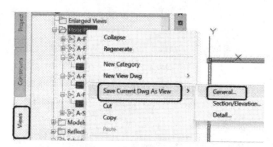

Open the Project Navigator.
Open the **Views** tab.
Right click on **Floor Plans**.
Select **Save Current Dwg As
View→General**.

21.

Name	A-FP01-2
Description	Level 1 Fire Rating Plan
Category	Views\Architectural\Floor Plans
File Name	A-FP01-2

For Name: Type **A-FP01-2**.
For Description:
Type **Level 1 Fire Rating Plan**.
Click **Next**.

22.

Level	Description	Division
Roof	Roof	☐
Level 1	First Floor	☑
Site	Site Datum Level	☐

Enable **First Floor**.

Click **Next**.

23.

Disable all selections except for 01 Core.

Click **Finish**.

24. Save and close all open drawings.

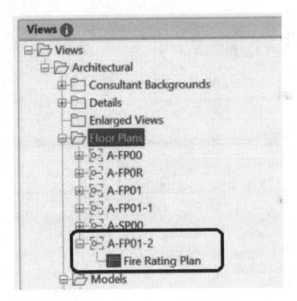

Exercise 8-11:
Applying a Fire Rating Line

Drawing Name: A-FP01-2.dwg
Estimated Time: 25 minutes

This exercise reinforces the following skills:

- Adding a Layout
- Layers
- Edit an External Reference In-Place
- Adding Fire Rating Lines
- Creating a tool on the tool palette

1.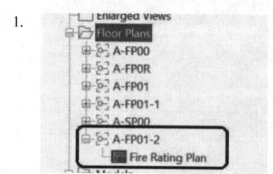

   Launch the Project Navigator.

   Open the **A-FP01-2** view drawing.

2.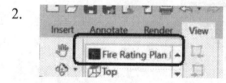

   Restore the **Fire Rating Plan** named view, if needed.

3.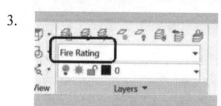

   Restore the **Fire Rating** layer state.

4.   Select an exterior wall.

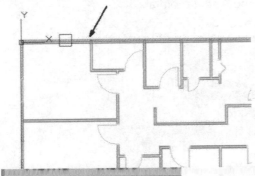

5.

Select **Edit Reference In-Place** from the ribbon.

6.

Click **OK**.

7.

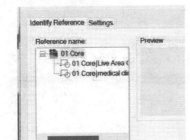

Launch the **Design Tools** palette from the Home ribbon if it is not available.

8.

Right click on the Design Tools palette title bar.

Enable **Document**.

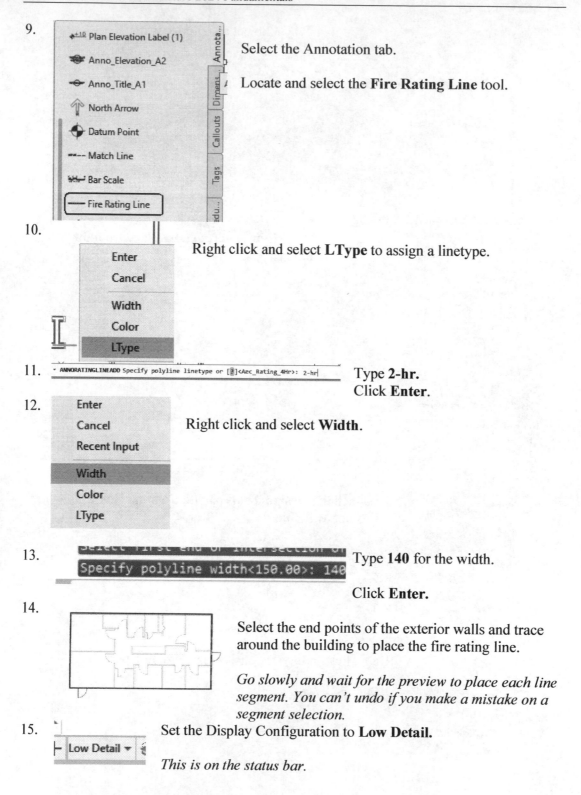

9. Select the Annotation tab.

    Locate and select the **Fire Rating Line** tool.

10. Right click and select **LType** to assign a linetype.

11. Type **2-hr.**
    Click **Enter**.

12. Right click and select **Width**.

13. Type **140** for the width.

    Click **Enter**.

14. Select the end points of the exterior walls and trace around the building to place the fire rating line.

    Go slowly and wait for the preview to place each line segment. You can't undo if you make a mistake on a segment selection.

15. Set the Display Configuration to **Low Detail.**

    This is on the status bar.

16.

Select the polyline that is the fire rating line.

Right click and select **Properties**.

Note that the line is placed on the F-Wall-Fire-Patt layer.

Click **ESC** to release the selection.

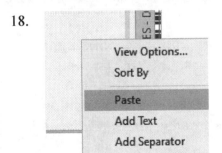

Change the View Scale to **1:100**.

You should be able to see the hatch pattern of the fire rating line.

17.

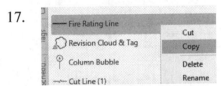

Select the **Fire Rating Line** tool.
Right click and select **Copy**.

18.

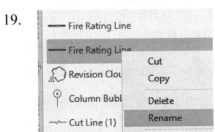

Right click on the palette.

Click **Paste**.

19.

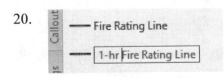

Select the duplicate Fire Rating Line tool.

Right click and select **Rename**.

20.

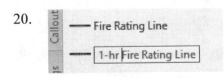

Rename **1-hr Fire Rating Line**.

21.

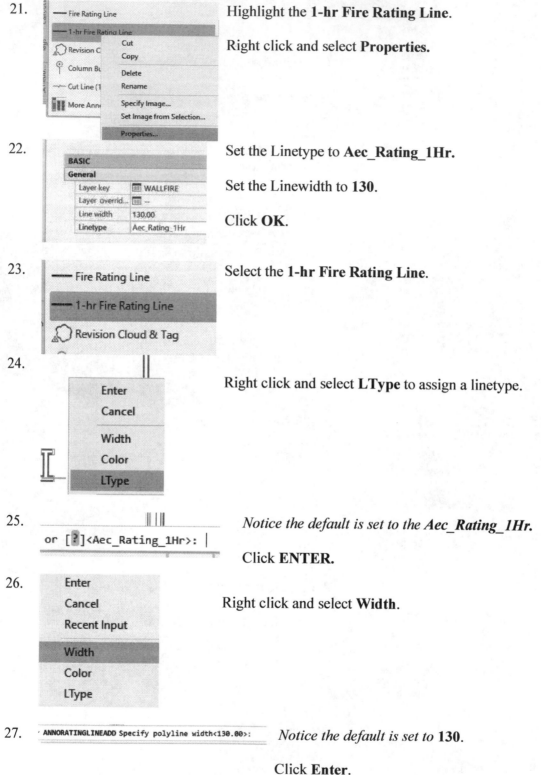

Highlight the **1-hr Fire Rating Line**.

Right click and select **Properties.**

22.

Set the Linetype to **Aec_Rating_1Hr.**

Set the Linewidth to **130**.

Click **OK**.

23.

Select the **1-hr Fire Rating Line**.

24.

Right click and select **LType** to assign a linetype.

25.

Notice the default is set to the Aec_Rating_1Hr.

Click **ENTER.**

26.

Right click and select **Width**.

27.

ANNORATINGLINEADD Specify polyline width<130.00>:

Notice the default is set to **130**.

Click **Enter**.

28.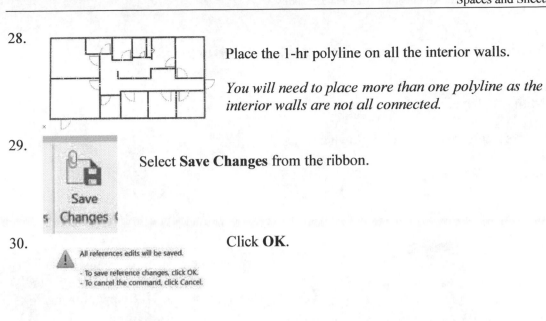

Place the 1-hr polyline on all the interior walls.

You will need to place more than one polyline as the interior walls are not all connected.

29. Select **Save Changes** from the ribbon.

Save
Changes

30. Click **OK**.

All references edits will be saved.

- To save reference changes, click OK.
- To cancel the command, click Cancel.

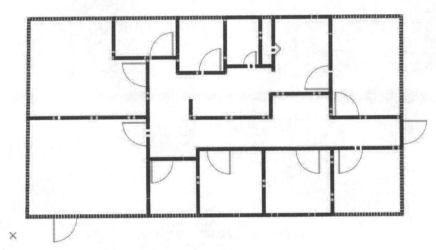

31. Save and close all drawings.

Exercise 8-12:
Assigning Fire Rating Tags to Wall Properties

Drawing Name: A-FP01-2.dwg
Estimated Time: 15 minutes

This exercise reinforces the following skills:

- ❑ Wall Properties
- ❑ Style Manager
- ❑ Edit Reference In-Place

1.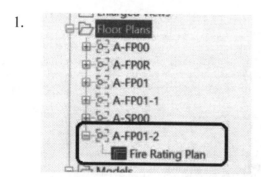

   Launch the Project Navigator.

   Open the **A-FP-01-2** view drawing.

2.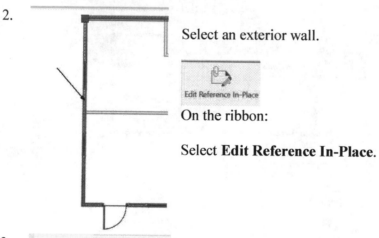

   Select an exterior wall.

   On the ribbon:

   Select **Edit Reference In-Place**.

3.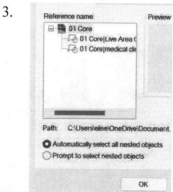

   Click **OK**.

4. Activate the Manage ribbon.

Select the **Style Manager**.

5. 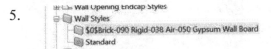 Locate and select the **exterior**Wall Style.

6.  On the General tab:

Select **Property Sets**.

Property sets control the values of parameters which are used in schedules and tags.

7. 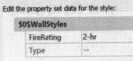 Type **2-hr** in the FireRating property.

Click **OK**.

When defining the attribute to use this value, we use the format WallStyles:FireRating. The first word is the property set and the second word is the property.
Close the Style Manager.

8.  Locate and select the **interior** Wall Style.

9.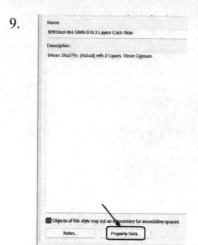

On the General tab:

Select **Property Sets**.

Property sets control the values of parameters which are used in schedules and tags.

10.

Type **1-hr** in the FireRating property.

Click **OK**.
Close the Style Manager.

11.

Select **Save Changes** from the ribbon.

12.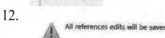

Click **OK**.

13. Double click outside the viewport to return to paper space.

Save and close all drawings.

Exercise 8-13:
Creating a Schedule Tag

Drawing Name: schedule tags_metric.dwg
Estimated Time: 15 minutes

This exercise reinforces the following skills:

- ❏ Wall Tags
- ❏ Attributes
- ❏ Create a Schedule Tag

1.   Create a new drawing called *schedule tags_metric.dwg*.

2.   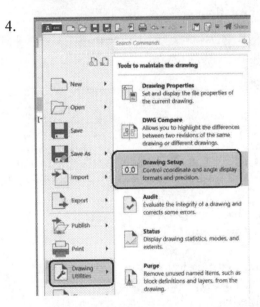   Activate the **Model** layout.

3.   Set the view scale to **1:1**

4.   On the Application Menu:

Go to **Drawing Utilities→Drawing Setup**.

5.   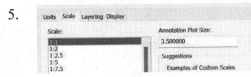   Select the Scale tab.
Highlight **1:1**.
Set the Annotation Plot Scale to. **3.5**

6.

Create a layer named **A-Anno-Wall-Iden.**
Set the **A-Anno-Wall-Iden** layer current.

This is the layer to be used by the new wall tag.

7.

Draw a 200 x 200 square.

Hint: Use the Rectangle tool. Click one corner and for the second corner, type @ 6, 6.

8.

Select the square.
Right click and select **Basic Modify Tools→Rotate.**

When prompted for the base point, right click and select **Geometric Center.**

Then select the center of the square.

Type **45** for the angle of rotation.

9.

Activate the Insert ribbon.

Select the **Define Attributes** tool.

10.

Type
WALLSTYLES:FIRERATING for the tag.
This is the property set definition used for this tag.
Type **Fire Rating** for the prompt.
Type **2-hr** for the Default value.

Set the Justification for the **Middle center.**
Set the Text style to **RomanS.**
Disable **Annotative.**
Set the Text Height to **2.**

Enable **Specify on-screen.**

Click **OK.**

11. Select the Geometric Center of the Square to insert the attribute.

WALLSTYLES:FIRERATING

12.

Launch the **Styles Browser** from the Home ribbon.

13.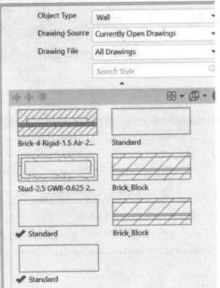

Import the Exterior and Interior wall styles used in the project.

Close the Styles Browser.

14.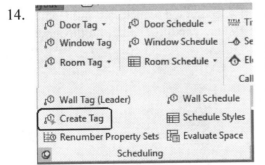

Activate the Annotate ribbon.

Select the **Create Tag** tool from the Scheduling panel in the drop-down area.

15.

Select the square and the attribute you just created.

Press **ENTER**.

16.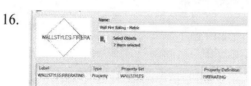

Name the new tag **Wall Fire Rating - Metric**.

You should see a preview of the tag and you should see the Property Set and Property Definition used by the tag based on the attribute you created.

Click **OK**.

17.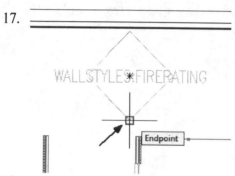

You will be prompted for the insertion point for the tag.

Select the bottom vertex of the rectangle.

18.

The tag will convert to a block and you will see the default attribute.

Notice that when it converts, the size will change.

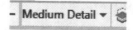

If you don't see the tag, check that the Display Property is set to Medium Detail on the status bar.

19. Save and close the drawing.

Adding a Schedule Tag Tool to the Document Palette

Drawing Name: A-FP01-2.dwg
Estimated Time: 10 minutes

This exercise reinforces the following skills:

- ❑ Wall Tags
- ❑ Create a Tool on the Tool Palette
- ❑ Property Sets

1. 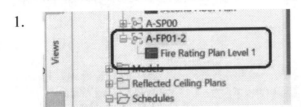 Launch the Project Navigator.

   Open the *A-FP01-2.dwg*.

2. 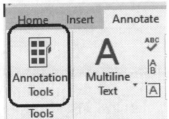 Switch to the Annotate ribbon.

   Launch the **Annotation Tools**.

3. 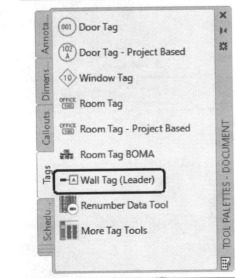 Click on the **Tags** tab of the Tool Palette.

   Locate the **Wall Tag (Leader) tool** on the Document tool palette.

4.  Highlight the **Wall Tag (Leader) tool**.
   Right click and select **Copy**.

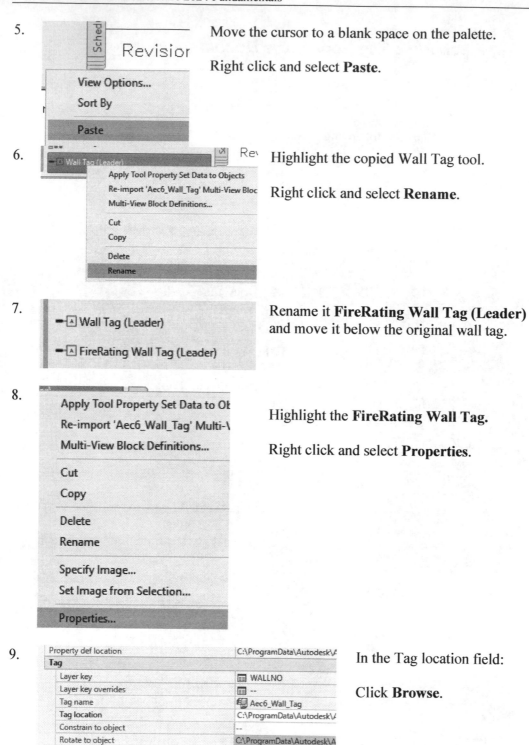

5.   Move the cursor to a blank space on the palette.

Right click and select **Paste**.

6.   Highlight the copied Wall Tag tool.

Right click and select **Rename**.

7.   Rename it **FireRating Wall Tag (Leader)** and move it below the original wall tag.

8.   Highlight the **FireRating Wall Tag.**

Right click and select **Properties**.

9.   In the Tag location field:

Click **Browse**.

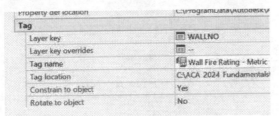

10. Set the Tag Location as *schedule_tags_metric.dwg*.
    Set the Tag Name to **Wall Fire Rating - Metric**.

    You may want to create a drawing to store all your custom schedule tags and then use that drawing when setting the tag location. In this case, I created the drawing and loaded the tag we created in the previous exercise.

    Click **OK**.

11.  Set the Display Configuration to **Medium Detail**.

12. 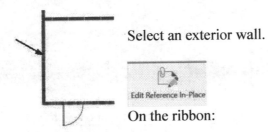 Select an exterior wall.

    On the ribbon:

    Select **Edit Reference In-Place**.

13. 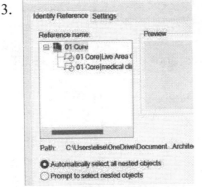 Click **OK**.

14. 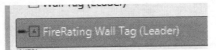 Select the **FireRating Wall Tag** tool and tag an exterior wall.

    Click **ENTER** to complete the placement.

15.

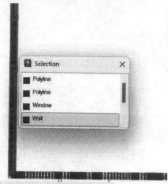

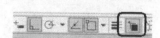

Hint: Enable Selection Cycling to make it easier to select a wall.

16.

Click **OK** to accept the data set.

17.

Left click to place the tag and then click **OK** to accept the properties.

You should see the updated tag.

If necessary, change the scale of the tag to .03.

18.

Go through the floor plan to tag all the walls.

Double click outside the viewport to switch to paper space.

19.

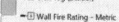

Select the **Wall Fire Rating Metric** tool and tag an interior wall.

Click **ENTER** to complete the placement.

20.

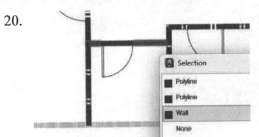

*Hint: Enable **Selection Cycling** to make it easier to select a wall.*

21.

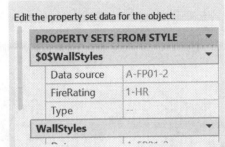

Click **OK** to accept the data set.

Repeat to tag additional interior walls.

22.

If the attribute doesn't fill in properly, select a wall.
Click **Edit Style** on the ribbon.

23.

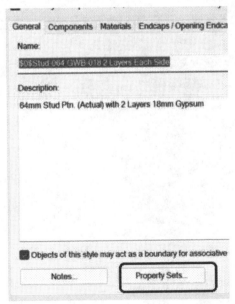

Click **Property Sets** on the General tab.

24.

WallStyles	
Type	--
FireRating	1-HR

0WallStyles	
FireRating	1-HR
Type	--

Edit the property set data for the style:

Type in the desired fire rating in both locations.

Click **OK**.

Click **OK**.

25.

Save Changes

Select **Save Changes** from the ribbon.

26.

⚠ All references edits will be saved.

- To save reference changes, click OK.
- To cancel the command, click Cancel.

Click **OK**.

27.

The drawing should appear like this.

If you have difficulty, open the xref and tag inside the xref. Save and then reload.

28. Save and close all drawings.

Exercise 8-15:

Modify a Dimension Style

Drawing Name: A-FP01.dwg
Estimated Time: 30 minutes

This exercise reinforces the following skills:

- ❑ Dimension Styles
- ❑ AEC Dimensions

1.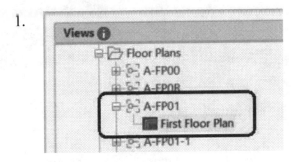

   Launch the Project Navigator.

   Open the **A-FP01.dwg** view.

   The view is empty except for elevation and section lines.

   There is a view title as well.

2.

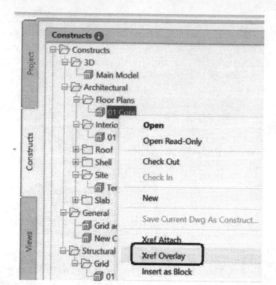

Open the Project Navigator.

Right click on the **01 Core** construct. Select **Xref Overlay**.

Double click the mouse wheel to **Zoom Extents**.

3.

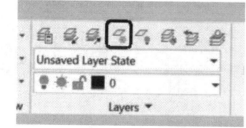

Position the xref so it is centered between the elevation and section lines.

4.

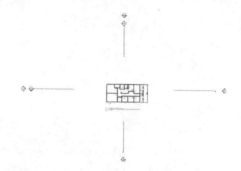

Select the **Freeze Layer** tool on the Home ribbon.

5.

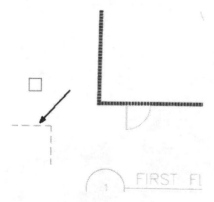

Select the grid lines indicated located at the lower left of the drawing.

Select the section and elevation lines.

6.

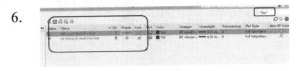

Launch the Layer Properties Manager.
Perform a search for layers with ***fire*** in their name.
Freeze those layers.

7.

Perform a search for layers with ***iden*** in their name.
Freeze those layers.

This freezes the wall tags layer.

Close the Layer Properties Manager.

8.

Go to the View ribbon.

Launch the **View Manager**.

9.

Highlight the **First Floor Plan**.

Click **Edit Boundaries**.

10.

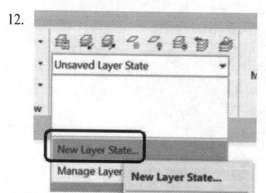

Window around the floor plan.

Click ENTER.

Click **OK** to close the View Manager.

11.

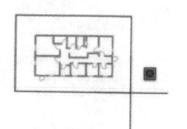

Set the Display Configuration to **Medium Detail**.

12.

On the Home ribbon/Layer panel:

Select **New Layer State**.

13.

New layer state name:

Floor Plan

Description

Floor Plan with dimensions

Type **Floor Plan** for the layer state name.

Type **Floor Plan with dimensions** for the description.

Click **OK**.

14.

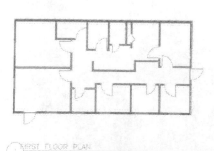

Move the view title to below the floor plan.

15. Use the grips to adjust the location of the view title and scale attributes.

FIRST FLOOR PLAN
:100

16. Activate the Annotate ribbon.

Select the **Dimension Style Editor** on the Dimensions panel.

17. Highlight **Annotative**.

18. Select **Modify**.

19. On the Lines tab,
set the Baseline spacing to **150**.

Extend beyond ticks: 0.00
Baseline spacing: 150.00

20. Set Extend beyond dim lines to **50**.
Set Offset from origin to **25**.

Extend beyond dim lines: 50.00
Offset from origin: 25.00

21. On the Symbols and Arrows tab,

Select **Architectural** tick for the arrowheads.
Set the Arrow size to **150**.

Arrow size:
150.00

22. On the Text tab,
set the Text height to **150**.

Text appearance
Text style: Arch-Dim
Text color: ByBlock
Fill color: None
Text height: 150.00
Fraction height scale: 0.75

23. Set the Offset from dim line to **3**.

Text placement
Vertical: Above
Horizontal: Centered
View Direction: Left-to-Right
Offset from dim line: 3.00

24. Set the Text alignment to **Horizontal.**

Text alignment
○ Horizontal
○ Aligned with dimension line
○ ISO standard

25.

Lines	Symbols and Arrows	Text	Fit	Primary Units
Linear dimensions				
Unit format:	Decimal			⌄
Precision	0			⌄

Activate the Primary Units tab.

Set the Unit format to **Decimal.**
Set the Precision to **0**.

26. Click **OK** and **Close**.

27. Save and close all drawings.

Exercise 8-16:

Dimensioning a Floor Plan

Drawing Name: A-FP01.dwg
Estimated Time: 30 minutes

This exercise reinforces the following skills:

- ❑ Drawing Setup
- ❑ AEC Dimensions
- ❑ Title Mark
- ❑ Attributes
- ❑ Design Center
- ❑ Dimension Styles

1.

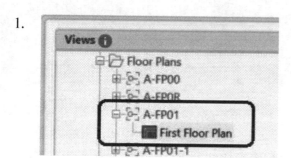

Launch the Project Navigator.

Open the Views tab.

Open the **A-FP01** view drawing.

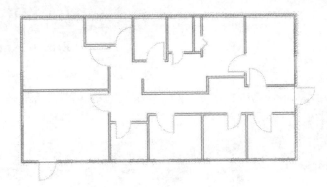

The floor plan should look like this.

Doors, windows, and walls should be visible in the viewport.

2.

Activate the Home ribbon.

In the Layers panel drop-down, select **Select Layer Standard**.

3.

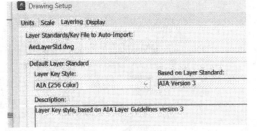

Set the Layer Key Style to **AIA (256 Color)**.

Click **OK**.

4.

On the Status Bar: Set the Global Cut Plane to **1400**.

That way you see the doors and windows in the floor plan. You can only set the global cut plane in model space.

5.

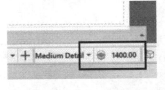

Left click on the Global Cut Plane icon.

Set the Cut Height to **1400**.

Click **OK**.

6.

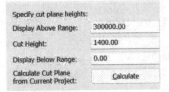

Verify that the **Add Annotative Scales** is enabled on the status bar.

7.
```
for use current <"use current">: A-Anno-Dims
```
Type **DIMLAYER**.

Type **A-Anno-Dims**.

This ensures any dimensions are placed on the A-Anno-Dims layer, regardless of which layer is current.

8. Launch the Design Center by typing **ADC**.

Or toggle ON from the View ribbon.

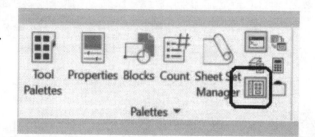

9.

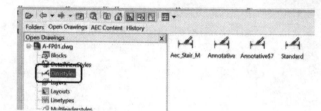

Click the **Open Drawings** tab.

Highlight **DimStyles** for the A-FP01.dwg.

You will see the available dimension styles in the drawing.

10.

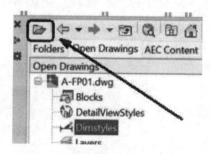

Click the **Open Drawing** tool.

11.

Open the *DimStyles.dwg* file located in the exercises folder.

12.

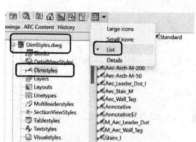

Highlight **DimStyles** in the DimStyles.dwg file.

Change the Views to **List**.

13. Highlight **AEC-Arch-M-100.**
Right click and select **Copy.**

14. Paste the dimension style into the A-FP-01.dwg file.

15. Close the Design Center.
Switch to the Annotate ribbon.

Set the Dimension Style to **AEC-Arch-M-100**.

16.  Select the **Linear Dimension** tool on the Home ribbon.

17. Click **ENTER** to select an object.

18.

Select the right side of the building.

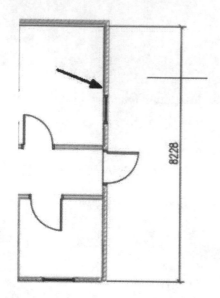
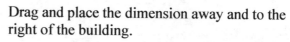

19.

Drag and place the dimension away and to the right of the building.

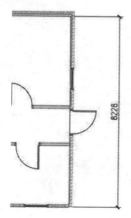

20.

Select the **Linear Dimension** tool on the Home ribbon.

21.

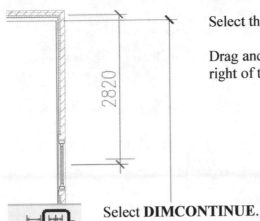

Select the upper right corner of the building.

Select the midpoint of the window.

Drag and place the dimension away and to the right of the building.

22.  Select **DIMCONTINUE**.

23.

Select the endpoint of the wall below the window.

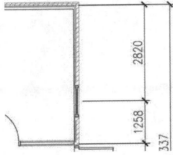

24.

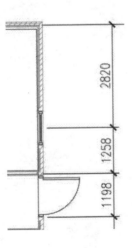

Select the wall below the door.

25.

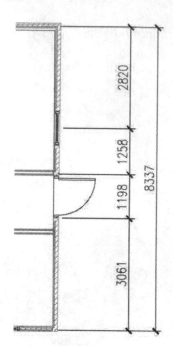

Select the endpoint of the wall below the door.

Use grips to align and position the dimensions.

It's OK if your dimensions are slightly different from mine.

26.

Hover over a dimension.

The dimension should automatically be placed on A-Anno-Dims because the DIMLAYER system variable was set to that layer.

27. Save and close the drawing.

28.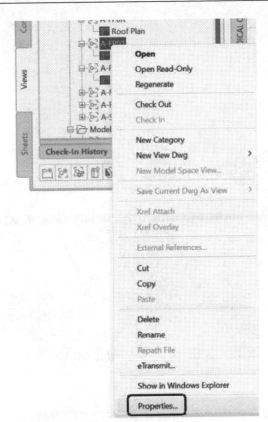

Open the Project Navigator.
Open the Views tab.
Highlight **A-FP01**.
Right click and select **Properties**.

29.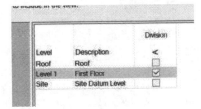

Click **Context**.
Enable **First Floor**.

30.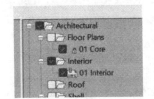

Click **Content**.

Enable **01 Core** and **01 Interior**.

Click **OK**.

31. Save and close the drawing.

Exercise 8-17:

Creating a Custom Title block

Drawing Name: Architectural Title Block.dwg
Estimated Time: 30 minutes

This exercise reinforces the following skills:

- ❏ Title blocks
- ❏ Attributes
- ❏ Edit Block In-Place
- ❏ Insert Image
- ❏ Insert Hyperlink

1.   Select the **Open** tool.

2.   Set the Files of type to *Drawing Template*.

3.   Browse to
*ProgramData\Autodesk\ACA 2024\enu\
Template\AutoCAD Templates*.

4.   Open the *Tutorial-iArch.dwt.*
Verify that *Files of type* is set to Drawing Template or you won't see the file.

5.   Perform a **File→Save as**.

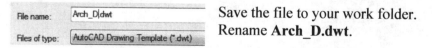

Save the file to your work folder.
Rename **Arch_D.dwt**.

6. 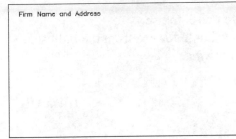 Enter a description and Click **OK**.

7. Firm Name and Address   Zoom into the Firm Name and Address rectangle.

8. A
   Multiline
   Text   Select the **MTEXT** tool from the **Annotation** panel on the Home ribbon.

   Draw a rectangle to place the text.

9. Set the Font to **Verdana**.

   Hint: If you type the first letter of the font, the drop-down list will jump to fonts starting with that letter.

   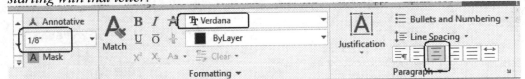

   Set the text height to ⅛″.
   Set the justification to **centered**.
   Enter the name and address of your college.

10.    Extend the ruler to change the width of your MTEXT box to fill the rectangle.

    LANEY COLLEGE
    800 FALLON STREET
    OAKLAND, CA

    Click **OK**.

11. Firm Name and Address   Use **MOVE** to locate the text properly, if needed.

    LANEY COLLEGE
    800 FALLON STREET
    OAKLAND, CA

12. Activate the **Insert** ribbon.

Select the arrow located on the right bottom of the Reference panel.

13. The External Reference Manager will launch.

Select **Attach Image** from the drop-down list.

Attach DWG...
Attach Image...
Attach DWF...

14. Locate the image file you wish to use.
There is an image file available for use from the publisher's website called *college logo.jpg*.
Click **Open**.

15. Click **OK**.

Name: college logo   Browse...

Preview

Path type
Full path

Scale
☑ Specify on-screen
1.0000

Insertion point
☑ Specify on-screen
X: 0"
Y: 0"
Z: 0"

Laney College

Rotation
☐ Specify on-screen
Angle: 0.00

16. Place the image in the rectangle.

Firm Name and Address

LANEY COLLEGE
800 FALLON STREET
OAKLAND, CA

Laney College

If you are concerned about losing the link to the image file (for example, if you plan to email this file to another person), you can use **INSERTOBJ** to embed the image into the drawing. Do not enable link to create an embedded object. The INSERTOBJ command is not available on the standard ribbon.

17. Select the image.
Right click and select **Properties**.

18. 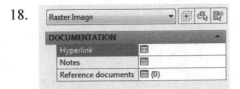 Select the **Extended** tab.
Pick the Hyperlink field.

19.  Type in the name of the school in the Text to display field.

Type in the website address in the *Type the file or Web page name* field.

Click **OK**.
Close the Properties dialog.

20. If you hover your mouse over the image, you will see a small icon. This indicates there is a hyperlink attached to the image. To launch the webpage, simply hold down the CTRL key and left click on the icon.

If you are unsure of the web address of the website you wish to link, use the Browse for **Web Page** button.

21. To turn off the image frame/boundary, type **IMAGEFRAME** at the command line. Enter **0**.

IMAGEFRAME has the following options:

        0: Turns off the image frame and does not plot

        1: Turns on the image frame and is plotted

        2: Turns on the image frame but does not plot

22. 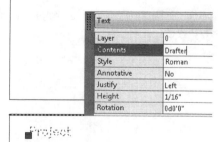   Select the titleblock.
          Right click and select **Edit Block in-place**.

23. Reference Edit   ×   Click **OK**.

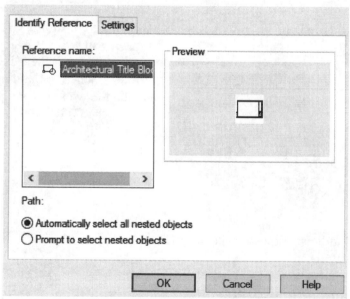

24. Change the text for Project to **Drafter**.
To change, simply double click on the text and an edit box will appear.

Text	
Layer	0
Contents	Drafter
Style	Roman
Annotative	No
Justify	Left
Height	1/16"
Rotation	0d0'0"

Project

25.   Select the **Define Attribute** tool from the Insert ribbon.

26. Select the **Field** tool.

27. Highlight Author and Uppercase.

Click **OK**.

28. In the Tag field, enter **DRAFTER**.
In the Prompt field, enter **DRAFTER**.
The Value field is used by the FIELD property.
  In the Insertion Point area:
    In the X field, enter **2' 6.5"**.
    In the Y field, enter **2-1/16"**.
    In the Z field, enter **0"**.
  In the Text Options area:
    Set the Justification to **Left**.
    Set the Text Style to **Standard**.
    Set the Height to **1/8"**.

Click **OK**.

29. Select the **Define Attributes** tool.

Define
Attributes

30. In the Tag field, enter **DATE**.
In the Prompt field, enter **DATE**.

  In the Insertion Point area:
    In the X field, enter **2' 6.5"**.
    In the Y field, enter **1-9/16"**.
    In the Z field, enter **0"**.
  In the Text Options area:
    Set the Justification to **Left**.
    Set the Text Style to **Standard**.
    Set the Height to **1/8"**.

31. Select the Field button to set the default value for the attribute.

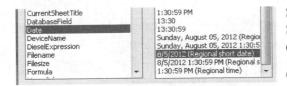

Select **Date**.
Set the Date format to **Regional short date**.

Click **OK**.

32. Select the **Define Attributes** tool.

33.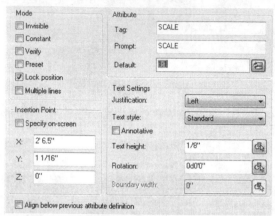

In the Tag field, enter **SCALE**.
In the Prompt field, enter **SCALE**.
  In the Insertion Point area:
    In the X field, enter **2′ 6.5″**.
    In the Y field, enter **1-1/16″**.
    In the Z field, enter **0″**.
  In the Text Options area:
    Set the Justification to **Left**.
    Set the Text Style to **Standard**.
Set the Height to **1/8″**.

34.

| Field category: | PlotScale: |
| Plot | 1:1 |

Field names:
Format:

DeviceName
Login
PageSetupName
PaperSize
PlotDate
PlotOrientation
PlotScale
PlotStyleTable

(none)
#:1
1:#
1" = #'
#" = 1'
#" = 1'-0"
Use scale name

Select the Field button to set the default value for the attribute.

Select **PlotScale**.
Set the format to **1:#**.

Click **OK**.

Click **OK** to place the attribute.

35. Select the **Define Attributes** tool.

36.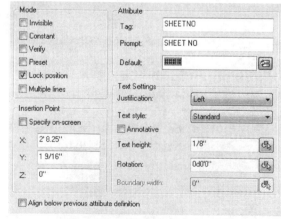

In the Tag field, enter **SHEETNO**.
In the Prompt field, enter **SHEET NO**.
  In the Insertion Point area:
    In the X field, enter **2′ 8.25″**.
    In the Y field, enter **1-9/16″**.
    In the Z field, enter **0**.
  In the Text Options area:
    Set the Justification to **Left**.
    Set the Text Style to **Standard**.
    Set the Height to **1/8″**.

37. Select the Field button to set the default value for the attribute.

Select **CurrentSheetNumber**. Set the format to **Uppercase**.

Click **OK**.

Click **OK** to place the attribute.

38. Select the **Define Attribute** tool.

39.

In the Tag field, enter **PROJECTNAME**.
In the Prompt field, enter **PROJECT NAME**.
In the Insertion Point area:
    In the X field, enter **2′ 7-9/16″**.
    In the Y field, enter **4″**.
    In the Z field, enter **0**.
In the Text Options area:
    Set the Justification to **Center**.
    Set the Text Style to **Standard**.
    Set the Height to **1/8″**.

40. Select the Field button to set the default value for the attribute.

Select **AEC Project** for the field category.
Set the Field Name to **Project Name**.
Set the format to **Uppercase**.

Click **OK**.

A common error for students is to forget to enter the insertion point. The default insertion point is set to 0,0,0. If you don't see your attributes, look for them at the lower left corner of your title block and use the MOVE tool to position them appropriately.

Your title block should look similar to the image shown.

If you like, you can use the MOVE tool to reposition any of the attributes to make them fit better.

41. Select **Save** on the ribbon to save the changes to the title block and exit the Edit Block In-Place mode.

42. Click **OK**.

AutoCAD ×

All references edits will be saved.

- To save reference changes, click OK.
- To cancel the command, click Cancel.

OK   Cancel

43. Save the file and go to **File→Close**.

Tips & Tricks

➤ Use templates to standardize how you want your drawing sheets to look. Store your templates on a server so everyone in your department uses the same titleblock and sheet settings. You can also set up dimension styles and layer standards in your templates.

➤ ADT 2005 introduced a new tool available when you are in Paper Space that allows you to quickly switch to the model space of a viewport without messing up your scale. Simply select the **Maximize Viewport** button located on your task bar to switch to model space.

➤ To switch back to paper space, select the **Minimize Viewport** button.

Exercise 8-18:
Adding Views

Drawing Name: A102 Floor Plans.dwg
Estimated Time: 20 minutes

This exercise reinforces the following skills:

- ❏ Project Browser
- ❏ Project Navigator
- ❏ Title blocks
- ❏ Layer States
- ❏ Named Views
- ❏ Project Views

1.

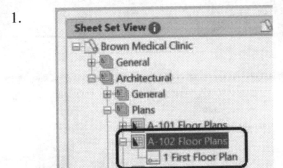

Set the **Brown Medical Clinic** project Current.

Launch the Project Navigator.
Open the Sheets tab.

Open the **A102 Floor Plans** sheet.

2.

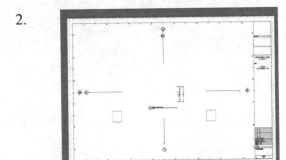

The layout tab opens to the sheet.

Select the title block.

Type **BREPLACE**.

3.

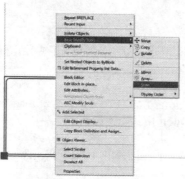

There will be a pause while ACA searches for blocks and brings up the dialog.

Select the *Architectual Title Block w Attributes.dwg*.

4.

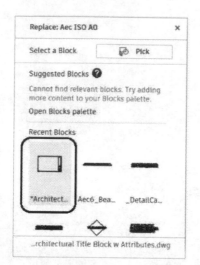

The title block is inserted.

Right click and select **Basic Modify Tools→Scale**.

5.

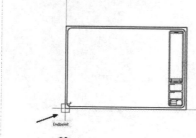

Select the lower left point of the title block as the base point.

6.

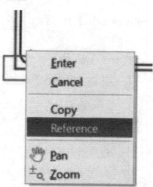

Right click and select **Reference**.

7.

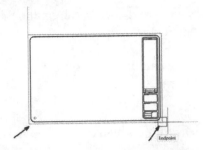

Select the lower left point of the title block.
Select the lower right point of the title block.

8.

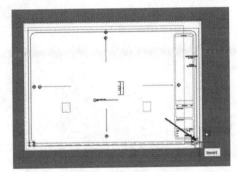

Select the lower right corner of the layout as the new size.

9.

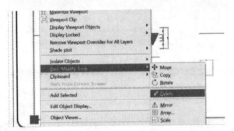

Select the viewport.
Right click and select **Basic Modify Tools→Delete.**

10.

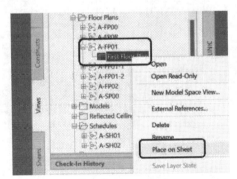

Open the Views tab of the Project Navigator.

Locate the **First Floor Plan** view under A-FP01.

Right click and select **Place on Sheet**.

11.

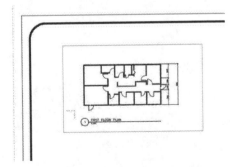

Place the view on the sheet.

You may need to adjust the size of the viewport.

12.

Switch to the View tab on the Project Navigator.

Drag and drop the **A-FP-01-1** view onto the sheet.

13.

The view is blank!

Switch to the Model tab.

14.

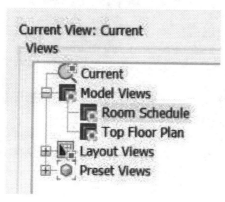

Move the spaces so they are positioned inside the model.

15.

Create a Named View for the building. Create a Named View for the Room Schedule.

16.

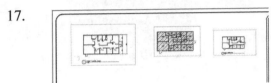

Create a layer state for the rooms.
Create a layer state for the fire rating plan.
Create a layer state for the dimensioned floor plan.

Hint: Use the Search feature in the Layers Property Manager to search for Wall, Glaz, Door, Fire, Spc.

17.

Drag **A-FP01-2** view onto the sheet.

Unlock the display on the viewport.
Adjust the view so it shows the fire rating plan.

18.

Add the Room Schedule to the sheet.

19.

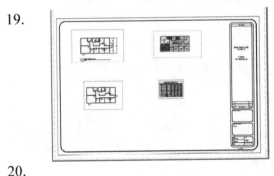

The sheet should look similar to this.

Hint: You can use GRIPS to move the MTEXT in the title block.

20.

Save and close all drawings.

Exercise 8-19:

Creating a Sheet in a Project

Drawing Name:
Estimated Time: 30 minutes

This exercise reinforces the following skills:

- ❑ Project Browser
- ❑ Project Navigator
- ❑ Title blocks

1.	Launch AutoCAD Architecture – US Imperial.
2. 	Set the **Tahoe Cabin** project Current. Open the **3D Model -Master** construct.
3.	Go to the Project Navigator. Select the Project tab. Click on **Edit Project**.
4.	Set the template to use the *Arch_D.dwt* template. 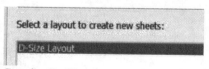 Set the **D-Size layout**.
5.	Click on the **Detailed Information** button.

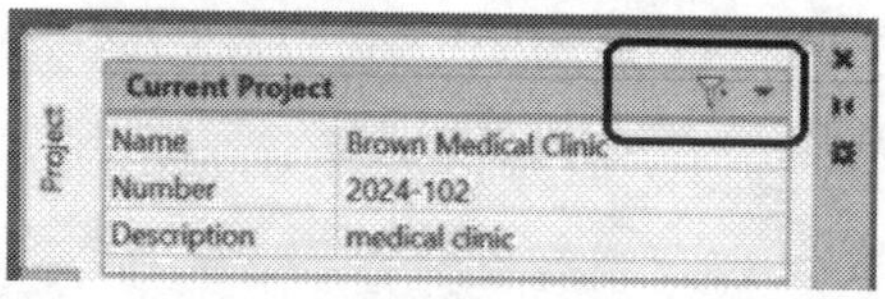

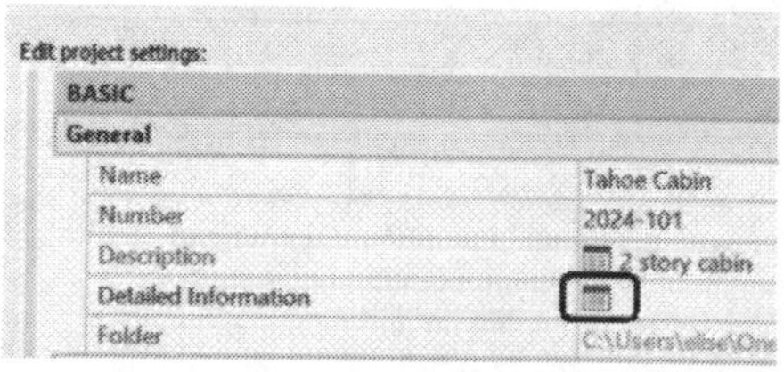

6.

Fill in as you like.

Why is the Detailed Information different from the metric project? Because we used a template for the metric project and it automatically created all those fields.

Click **OK**.

7.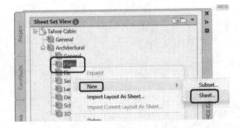

Open the Project Navigator.
Open the Sheets tab.
Highlight **Plans**.
Right click and select **New→Sheet**.

8.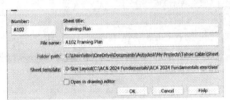

Type **A102** for the Number.
Type **Framing Plan** for the Sheet Title.

Click **OK**.

9.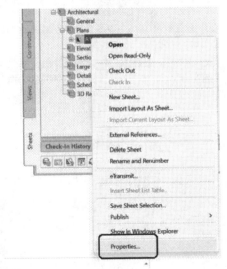

Highlight the **A102 Framing Plan.**
Right click and select **Properties**.

10.

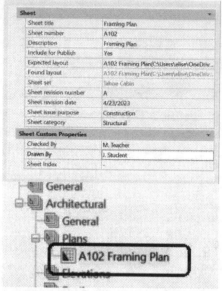

Fill in the fields.

These fields can be utilized by the title block attributes.

Click **OK**.

11.

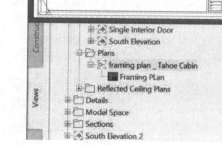

The sheet is created.

Highlight the **A102 Framing Plan**.
Right click and select **Open**.

12.

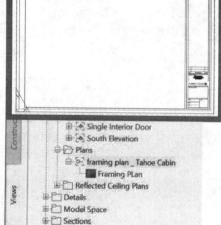

The sheet opens to the layout tab.
The viewport is empty.

Select the viewport and delete.

13.

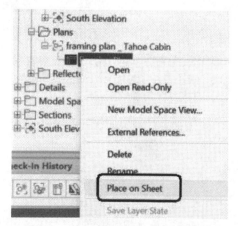

Switch to the Views tab on the Project Navigator.
Locate the **Framing Plan** view.

14.

Right click and select Place on Sheet.

15.

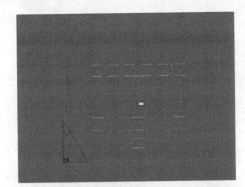

If you zoom out, you will see that the viewport is a lot bigger than the sheet.

Scale the viewport so it fits onto the sheet.

16.

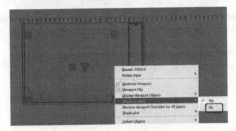

Select the viewport.
Right click and select **Display Locked→No**.

17.

Display locked	No
Annotation scale	1/8" = 1'-0"
Standard scale	1/8" = 1'-0"
Custom scale	0"

Set the viewport Standard scale to **1/8" = 1'-0"**.
Set the Annotation scale to **1/8" = 1'-0"**.

18.

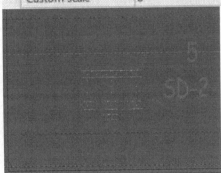

The annotation tags will appear huge.

Select the tags and rescale so that they fit better with the view.

To see the sheet better: Go to Options, Display, Colors and set the Sheet Background to White.

The view should look similar to this.

19.

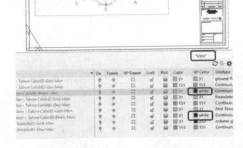

You can change the colors of layers in the viewport to make the labels easier to read.

20.

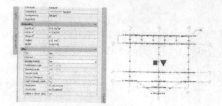

Set the viewport scale to **1/8" = 1'-0"**.
Set the annotation scale to **1/8" = 1'-0"**.

Lock the display.

21.

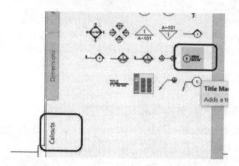

Launch the Document tool palette.
Open the Callouts tab.
Select the **Title Mark** tool.

22.

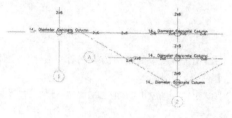

Modify the attributes for the view title.

23.

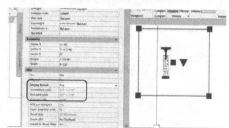

Locate the column detail view in the Project Navigator.

Drag and drop it onto the sheet.

24.

Unlock the viewport display.
Zoom into the named view for the column detail.
Set the viewport scale to **1/2" = 1'-0"**.
Set the annotation scale to **1/2" = 1'-0"**.
Lock the display.

25. 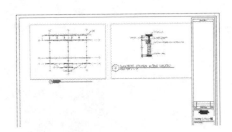 Your sheet should look similar to this.

26. Save and close your drawings.

Exercise 8-20:

Creating a Framing Schedule

Drawing Name: A102 Framing Plan.dwg
Estimated Time: 30 minutes

This exercise reinforces the following skills:

- ❑ Project Browser
- ❑ Project Navigator
- ❑ Style Manager
- ❑ Property Set Definitions
- ❑ Schedules

1. Launch **AutoCAD Architecture – US Imperial**.

2.  Set the **Tahoe Cabin** project Current.

   Open the **3D Model -Master** construct.

3. 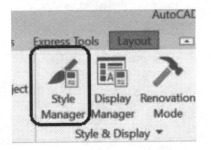 Go to the Project Navigator.
   Select the Sheets tab.
   Click on **A102 Framing Plan**.
   Right click and select **Open**.

4. 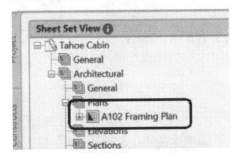 Go to the Manage ribbon.

   Click **Style Manager**.

5.

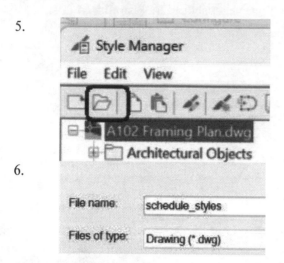

Click **Open**.

6.

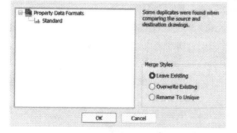

Open *schedule_styles.dwg*.

7.

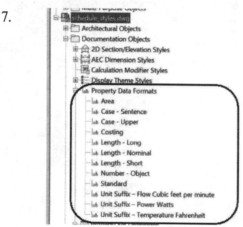

Expand the Documentation Objects category.

Highlight **Property Data Formats**.

Copy and paste the Property Data Formats from *schedule_styles.dwg* to *A102 Framing Plan.dwg*.

Hint: Go to the source drawing. Highlight Property Data Formats. Right click and select Copy. Go to the target drawing. Highlight Property Data Formats. Right click and select Paste.

8.

Enable **Leave Existing**.

Click **OK**.

9.

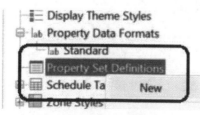

Expand the Documentation Objects category.

Highlight **Property Set Definitions**. Right click and select **New**.

10.

General   Applies To   Definition

Name:

Structural Members

Description:

Select the General tab.

Type **Structural Members** in the Name field.

11.

5-View Block Reference   ☐ Slab
5-View Part   ☐ Slice
ning   ☐ Solid (2D)
l   ☐ Solid (3D)
Underlay   ☐ Space
                          ☐ Spline
Custom Fitting   ☐ Stair
Fitting   ☑ Structural Member
Flex   ☐ Surface
bing Fitting   ☐ Surface (Extrusion)
bing Line   ☐ Surface (Planar)
t   ☐ Surface (Revolve)

Select the Applies To tab.

Enable **Structural Member.**

12.

Click **Automatic Definition**.

13.

Hyperlink   ☐
Justification   ☐
Layer   ☐
Length   ☑
Length along Baseline ☐
Linetype   ☐
Location   ☐
Location X   ☐
Location Y   ☐
Location Z   ☐
Member Type   ☑
Notes   ☐
Notes from Style   ☐
Object ID   ☐
Object Type   ☐
Roll   ☐
Rotation   ☐
Shape Designation - ☐
Shape Designation - ☐
Start Offset   ☐
Style   ☑
Trim Automatically ☐
Volume   ☐

Enable:
- Length
- Member Type
- Style

Click **OK**.

14.

General   Applies To   Definition

Name	Description	Type	Source	Default	Units	Format
Length	Length	Automatic	Structural Member:Length			Length - Short
MemberType	Member Type	Automatic	Structural Member:Member ...			Case - Sentence
Style	Style	Automatic	Structural Member:Style			Case - Sentence

Set the Format for Length to **Length-Short**.

15. Set the Format for Member Type and Style to **Case – Sentence**. Click **Apply**.

16.

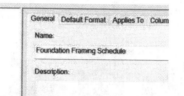

Expand the Documentation Objects category.

Highlight **Schedule Table Styles**. Right click and select **New**.

17.

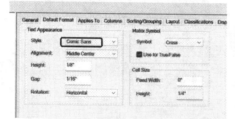

Select the General tab.

Type **Foundation Framing Schedule** in the Name field.

18.

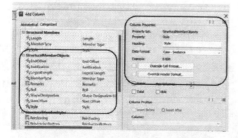

Select the Default Format tab.

Set the Text Style to **Comic Sans**.

If you don't like that font, assign your preferred text style.

19.

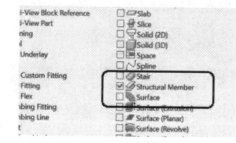

Select the Applies To tab.

Enable **Structural Member**.

20.

Select the Columns tab.

Click **Add Column**.

21.

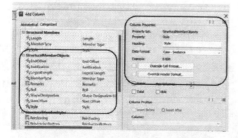

Look under StructuralMemberObjects.

Highlight **Style**.
Change the Heading to **Style**.
Set the Data Format to **Case – Sentence**.
Click **OK**.

22. 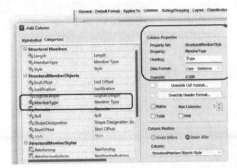 Click **Add Column**.

(Note: above image id refers to the Add Column button for step 22.)

23.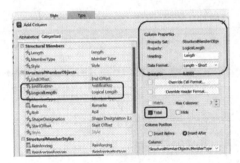

Look under StructuralMemberObjects.

Highlight **Member Type**.
Change the Heading to **Type**.
Set the Data Format to **Case – Sentence**.

Enable **Insert After**.
Click **OK**.

24. Click **Add Column**.

25.

Look under StructuralMemberObjects.

Highlight **LogicalLength**.
Enable **Total**.
Change the Heading to **Length**.
Set the Data Format to **Length - Short**.

Enable **Insert After**.
Click **OK**.

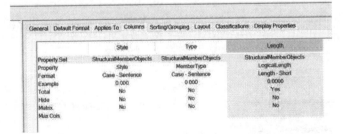

The columns tab should look like this.

	Style	Type	Length
Property Set	StructuralMemberObjects	StructuralMemberObjects	StructuralMemberObjects
Property	Style	MemberType	LogicalLength
Format	Case - Sentence	Case - Sentence	Length - Short
Example	0.000	0.000	0.0000
Total	No	No	Yes
Hide	No	No	No
Matrix	No	No	No
Max Cols			

26.

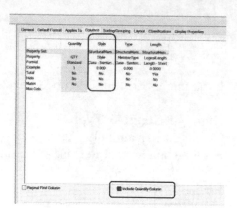

Highlight the Style column.

Enable **Include Quantity Column**.

A column for quantity will be added.

27.

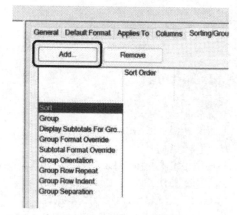

Select the Sorting/Grouping tab.

Highlight **Sort**.

Click **Add**.

28.

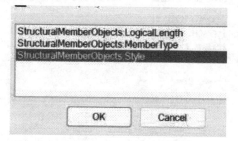

Highlight **Style**.

Click **OK**.

29.

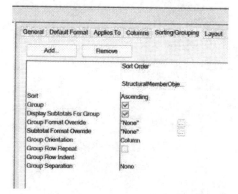

Enable **Group**.
Enable **Display Subtotals for Group**.

Click **Apply**.

30.

Select the Layout tab.

Change the Table Title to:

Foundation Framing Schedule.

Click **OK**.

Save the **A102 Framing Plan** drawing.

31.

Launch the Annotation tool palette.

Open the **Scheduling** tab.

32.

Open the **Style Manager**.

33.

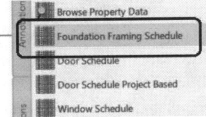

Drag and drop the **Foundation Framing Schedule** onto the tool palette.

34.

Foundation Framing Schedule		
Structural Member	Member Type	Length

Open the Model layout.

Click the **Foundation Framing Schedule** tool.

Click ENTER.

Click to place the schedule.

The schedule will show the headings only, no data.

35.

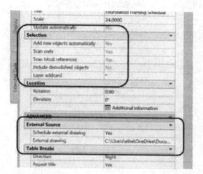

Select the schedule.
Right click and select **Properties**.

Set Schedule external drawing to **Yes**.
Browse to the location of the current drawing and select it.

Note that the Selection will change to scan xrefs and block references.

36.

Click **Update** on the ribbon.

Foundation Framing Schedule		
Structural Member	Member Type	Length
	?	?
	?	?
	?	?

The schedule displays a series of ?s, but no data.

This is because the external references also need to have their data property sets updated.

37. Type **XREF** to bring up the external reference manager.

The external reference files are listed.

38.

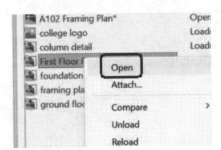

Highlight **First Floor Foundation – Tahoe Cabin**.

Right click and select **Open**.

39.

Go to the Manage ribbon.

Launch the **Style Manager**.

40.

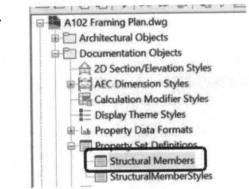

Locate **Structural Members** under the *A102 Framing Plan.dwg*.

Copy and paste to the *First Floor Foundation – Tahoe Cabin.dwg*.

Save the changes and close the Style Manager.

41. Save and close the *First Floor Foundation – Tahoe Cabin.dwg*.

42.

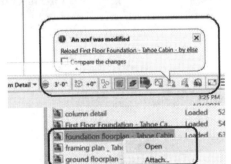

Click to reload the external reference.

43.

Return to the Xref Manager palette.

Highlight **Foundation floorplan– Tahoe Cabin**.

Right click and select **Open**.

44.

Go to the Manage ribbon.

Launch the **Style Manager**.

45.

Locate **Structural Members** under the *A102 Framing Plan.dwg*.

Copy and paste to the *Foundation floorplan– Tahoe Cabin.dwg*.

Save the changes and close the Style Manager.

46. Save and close the *Foundation floorplan – Tahoe Cabin.dwg*.

47. Click to reload the external reference.

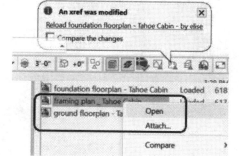

48. Return to the Xref Manager palette.

Highlight **framing plan– Tahoe Cabin**.

Right click and select **Open**.

49.

Go to the Manage ribbon.

Launch the **Style Manager**.

50.

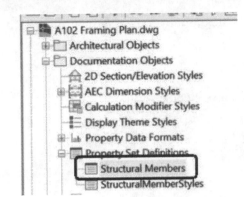

Locate **Structural Members** under the *A102 Framing Plan.dwg.*

Copy and paste to the *framing plan–Tahoe Cabin.dwg.*

Save the changes and close the Style Manager.

51.   Save and close the *framing plan – Tahoe Cabin.dwg.*

52.   Click to reload the external reference.

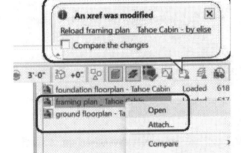

53.   Return to the Xref Manager palette.

Highlight **ground floorplan– Tahoe Cabin**.

Right click and select **Open**.

54.   Go to the Manage ribbon.

Launch the **Style Manager**.

55.

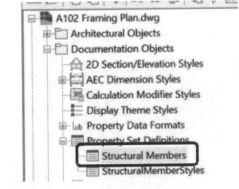

Locate **Structural Members** under the *A102 Framing Plan.dwg.*

Copy and paste to the *framing plan–Tahoe Cabin.dwg.*

Save the changes and close the Style Manager.

56.   Save and close the *framing plan – Tahoe Cabin.dwg.*

57.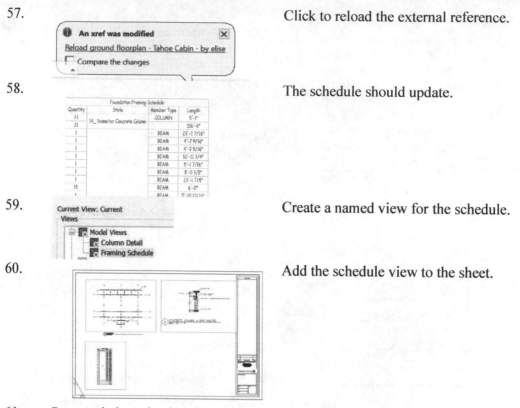

Click to reload the external reference.

58.

The schedule should update.

59.

Create a named view for the schedule.

60.

Add the schedule view to the sheet.

61. Save and close the drawing.

Sheet Set Manager

The Sheet Set Manager provides a list of layout tabs available in a drawing or set of drawings and allows you to organize your layouts to be printed.

The Sheet Set Manager allows you to print to a printer as well as to PDF. Autodesk does not license their PDF creation module from Adobe. Instead, they have a partnership agreement with a third-party called Bluebeam. This means that some fonts may not appear correct in your PDF print.

Exercise 8-21:

Adding a Sheet

Drawing Name: new (Tahoe Cabin Project)
Estimated Time: 15 minutes

This exercise reinforces the following skills:

- ❑ Creating Views
- ❑ Add Sheet
- ❑ Add Elevation
- ❑ Xref Manager

1. Launch **AutoCAD Architecture – US Imperial**.

2.  Set the **Tahoe Cabin** project Current.

3. 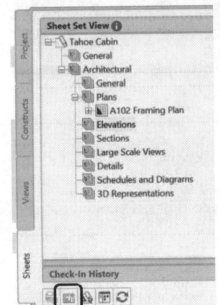 Launch the Project Navigator.
Open the Sheets tab.

   Click **Add Sheet** at the bottom of the palette.

4.  Type **A201** in the Number field.
Type **Exterior Elevations** in the Sheet Title.
Enable **Open in drawing editor**.
Click **OK**.

5.

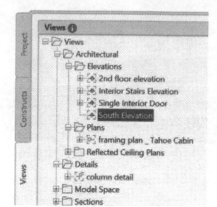

Open the Views tab on the Project Navigator.

Locate the **South Elevation** view.

Drag and drop it onto the sheet.

6.

Select the viewport.
Right click and change the Display Locked to **No**.

7.

Switch to Model space and zoom extents.
You will see the south elevation.

Create a named view for the south elevation.

8.

Return to the layout sheet.

Double click inside the viewport and set the viewport to the named view for the south elevation.

9.

Select the viewport.
Right click and select **Properties**.
Set the Annotation and Standard scales to **1/8" = 1'-0"**.
Lock the display.

10.

Add a view title.
Change the view title as shown.
The title should read **South Exterior Elevation**.
The scale should read Scale: **1/8" = 1'-0"**.

11.

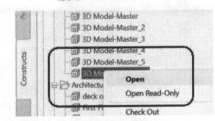

Can you create North/West/Exterior elevation views and add them to the sheet?

Hint: Open the Master 3D construct.
Create the elevation views and then add as new view drawings.

12.

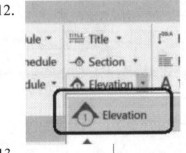

Select the **Elevation** tool from the Annotate ribbon.

13.

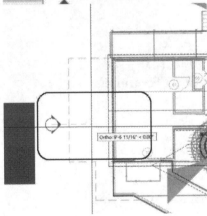

Place the elevation at the desired location.

Point it towards the building model.

14.

Give the elevation a name.

Click Current Drawing.

15.

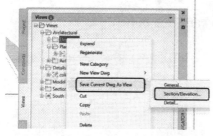

Window around the entire building.
Click to place the view in the drawing.

16.

Update the attributes for the view title.

17.

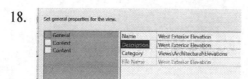

Save the **Current Dwg As
View→Section/Elevation**.

18.

Provide a name and Description.

Click Next.

19.

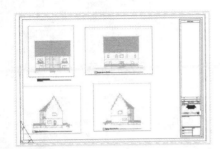

Enable all the levels.

Click **Next**.

Click **Finish**.

Save and close the view drawing.

20.

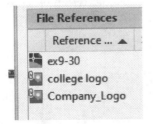

Return to the **A201 Exterior Elevations** sheet.
Add the exterior view to the sheet.
Repeat for the remaining exterior views.

21.

Save and close all drawings.

Exercise 8-22:

Creating a PDF Document

Drawing Name: A201 Exterior Elevations.dwg
Estimated Time: 5 minutes

This exercise reinforces the following skills:

- ❑ Printing to PDF
- ❑ Sheet Set Manager
- ❑ Xref Manager

1.   Open *A201 Exterior Elevations.dwg*.

2.

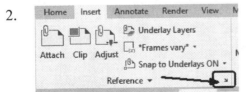

Switch to the **Insert** ribbon.
Launch the XREF Manager.

The file has a couple of image files used in the titleblock.

We need to re-path the image files since you downloaded the files and need to point the reference to the correct file locations.

3.

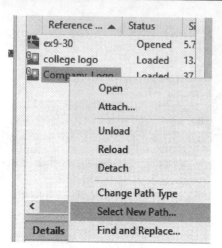

Highlight the image file.
Right click and click **Select New Path**.

4.

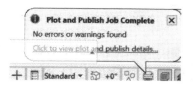

Locate the file in your downloaded folder.

Click **Open**.

Close the XREF Manager.

5.

Launch the Project Navigator.

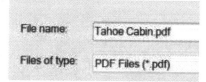

Open the Sheets tab.

Highlight Tahoe Cabin.

Right click and select
Publish→Publish to PDF.

6.

Browse to where you want to save the pdf file.

Click **Select**.

7.

All sheets that exist in the project will be converted
to PDF. You will see a notification when the plot is
complete.
Click the link.

The sheets that were plotted are listed.

8. Locate the pdf and open it to review.

Extra:

1) Update the view titles with the correct sheet numbers based on the sheets where the views are placed.

2) Create additional elevation and section views for the kitchen, the second bathroom, the exterior elevations, and windows.

3) Add the new views to existing sheets or new sheets.

4) Add window tags to all the windows.

5) Create a window schedule.

6) Create a window legend.

7) Create a detail view for window framing.

8) Create a callout to the window framing detail view.

9) Create a section elevation view of an interior wall.

10) Add keynotes to the interior wall section elevation view.

11) Update the attributes on each title block with the correct sheet number and title information.

Notes:

QUIZ 8

True or False

1. The length of a structural member is its logical length, not the actual length.

2. Sheets are used to plot view drawings of your building project.

3. Once you create a 2D Elevation, it cannot be updated.

4. You can import space styles from the Style Browser.

5. You can use a 2D Section/Elevation style to control the display of an elevation.

Multiple Choice

6. A sheet is a _____ file:

   A. DWG
   B. PDF
   C. DWF
   D. DWT

7. Sheets are created after:

   A. Construct drawings
   B. View drawings
   C. Shedules
   D. Detail Views

8. Space Styles: (select all that apply)

   A. Control the hatching displayed
   B. Control the name displayed when spaces are tagged
   C. Cannot be modified
   D. Have materials assigned

9. External references:

    A. should be managed by the Project Navigator
    B. should be managed by the Xref Manager
    C. should never be used
    D. may require extra care

10. An element in the building project:

    A. is a multi-view block
    B. is a generic building block for multiple use in different levels or divisions
    C. is a wall, window, or other object used in a building
    D. is a fixture or piece of furniture used in a building

ANSWERS:

1) T; 2) T; 3) F; 4) T; 5) T; 6) A; 7) B; 8) A, B, C, and D; 9) A; 10) B

Lesson 9:
Rendering & Materials

AutoCAD Architecture divides materials into TWO locations:

- Materials in Active Drawing
- Autodesk Library

You can also create one or more custom libraries for materials you want to use in more than one project. You can only edit materials in the drawing or in the custom library—Autodesk Materials are read-only.

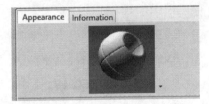

Materials have two asset definitions—Appearance and Information. Materials do not have to be fully defined to be used. Each asset contributes to the definition of a single material.

You can define Appearance independently without assigning/associating it to a material. You can assign the same assets to different materials. Assets can be deleted from a material in a drawing or in a user library, but you cannot remove or delete assets from materials in locked material libraries, such as the Autodesk Materials library or any locked user library. You can delete assets from a user library – BUT be careful because that asset might be used somewhere!

You can only duplicate, replace or delete an asset if it is in the drawing or in a user library. The Autodesk Asset Library is "read-only" and can only be used to check-out or create materials.

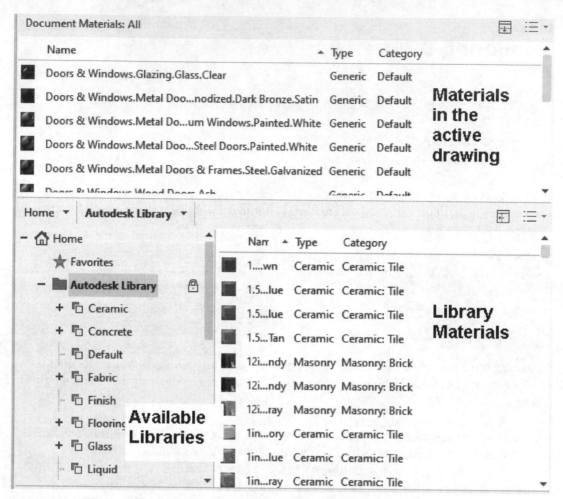

The Material Browser is divided into three distinct panels.

The Document Materials panel lists all the materials available in the drawing.

The Material Libraries list any libraries available.

The Library Materials lists the materials in the material libraries.

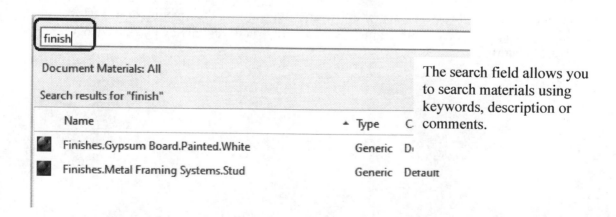

The search field allows you to search materials using keywords, description or comments.

The View control button allows you to control how materials are sorted and displayed in the Material Libraries panel.

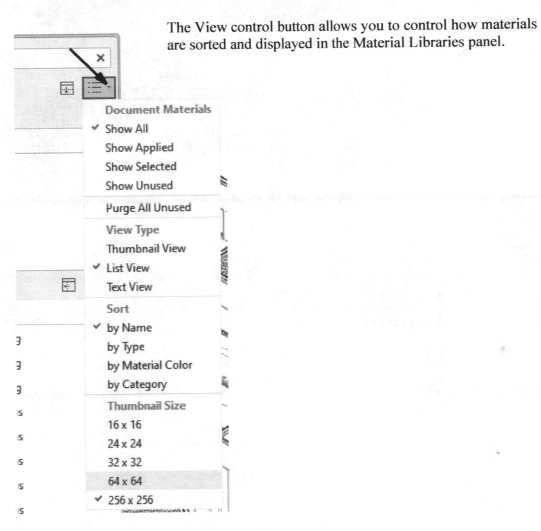

Material Editor

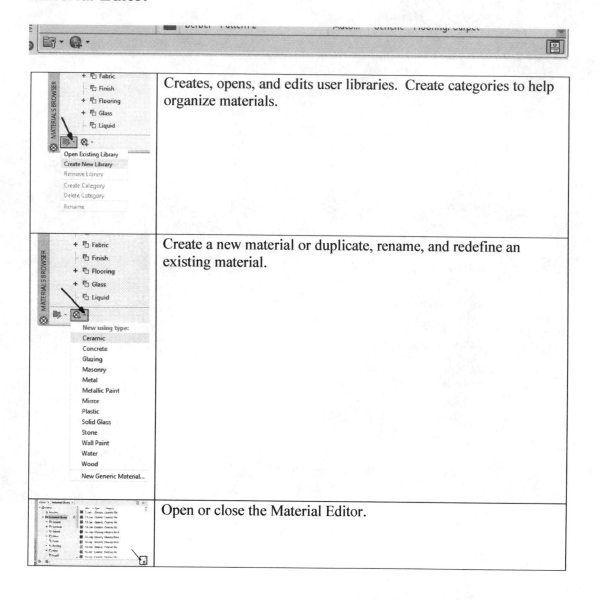

	Creates, opens, and edits user libraries. Create categories to help organize materials.
	Create a new material or duplicate, rename, and redefine an existing material.
	Open or close the Material Editor.

Exercise 9-1:
Modifying the Material Browser Interface

Drawing Name: material_browser.dwg
Estimated Time: 10 minutes

This exercise reinforces the following skills:

❑   Navigating the Material Browser and Material Editor

1.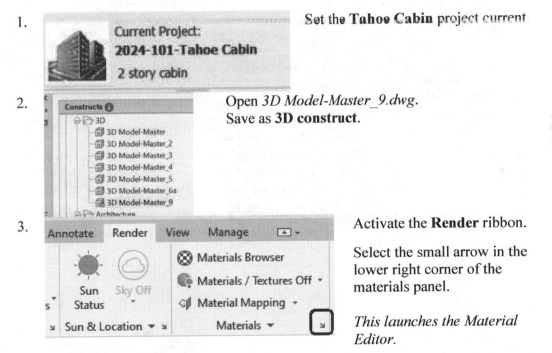
Set the **Tahoe Cabin** project current

2. Open *3D Model-Master_9.dwg*.
Save as **3D construct**.

3. Activate the **Render** ribbon.

Select the small arrow in the lower right corner of the materials panel.

This launches the Material Editor.

4.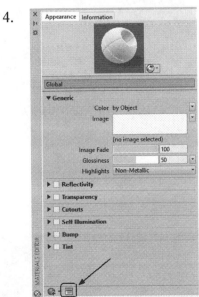
Launch the Material Browser by selecting the button on the bottom of the Material Editor dialog.

5.

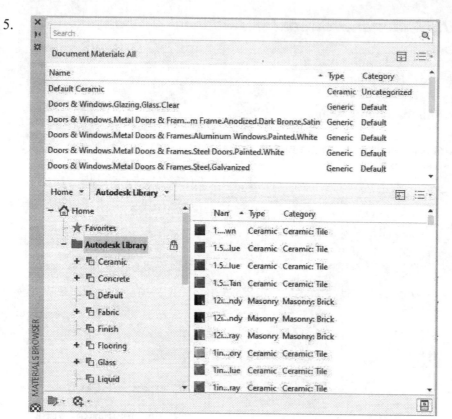

Make the
Materials
Browser
appear as
shown.

6.

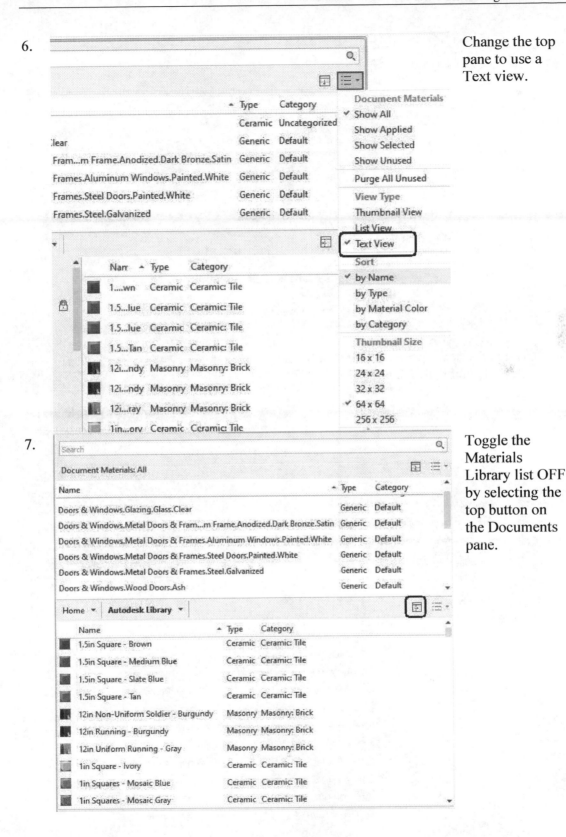

Change the top pane to use a Text view.

7.

Toggle the Materials Library list OFF by selecting the top button on the Documents pane.

8.

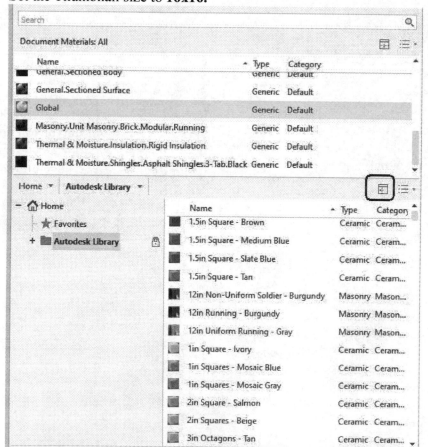

Change the upper pane to a **List View**.
Set the Thumbnail size to **16x16**.

9.

Toggle the
Library tree
back on.

10. 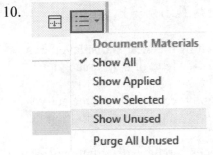   In the Document Materials panel, set the view to **Show Unused**.

Scroll through the list to see what materials are not being used in your drawing.

11. 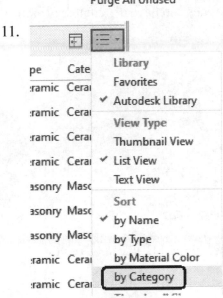   In the Material Library pane, change the view to sort **by Category**.

12. Close the Material dialogs.

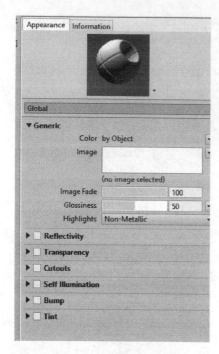

The Appearance Tab

Appearance controls how material is displayed in a rendered view, Realistic view, or Ray Trace view.

A zero next to the hand means that only the active/selected material uses the appearance definition.

Different material types can have different appearance parameters.

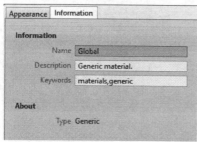

The Information Tab

The value in the Description field can be used in Material tags.

Using comments and keywords can be helpful for searches.

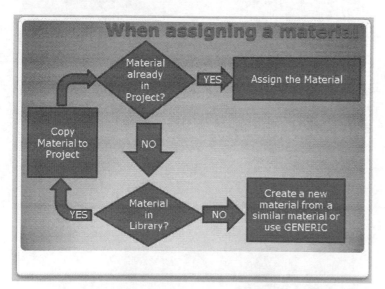

Use this flow chart as a guide to help you decide whether or not you need to create a new material in your project.

Exercise 9-2:

Copy a Material from a Library to a Drawing

Drawing Name: 3D-Model Master_9.dwg
Estimated Time: 10 minutes

In order to use or assign a material, it must exist in the drawing. You can search for a material and, if one exists in the Autodesk Library, copy it to your active drawing.

This exercise reinforces the following skills:

- ❏ Copying materials from the Autodesk Materials Library
- ❏ Modifying materials in the drawing

1. 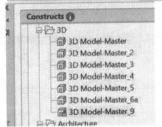   Set the **Tahoe Cabin** project current.

2.    Open *3D Model-Master_9.dwg.*

3. Go to the **Render** ribbon.

4. ![Render ribbon with Materials Browser highlighted]   Select the **Materials Browser** tool.

5. ![Search field with yellow typed]   Type **yellow** in the search field.

    There are no yellow materials in the drawing.

6.

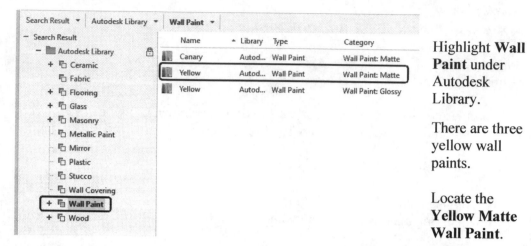

Highlight **Wall Paint** under Autodesk Library.

There are three yellow wall paints.

Locate the **Yellow Matte Wall Paint**.

7.

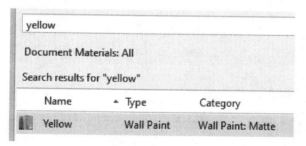

Select the up arrow to copy the material up to the drawing.

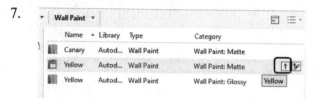

The yellow paint material is now listed in the Document Materials.

8. Type **blue** in the search field.

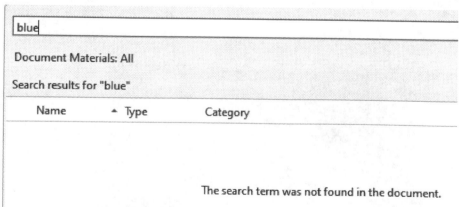

9. There are no blue materials in the drawing, but the Autodesk Library has several different blue materials available.

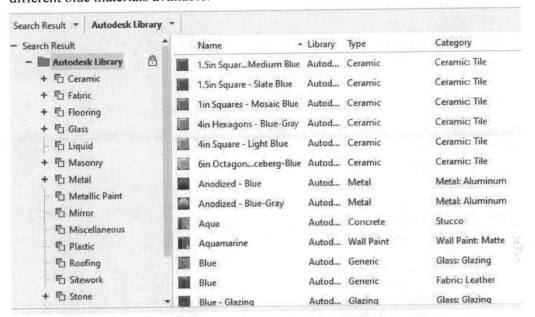

10. Highlight the Wall Paint category in the Library pane.

    This confines the search to blue wall paints.

11. Locate the **Periwinkle Matte** Wall Paint and then copy it over to the drawing.

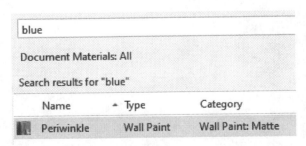

    Notice the lock symbol next to the material name. Materials in the library cannot be modified. However, once you import a material into your document, it can be edited.

    The blue wall paint is now available in the drawing.

    You have added material definitions to the active drawing ONLY.

12. Close the Materials dialog.

13. Save the drawing.

Adding Color to Walls

Drawing Name: *3D Model-Master_9.dwg.*
Estimated Time: 30 minutes

This exercise reinforces the following skills:

- ❑ Edit External Reference
- ❑ Materials
- ❑ Colors
- ❑ Display Manager
- ❑ Style Manager

1.   Open *3D Model-Master_9.dwg..*

2.   Switch to a **SE Isometric** view.

3.   Switch to the Home ribbon.

   Set the layer state to **main floor – Tahoe Cabin_floorplan.**

4.

   Select a wall.

   Click **Edit Reference In-Place** on the ribbon.

5.

Click **OK**.

6.

Go to the Manage ribbon.

Launch the **Style Manager**.

7.

Locate the Stud-3.5 GWB… wall style.

Create a copy of the wall style.

8.

Rename the copy **3.5 GWB…Blue-Yellow**.

9.

Open the Components tab.

Rename the exterior components **GWB-Yellow** and **GWB-Blue**.

Just click in the field to modify the names.

10. Activate the Materials tab.

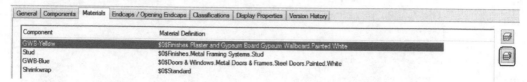

Highlight the **GWB-Yellow** component.
Select the **New** button.

11.

New Name: Finishes.Wall Paint.Yellow

OK   Cancel

Type **Finishes.Wall Paint.Yellow.**

Click **OK.**

12.

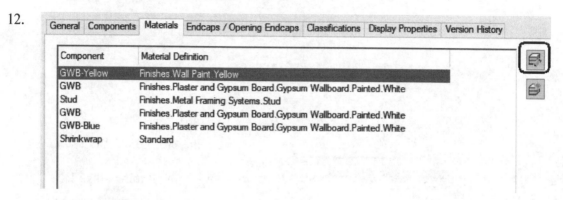

Highlight the **GWB-Yellow** component.
Select the **Edit** button.

13.

Activate the **Display Properties** tab.
Select the **Edit Display Properties** button.

14.

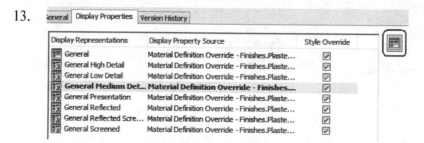

Activate the **Layer/Color/ Linetype** tab.

Highlight **3D Body**.

Left pick on the color square.

15.

ByLayer   ByBlock

Color:
yellow

Select the color **yellow**.

Click **OK.**

16. 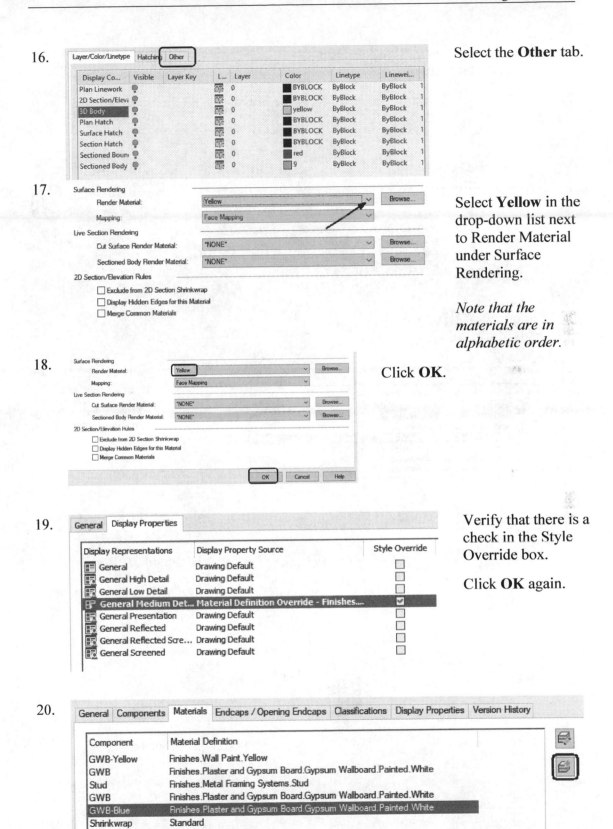 Select the **Other** tab.

17. Select **Yellow** in the drop-down list next to Render Material under Surface Rendering.

    Note that the materials are in alphabetic order.

18. Click **OK**.

19. Verify that there is a check in the Style Override box.

    Click **OK** again.

20.

Highlight the **GWB-Blue** component.
Select the **New** button.

21.

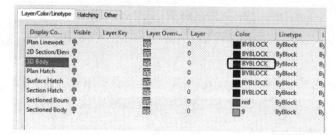

Type **Finishes.Wall Paint.Periwinkle**.

Click **OK**.

22.

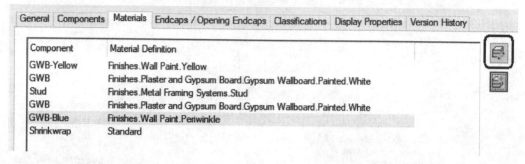

Highlight the **GWB-Blue** component.
Select the **Edit** button.

23.

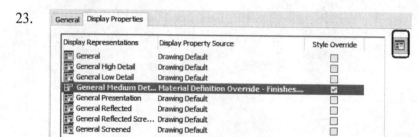

Activate the **Display Properties** tab.
Select the **Edit Display Properties** button.

24.

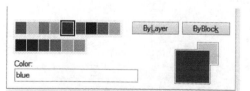

Activate the **Layer/Color/ Linetype** tab.

Highlight **3D Body**.

Left pick on the color square.

25.

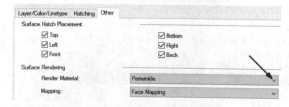

Select the color **blue**.

Click **OK**.

26.

Select the **Other** tab.

Use the drop-down list to select the color Periwinkle.

Click **OK**.

27.

Verify that there is a check next to the Style Override.

Click **OK**.

Display Representations	Display Property Source	Style Override
General	Drawing Default	☐
General High Detail	Drawing Default	☐
General Low Detail	Drawing Default	☐
General Medium Det...	Material Definition Overrid...	☑
General Presentation	Drawing Default	☐
General Reflected	Drawing Default	☐
General Reflected Scre...	Drawing Default	☐
General Screened	Drawing Default	☐

28.

Your components should show two different material definitions for the GWB components.

Click **OK**.

Component	Material Definition
GWB-Yellow	Finishes.Wall Paint.Yellow
GWB	Finishes.Plaster and Gypsum Board.Gypsum Wallboard.Painted.White
Stud	Finishes.Metal Framing Systems.Stud
GWB	Finishes.Plaster and Gypsum Board.Gypsum Wallboard.Painted.White
GWB-Blue	Finishes.Wall Paint.Periwinkle
Shrinkwrap	Standard

29.

Select a wall.
Right click and select **Properties**.
Change the wall style to use the **Blue-Yellow** wall style.

Click **ESC** to release the selection.

30. −][Custom View][Realistic]

On the View ribbon,
change the display to **Realistic**.

31. Inspect the wall.

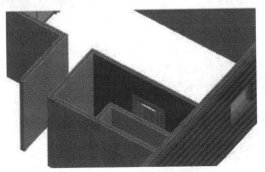

Notice that all walls using the new wall style have blue on one side and yellow on the other.

In order to apply different colors to different walls, you would have to create a wall style for each color assignment.

32.  To flip the color so walls are yellow/blue, simply use the direction arrows.

33.

Click **Save Changes** on the ribbon to save the changes to the external reference.

Click **OK** to complete the changes.

34. Save the drawing.

Elements

An element is a generic building block for multiple use. For example, you can create an element for a repeating design object like a desk/chair combination to place in a number of cubicles. You can also create an element for a typical bathroom layout and reference it multiple times into one or more constructs. Because you can annotate individual instances of an external reference, you can use the same element and annotate it differently in different locations. Unlike constructs, elements are not assigned to a level/division in the building.

Elements are dwg files. When you convert or add a dwg file as a project element, an additional XML file with the same name is created. The accompanying XML file contains information to connect the drawing file to the project.

Note: The XML file is created and updated automatically. You do not need to edit it, but be careful not to accidentally delete it in File Explorer.

Exercise 9-4:
Adding Elements

Drawing Name: woman1.dwg
Estimated Time: 15 minutes

This exercise reinforces the following skills:

□ Project Navigator
□ Elements

1.  Set the **Tahoe Cabin** project current.

2. 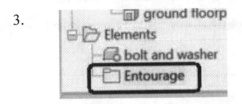 In the Project Navigator, open the Constructs tab.
Highlight **Elements**.
Right click and select **New→Category**.

3. 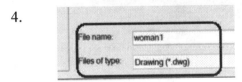 Name the category: **Entourage**.

4. 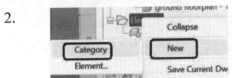 Open *woman1.dwg.*

I have included several 3D drawings of men, women, a dog, etc. in the exercise files that can be used as elements in your projects.

5.  Locate 0,0,0 in the drawing.

I locate this by drawing a line from 0,0,0.

6.

Move the figure so it is standing at the 0,0,0 location.

This will make it easier to insert and position the block.

Save the file.

7.

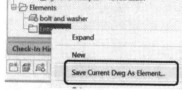

Highlight the **Entourage** category in the Project Navigator.

Right click and select **Save Current Dwg As Element**.

8.

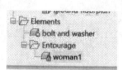

Type a Description for the element:
3D woman figure standing.

Click **OK**.

The element is now listed in the Project Navigator.
Close the file.

9.

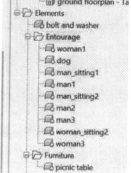

Add additional files to the Project Navigator as elements.

Camera

You can place a camera in a drawing to define a 3D view.

You can turn a camera on or off in a drawing and use grips to edit a camera's location, target, or lens length. A camera is defined by a location *XYZ* coordinate, a target *XYZ* coordinate, and a field of view/lens length, which determines the magnification, or zoom factor.

By default, saved cameras are named sequentially: Camera1, Camera2, and so on. You can rename a camera to better describe its view. The View Manager lists existing cameras in a drawing as well as other named views.

Exercise 9-5:

Camera View

Drawing Name: 3D Model-Master_9a.dwg
Estimated Time: 20 minutes

This exercise reinforces the following skills:

- ❑ Camera
- ❑ Display Settings
- ❑ Navigation Wheel
- ❑ Saved Views
- ❑ Adjust Camera

1.   Open 3D Model-Master_9a.dwg.

2.   Use the View Cube to orient the floor plan to a Top View with North in correct position. *You can also use the right click menu on the upper left of the display window.*

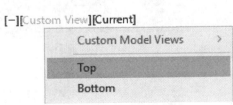

3.

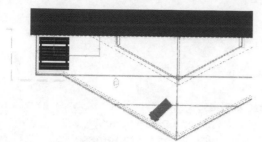

Zoom into the outside deck area.

There should be a BBQ, picnic table, and person.

4.

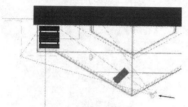

Activate the Render ribbon.

On the Camera panel,

select **Create Camera**.

5.

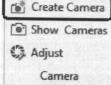

Place the Camera so it is on the rear deck and looking toward the southwest corner of the building.

Click ENTER.

6.

Click Height and set the Height of the camera to **8'-0"**.

7.

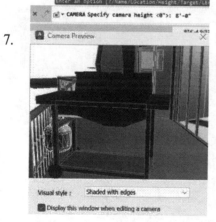

Select the camera so it highlights.

A Camera Preview window will open.

If you don't see the Camera Preview window, left click on the camera and the window should open.

Set the Visual Style to **Shaded with Edges**.

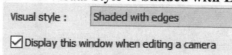

8.

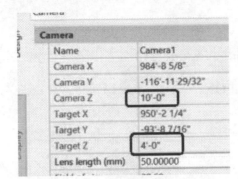

Right click on the camera and select **Properties**.

Change the Camera Z value to **10'-0"**.

Change the Target Z to **4'-0"**.

This points the camera slightly down.

This sets the camera at 4' 0" above floor level.

9.

Note that the Camera Preview window updates.

Close the preview window.

Click ESC to release the camera.

10.

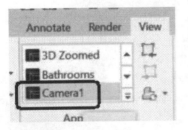

Activate the View ribbon.

Notice the Camera1 view has been added to the view lists.

Left click to activate the Camera1 view.

11.

Switch to the Render ribbon.

Click **Adjust**.
Click ENTER to see a list of cameras.

12.

Highlight **Camera1**.

Click **OK**.

13.

You can use the dialog to adjust the view to your liking.

Click **OK**.

14.

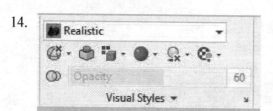

Switch to the View ribbon.

Change the Display Style to **Realistic**.

15.

Set Materials/Textures **On**.

16.

Set **Full Shadows** on.

17.

Switch to the View ribbon.

Set **Realistic Face Style** on.

18.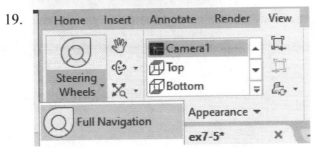

The Camera view should update as the different settings are applied.

19.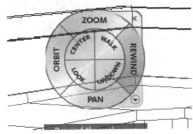

Select **Full Navigation**.

20.

Use the controls on the navigation wheel to adjust the view.

Try LOOK to look up and down.

Click the small x in the upper right to close when you are satisfied with the view.

21.

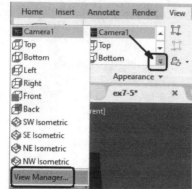

Click the bottom arrow on the view list.

Select **View Manager** located at the bottom of the view list.

22.

Select **New**.

23.

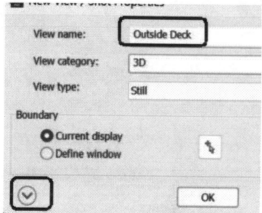

Name the view **Outside Deck.**

Select the down arrow to expand the dialog.

24.

Set the Visual style to **Realistic**.

Set the background to **Sun & Sky.**

25.

The saved view is now listed under Model Views.

Click **OK**.

Save as *ex9-4.dwg*.

Rendering

Rendering is the process where you create photo-realistic images of your model. It can take a lot of trial and error to set up the lighting and materials as well as the best camera angle before you get an image you like. Be patient. The more you practice, the easier it gets. Make small incremental changes – like adding more or less lighting – turning shadows off and on – changing materials – and render between each change so you can understand which changes make the images better and which changes make the images worse.

If you have created a rendering that is close to what you want, you may be able to use the built-in image editor to make it lighter/darker or add more contrast. That may be enough to get the job over the finish line.

Exercise 9-6:
Create Rendering

Drawing Name: rendering.dwg
Estimated Time: 5 minutes

This exercise reinforces the following skills:

- ❑ Render
- ❑ Display Settings
- ❑ Set Location

You need an internet connection in order to perform this exercise.

1.   Open *rendering.dwg*.

2.   Set the Top view active.

3.   Activate the Render ribbon.

4.   Select **Sky Background and Illumination**.

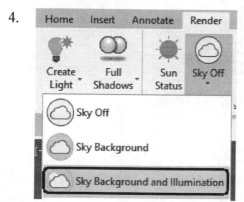

5.

In the drop-down panel under Sky Background and Illumination, select

Set Location→From Map.

6.

Click **Yes** if you have internet access.

7.

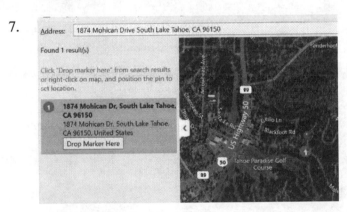

Type in an address in the search field where your project is located.

Once it is located on the map, select **Drop Marker here.**

Click **Next**.

8.

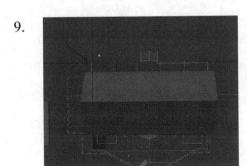

Select the coordinate system to be applied to the project.

Highlight **CA-III**, then Click **Next**.

9.

Select the upper left corner of the building as the site location.

Click ENTER to accept the default North location.

Your building model will appear on a satellite image of the address you provided.

This is useful to provide an idea of how the model will look in the area where it will be constructed.

10.

Switch back to the Render ribbon.

Select the small down arrow at the bottom right of the Sun & Location panel.

11.

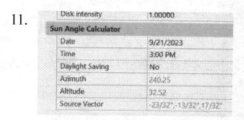

Scroll to Sun Angle Calculator.

Set the time to **3:00 PM**.

Close the palette.

12.

Activate the View ribbon.

Set the view to **Outside Deck**.

13.

Full Shadows should be set to **ON**.

14. 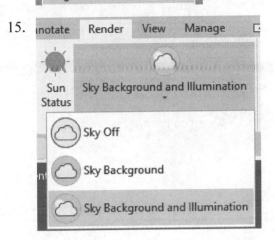 Materials and Textures should be set to **ON**.

15. Switch to the Render ribbon.

Set the **Sky Background and Illumination on**.

You should see a slight change in the view.

16. On the Render to Size drop-down, select the **More Output Settings**.

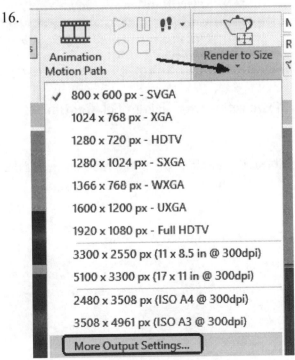

17.

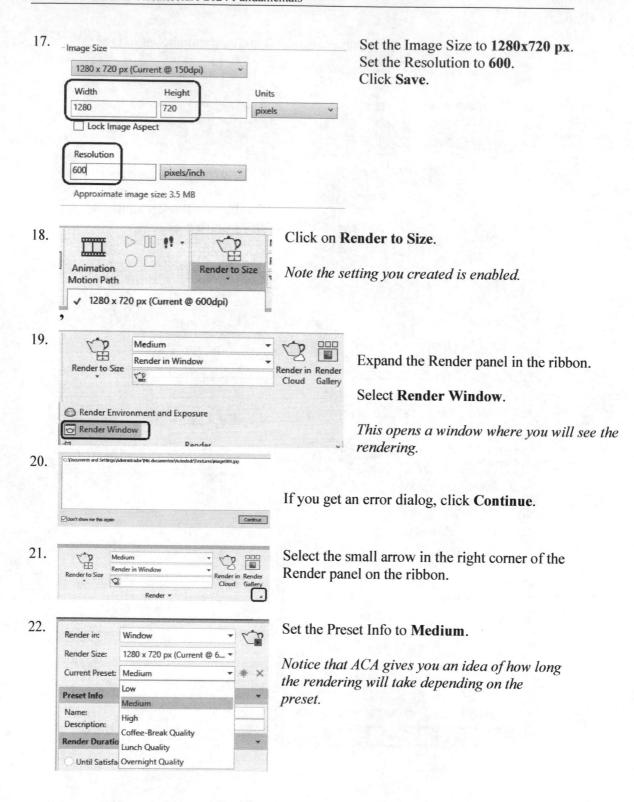

Set the Image Size to **1280x720 px**.
Set the Resolution to **600**.
Click **Save**.

18.

Click on **Render to Size**.

Note the setting you created is enabled.

19.

Expand the Render panel in the ribbon.

Select **Render Window**.

This opens a window where you will see the rendering.

20.

If you get an error dialog, click **Continue**.

21.

Select the small arrow in the right corner of the Render panel on the ribbon.

22.

Set the Preset Info to **Medium**.

Notice that ACA gives you an idea of how long the rendering will take depending on the preset.

23. 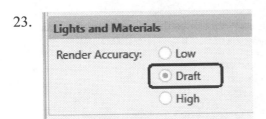 Set the Render Accuracy for the Lights and Materials to **Draft.**

24. 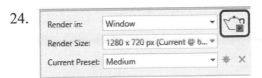 Click on the **Render** button.

25. 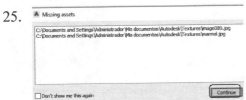 If a dialog comes up stating some material images are missing, Click **Continue**.

26. A window will come up where the camera view will be rendered.

    If you don't see the window, use the drop-down to open the Render Window.

27. 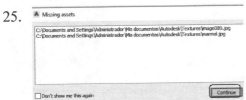 Select the **Save** icon to save the rendered image.

28. Save the image file to the exercise folder.

29.

PNG Image Options

Color
- ○ Monochrome
- ○ 8 Bits(256 Grayscale)
- ○ 8 Bits(256 Color)
- ○ 16 Bits(65,536 Grayscale)
- ○ 24 Bits(16.7 Million Colors)
- ⦿ 32 Bits(24 Bits + Alpha)

☐ Progressive

Dots Per Inch: 600

OK   Cancel

Click **OK**.

Close the Render Window.

30. Save as *ex9-6.dwg*.

Exercise 9-7:
Render in Cloud

Drawing Name: cloud_render.dwg
Estimated Time: 5 minutes

This exercise reinforces the following skills:

- ❑ Render in Cloud
- ❑ Design Center
- ❑ Blocks

Autodesk offers a service where you can render your model on their server using an internet connection. You need to register for an account. Educators and students can create a free account. The advantage of using the Cloud for rendering is that you can continue to work while the rendering is being processed. You do need an internet connection for this to work.

1.   Open *cloud_render.dwg*.

2.

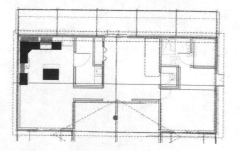

Switch to a top view.

Use the Layer Properties Manager to make visible the kitchen appliances and counter tops.

3. 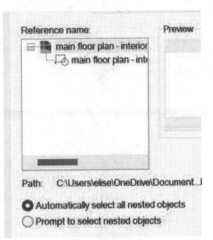 Select the kitchen cabinets so they are highlighted.

Click **Edit Reference In-Place** on the ribbon.

4.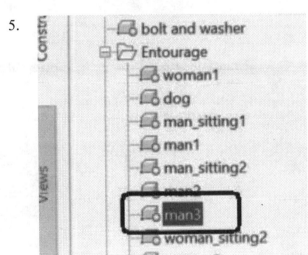

Click **OK**.

You can only add elements to constructs, so you need to open an external reference for editing prior to inserting the element.

5. Locate an entourage element to add to the kitchen area.

I selected man3.

6.

Highlight **man2** (or your preferred element).

Right click and select **Insert as Block**.

7.

Click the Recent tab on the Blocks palette.

You will see a list of elements you defined in your project.

Click on *man3.dwg*.

Place the block inside the kitchen area.

8.

Select the block.
Right click and select Properties.
Change Position Z to **4'-0"**.

Change the scale to **254.**

9.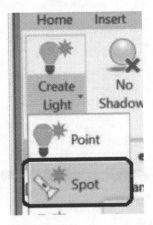

Open the Render ribbon.

Select **Create Light→Spot**.

10.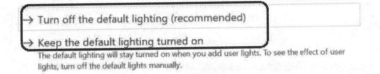

Click **Turn off the default lighting**.

11.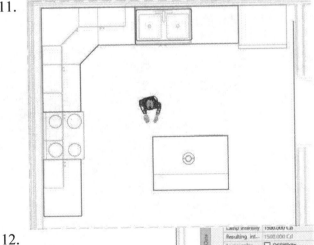

Place the light above the kitchen island.

Click **ENTER** to accept the default values.

Select the light.
Right click and select **Properties**.

12.

Change the Position Z to **8'-0"**.
This locates the light at the ceiling.

Change the Target Z to **4'-0"**.
This positions the light to fall to the kitchen island.

Click **ESC** to release the selection.

13. 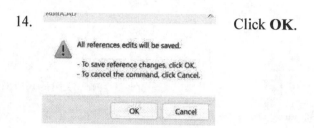 Select **Save Changes** from the ribbon.

14.  Click **OK**.

15. Select the **Create Camera** tool on the Render ribbon.

16. 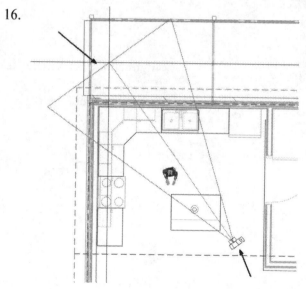 Position the camera so it is looking into the kitchen towards the upper left corner of the building.

17.

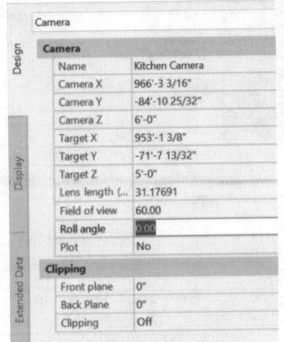

Select the camera.
Right click and select **Properties**.
Change the Name to **Kitchen Camera**.
Change the Camera Z to **6'-0"**.
Change the Target Z to **5'-0"**.
Change the Field of view to **60.0**.
This widens the angle of the view.

Click **ESC** to release the selection.

18.

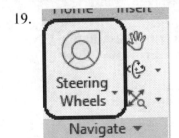

Go to the View ribbon.

Select **Kitchen Camera**.

19.

Use the Full Navigation Wheel to adjust the view as needed.

20.

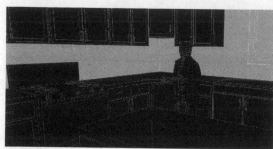

You should see the 3D figure inside the kitchen.

Change the visual style to **Hidden** to make it easier to see.

In order to render using Autodesk's Cloud Service, you need to sign into an Autodesk account. Registration is free and storage is free to students and educators.

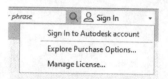

Sign into your Autodesk account.

You can create an account for free if you are an educator, student or if you are on subscription.

21.

Enter your log-in information and Click **Sign in**.

22.

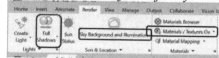

Your user name should display once you are signed in.

23. Save the file as *ex9-6.dwg*.

You must save the file before you can render.

24. Activate the Render ribbon.
Enable **Full Shadows**.
Set **Sky Background and Illumination** ON.
Enable **Materials/Textures On**.
Save the drawing.

25. Select **Render in Cloud**.

26. Your free preview renderings will appear in your Autodesk Rendering Account online. From there you can view, download, and share, or fine tune render settings to create higher resolution final images, interactive panoramas, and more.

Set the Model View to render the current view.

Click the **Start Rendering** model.

27.

A small bubble message will appear when the rendering is completed.

28.

You will also receive an email when the rendering is complete. The email is sent to the address you used to sign into your account.

29. **Render Gallery**

Select **Render Gallery** on the Render ribbon to see your renderings.

30. **MY RENDERINGS**

Click on the My Renderings tab on the browser to review your renderings.

31. You can also select View completed rendering from the drop-down menu under your sign in.

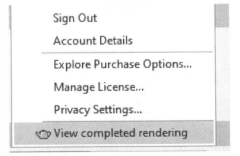

You may need to sign into the gallery in order to see your rendering.

Compare the quality of the rendering performed using the Cloud with the rendering done in the previous exercise.

I have found that rendering using the Cloud is faster and produces better quality images. It also allows me to continue working while the rendering is being processed.

Walk-Through

You can create 3D preview animations and adjust the settings before you create a motion path animation.

Preview animations are created with the controls on the Animation panel found on the ribbon and the 3D navigation tools. Once a 3D navigation tool is active, the controls on the Animation panel are enabled to start recording an animation.

Walk-through animations are useful when reviewing building sites or discussing building renovations.

The easiest way to create a walk-through is to sketch out the path using a polyline. Then, attach a camera to follow the desired path.

Exercise 9-8:
Create an Animation

Drawing Name: animation.dwg
Estimated Time: 5 minutes

This exercise reinforces the following skills:

❑   Walk-Through

> *It will take the software several minutes to compile the animation. It is a good idea to do this exercise right before a 30-minute break to allow it time to create the animation during the break.*

1.   Open *animation.dwg.*

2.   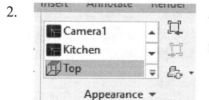 Activate the View ribbon.

Switch to a **Top** View.

3. 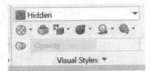 Set the view display to **Hidden**.

4.  Activate the Home ribbon.

Set the layer named **Walkthrough** Current.

This layer is set so it doesn't plot.

5.  Select the **Polyline** tool located under the Line drop-down list.

6.  Draw a polyline that travels through the floor plan to view different rooms.

7. Select the polyline.

Right click and select **Properties**.

8.

Geometry	
Current Vertex	1
Vertex X	34'-6"
Vertex Y	24'-9 27/32"
Start segment wid...	0"
End segment width	0"
Global width	0"
Elevation	5'-3"

Set the elevation of the polyline to **5' 3"**.

If you don't raise the line, your camera path is going to be along the floor.

Click **ESC** to release the selected polyline.

9. [−][SW Isometric][Realistic]

Switch to a **SW Isometric Realistic** view.

10.

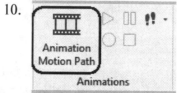

Activate the Render ribbon.

Select **Animation Motion Path**.

11.

Animation settings

Frame rate (FPS): 10

Number of frames: 3000

Duration (seconds): 300.00

Visual style: Realistic

Format: AVI   Resolution: 1024 x 768

☑ Corner deceleration   ☐ Reverse

Set the Frame Rate to **10**.

Set the Number of frames to **3000**.

Set the Duration to **300**.

Set the Visual Style to **Realistic**.

Set the Format to **AVI**.
This will allow you to play the animation in Windows Media Player.

Set the Resolution to **1024 x 768**.

12.

Camera

Link camera to:

○ Point   ● Path   [select button]

For Camera:

Enable **Path**.

Left click on the select button.

13.

Name: Path1

OK

Select the polyline.

Name the camera path.

Click **OK**.

14.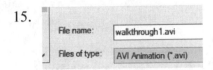

For Target:

Enable **Path.**

Left click on the select button.

Select the polyline.

Name the camera path.

Click **OK.**

15.

File name: walkthrough1.avi

Files of type: AVI Animation (*.avi)

Click **OK.**

Browse to your classwork folder.

Name the file to create *walkthrough1.avi.*

Click **Save**.

16.

Frame number - 468 of 18000

22 Minutes, 0 Seconds Remaining

You will see the animation preview as it processes.

This will take several minutes.

17.

When it is complete, open Windows File Explorer.

Browse to the folder where it is saved.

18.

Right click on it and select **Open with Windows Media Player.**

19. Try creating different walkthroughs using different visual styles.
    You can use different video editors to edit the files you create.

20. Save as *ex9-7.dwg.*

Exercise 9-9:
Using the Sun & Sky

Drawing Name: sun_study.dwg
Estimated Time: 15 minutes

This exercise reinforces the following skills:

- ❑ Create a saved view
- ❑ Apply sunlight
- ❑ Add an image as background
- ❑ Adjust background

1.   Open *sun_study.dwg*.

2.   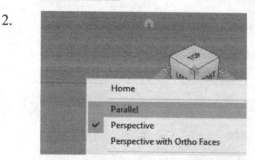   Right click on the ViewCube and select **Parallel**.

    How does the view shift?

3.   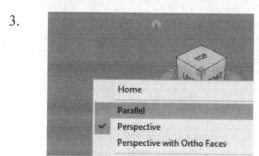   Right click on the ViewCube and select **Perspective**.

    How does the view shift?

4.      Double click the wheel on the mouse to Zoom Extents.

    Adjust the view so that the back edge of the slab is horizontal in the view.

    Hint: Switch to a Front view then tip the view slightly forward.

5. Switch to the View tab.

Launch the **View Manager**.

6. Click **New**.

7. Type **Front Isometric** for the View Name.

8. 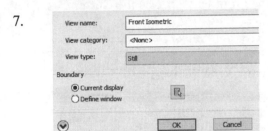 Expand the lower pane of the dialog.

Under Background:

Select **Image**.

9. Set image to *sky.JPG*.

You may need to change the Files of type to JFIF. This file is included in the downloaded exercise files.

Click **Open**.

10.

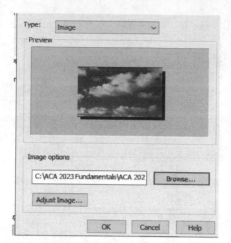

Click **OK**.

11.

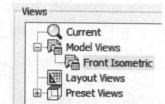

Click **OK**.

Set the Front Isometric as the Current View.

You have now saved the view.

Close the View Manager.

12.

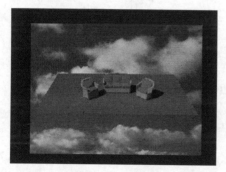

Type **Render** to render the view.

A preview window will open and the view will render.

Close the preview window.

13.

Switch to the Render tab.

Enable **Sky Background and Illumination**.

14.  Enable **Materials/Textures On**.

15. 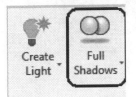 Type **Render** to render the view.

A preview window will open and the view will render.

Close the preview window.

16.  Enable **Full Shadows**.

17. 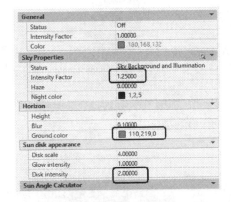 Click the small arrow in the lower right corner of the Sun & Location panel.

18. 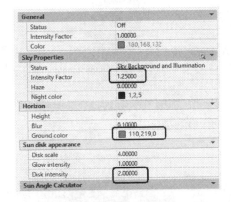 Set the Intensity Factor to **1.25**.
Set the Ground color to **110,219,0**.
(This is a light green.)
Set the Disk Intensity to **2**.

Close the dialog.

19.  Type **Render** to render the view.

A preview window will open and the view will render.

Close the preview window.

20. Save as *ex9-8.dwg*.

This exercise lets you try different settings and compare the rendering results. Try applying different backgrounds and lighting to see how the rendering changes.

Exercise 9-10:
Applying Materials

Drawing Name: materials.dwg
Estimated Time: 45 minutes

This exercise reinforces the following skills:

- ❑   Navigate the Materials Browser
- ❑   Apply materials to model components
- ❑   Edit Styles
- ❑   Change environment
- ❑   View Manager
- ❑   Render

1.   Open *materials.dwg*.

2.   [−][SW Isometric][Realistic]   Set the display to **Realistic**.

      If you don't set the display to Realistic, you won't see the material changes.

3.   Switch to the Render ribbon.

      Click on **Materials Browser**.

4.   Locate **Beechwood – Honey** and load into the drawing.

      This is for the flooring.

5.   Locate **Glass Clear - Light** and load into the drawing.

      This is for the window glazing.

6.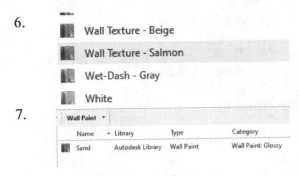

Locate **Wall Texture - Salmon** and load into the drawing.

This is for the walls.

7.

Locate **Sand Wall Paint** and load into the drawing.

This is for the wall and window trim.

8.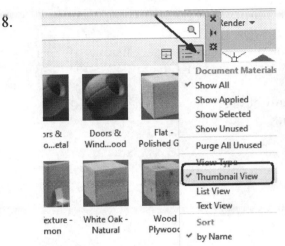

Change the Material Browser View to use a Thumbnail View.

Now you can see swatches of all the materials available in your drawing.

Close the Materials Browser.

9. Select a wall.

On the ribbon:
Select **Edit Style**.

10. Click the Materials tab.

Highlight the top component for the wall.
Click on the **Edit Material** button.

11. Click the Display Properties tab.

Click on the override button.

12.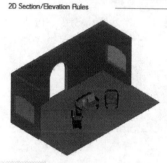

Click the Other tab.

Set the Render Material to **Wall Texture – Salmon**.

Click **OK**.

Close all the dialogs.

Click ESC to release the selection. You should see the new material on the walls.

If you don't see the material change, verify that your display is set to **Realistic**.

This only changes the wall style used in this drawing.

13.

Click on a window.
On the ribbon:
Select **Edit Style**.

14.

Click the Materials tab.

Component	Material Definition
Frame	Doors & Windows.Metal Doors & Frames.
Sash	Doors & Windows.Metal Doors & Frames..
Glass	Doors & Windows.Glazing.Glass.Clear
Muntins	Standard

Highlight the **Frame Component**.
Click on **New Material**.

15.

New Name: Sand Paint

OK

Create a new material called **sand paint**.

Click **OK**.

16.

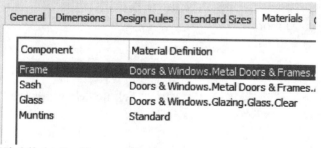

Component	Material Definition
Frame	Sand Paint
Sash	Doors & Windows.Metal Doors & Frames.
Glass	Doors & Windows.Glazing.Glass.Clear
Muntins	Standard

The Frame is now assigned the new material.

17. Highlight the Glass Component.
Click on **New Material**.

18. New Name: Glass - Clear Light

Create a new material called **Glass – Clear Light**.

OK

Click **OK**.

19.
Component	Material Definition
Frame	Sand Paint
Sash	Doors & Windows.Metal
Glass	Glass - Clear Light
Muntins	Standard

The Glass is now assigned the new material.

20.
General Dimensions Design Rules Standard Sizes Materials Classifications Display Properties Version History

Component	Material Definition
Frame	Sand Paint
Sash	Doors & Windows.Metal Doors & Frames.Aluminum Windows.Painted.White
Glass	Glass – Clear Light
Muntins	Standard

Highlight the Frame and select **Edit Material**.

21.
General Display Properties

Display Representations	Display Property Source	Style Override
General	Drawing Default	
General High Detail	Drawing Default	
General Low Detail	Drawing Default	
General Medium Det...	Drawing Default	
General Presentation	Drawing Default	
General Reflected	Drawing Default	
General Reflected Scre...	Drawing Default	
General Screened	Drawing Default	

Place a Check next to the General Medium Detail to enable Style Override.

22.
Layer/Color/Linetype Hatching Other

Surface Hatch Placement

☑ Top   ☑ Bottom
☑ Left   ☑ Right
☑ Front   ☑ Back

Surface Rendering

Render Material: Sand

Mapping: Face Mapping

Live Section Rendering

Cut Surface Render Material: *NONE*

Sectioned Body Render Material: *NONE*

2D Section/Elevation Rules

Click on the Other tab.
Set the Render Material to the **Sand** material.

Click **OK**.

Close the Display Properties dialog.

23.
General Dimensions Design Rules Standard Sizes Materials Classifications Display Properties Version History

Component	Material Definition
Frame	Sand Paint
Sash	Doors & Windows.Metal Doors & Frames.Aluminum Windows.Painted.White
Glass	Glass - Clear Light
Muntins	Standard

Highlight the Glass and select **Edit Material**.

24.
General Display Properties

Display Representations	Display Property Source	Style Override
General	Drawing Default	
General High Detail	Drawing Default	
General Low Detail	Drawing Default	
General Medium Det...	Drawing Default	
General Presentation	Drawing Default	
General Reflected	Drawing Default	
General Reflected Scre...	Drawing Default	
General Screened	Drawing Default	

Place a Check next to the General Medium Detail to enable Style Override.

25.

Surface Rendering	
Render Material:	Clear - Light
Mapping:	Face Mapping
Live Section Rendering	
Cut Surface Render Material:	*NONE*
Sectioned Body Render Material:	*NONE*
2D Section/Elevation Rules	

Click on the Other tab.
Set the Render Material to the **Clear - Light** material.

Click **OK**.

Close the Display Properties dialog.

26.

Component	Material Definition
Frame	sand paint
Sash	Doors & Windows.Metal Doors & Frames.Aluminum Windows.Painted.White
Glass	Glass - Clear-Light
Muntins	Standard

General | Dimensions | Design Rules | Standard Sizes | Materials | Classifications | Display Properties

Verify that:

Set the Frame to **sand paint**.

Set the Glass to **Glass – Clear Light**.

Close all the dialogs.

Click ESC to release the selection.
Select the floor slab.

27.

28. On the ribbon:
Select **Edit Style**.

Edit Style

29.

General | Components | Materials | Classifications | Display Properties

Total Thickness: 2"

Index	Name	Thickness	Thickness Offset
1	concrete slab	2"	-2"

On the Components tab:

Click **Add component**.

30.

Index	Name
1	wood flooring
2	concrete slab

Rename **wood flooring**.

31.

General | Components | Materials | Classifications | Display Properties

To

Index	Name	Thickness	Thickness Offset
1	wood flooring	2"	0"
2	concrete slab	2"	-2"

Make sure the Thickness Offset is set to 0" or you won't see the new material.

32. 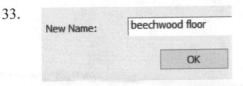 Switch to the Materials tab.

Highlight the wood flooring Component.
Click on **New Material**.

33. Name the new material **beechwood floor**.

New Name: beechwood floor

OK

Click **OK**.

34. Highlight the wood flooring component and select **Edit Material**.

Component	Material Definition
wood flooring	beechwood floor
concrete slab	concrete
Shrink Wrap Body	Standard

35. Place a Check next to the General Medium Detail to enable Style Override.

Display Representations	Display Property Source	Style Override
General	Drawing Default	
General High Detail	Drawing Default	
General Low Detail	Drawing Default	
General Medium Det...	Drawing Default	
General Presentation	Drawing Default	
General Reflected	Drawing Default	
General Reflected Scre...	Drawing Default	
General Screened	Drawing Default	

36. Click on the Other tab.
Set the Render Material to the **Beechwood - Honey** material.

Surface Rendering

Render Material: Beechwood - Honey

Mapping: Face Mapping

Click **OK**.

Close the Display Properties dialog.
Click **ESC** to release the selection.

37. 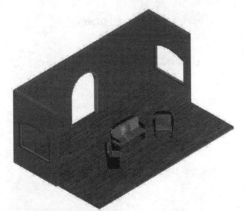 Verify that you see all the material changes that you made. The walls should appear salmon colored, the windows should have a light brown frame, and the floor should have a wood surface.

38. Switch to the Render tab.

Enable **Sun Status**.

39.

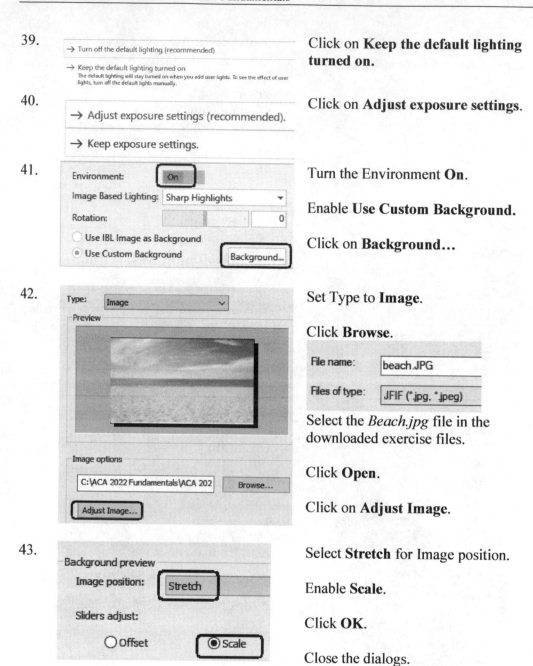

Click on **Keep the default lighting turned on.**

40.

Click on **Adjust exposure settings.**

41.

Turn the Environment **On.**

Enable **Use Custom Background.**

Click on **Background...**

42.

Set Type to **Image.**

Click **Browse.**

Select the *Beach.jpg* file in the downloaded exercise files.

Click **Open.**

Click on **Adjust Image.**

43.

Select **Stretch** for Image position.

Enable **Scale.**

Click **OK.**

Close the dialogs.

44.

Orbit, pan and zoom to position the model on the beach so it appears proper.

45.

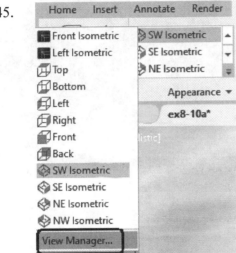

Switch to the View tab.

Launch the **View Manager**.

46.

Click **New**.

47.

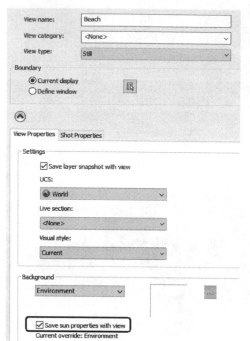

Save the view as **Beach**.

Enable **Save sun properties with view**.

Click **OK**.

48.

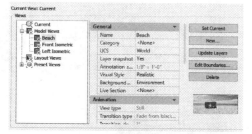

Click **OK**.

49.

Type **RENDER**.

The preview window will appear and the view will render.

Save as *ex9-9.dwg*.

Exercise 9-11:
Applying Materials to Furniture

Drawing Name: Wicker chair and sofa.dwg
Estimated Time: 15 minutes

This exercise reinforces the following skills:

- ❑ Navigate the Materials Browser
- ❑ Apply materials to model components
- ❑ Customize materials

1.
   | File name: | Wicker chair and sofa.dwg |
   | Files of type: | Drawing (*.dwg) |

   Open the *Wicker chair and sofa* file in the exercises folder.

2.
   | File name: | Wicker chair and sofa - Sea Fabric.dwg |
   | Files of type: | AutoCAD 2018 Drawing (*.dwg) |

   Save the file with a new name - *Wicker chair and sofa – Sea Fabric.*

3.
   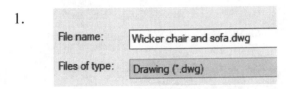

   Use explode to explode the blocks until you see solids when you hover over the elements.

   Type **EXPLODE, ALL, ENTER**.

   Repeat until you see 3D solid when you hover over each item.

   3D Solid
   Color   ☐ White
   Layer   None
   Linetype Continuous

4.
   Annotate   Render   View   Manage

   Sun Status   Sky Background and Illumination

   ⊗ Materials Browser
   ⊗ Materials On / Textures Off
   ◁ Material Mapping

   Sun & Location   Materials

   Switch to the Render ribbon.

   Click on **Materials Browser**.

5.
   Pebbled - Black
   Pebbled - Light Brown
   Pebbled - Mauve

   Locate **Pebbled – Mauve** and load into the drawing.

   This is for the chair fabric.

6.

Locate **Chrome - Satin** and load into the drawing.

This is for the chair frame.

7.

Highlight the Pebbled – Mauve fabric.

Right click and select **Duplicate**.

8.

Right click on the duplicate material.

Select **Rename**.

9.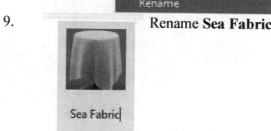

Rename **Sea Fabric**.

10.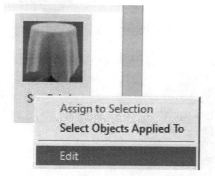

Select the Sea Fabric material.

Right click and select **Edit**.

11.

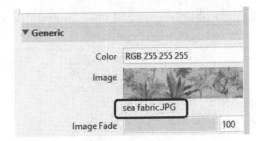

Click on the image name and set to *sea fabric.jpg*.
This file is available as part of the exercise downloads.

Change the Color to **255,255,255**.
This is white.

Close the Materials Editor.

12.

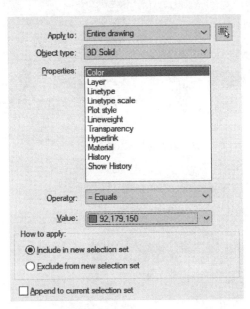

Select one of the cushions.

Right click and select **Properties**.

13.

3D Solid	▼
General	
Color	92,179,150
Layer	None
Linetype	Continuous
Linetype scale	1.0000

On the Properties palette:

Note the color is set to **92, 179,150**.

Click **ESC** to release the selection.

14.

Apply to:	Entire drawing
Object type:	3D Solid
Properties:	Color
	Layer
	Linetype
	Linetype scale
	Plot style
	Lineweight
	Transparency
	Hyperlink
	Material
	History
	Show History
Operator:	= Equals
Value:	92,179,150

How to apply:
◉ Include in new selection set
○ Exclude from new selection set

☐ Append to current selection set

Type **QSELECT** for quickselect.

Set the Object type to **3D Solid**.

Set the Color Equals to **92, 179, 150**.

Click **OK**.

All the chair cushions are selected.

15.

Highlight the Sea Fabric material in the Materials Browser.

Right click and select **Assign to Selection**.

Click ESC to release the selection.

If the view is set to Realistic, you should see the material as it is applied.

16.

Select one of the chair frame elements.

General	
Color	◧ White
Layer	None
Linetype	———— Continuous
Linetype scale	1.0000

The Color is set to White.

Click ESC to release the selection.

17.

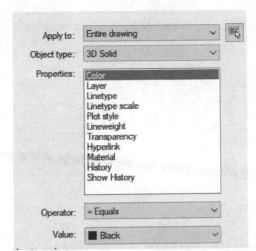

Type **QSELECT** for quickselect.

Set the Object type to **3D Solid**.

Set the Color Equals to **White**.

Click **OK**.

The chair frames are selected.

18.

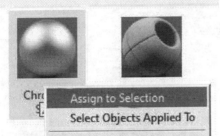

Highlight the **Chrome - Satin** material.

Right click and select **Assign to Selection**.

Click **ESC** to release the selection.

19.

The drawing has been redefined with new materials.

Save as *Wicker chair and sofa – Sea Fabric.dwg*

Exercise 9-12:
Replace a Block

Drawing Name: replace_block.dwg
Estimated Time: 10 minutes

This exercise reinforces the following skills:

- ❑ Insert a block
- ❑ Replace a block
- ❑ Render

In order to sync blocks, you need to have access to the internet and you need to be signed in to your Autodesk account.

1. ⬜📂🖫🖫🖳   Open *replace_block.dwg*.

2. Type **INSERT**.

    Click ENTER.

3.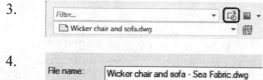
    Select the **Browse** button next to the Filter box.

4.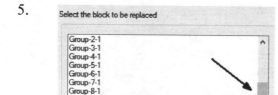
    Locate the *Wicker chair and sofa – Sea Fabric.dwg*.

    Click **Open**.

    Click a point in the view to place the block away from the room.

    Click ENTER when prompted for a rotation angle.

5. ![Select the block to be replaced dialog]
    Type **BLOCKREPLACE**.

    Scroll down and select *Wicker chair and sofa*.

    Click **OK**.

6.

Scroll down and select the *Wicker chair and sofa – Sea Fabric* block.

Click **OK**.

7.

Click ENTER.

8.

The block is replaced using the same location and orientation.

Delete the block that was inserted and isn't needed any more.

Note that you had to fully insert the new block into the drawing before you could replace the block.

9.

Type RENDER.

The preview window will appear and the view will render.

Save as *ex9-11.dwg*.

Exercise 9-13:
Using Artificial Light

Drawing Name: lights.dwg
Estimated Time: 15 minutes

This exercise reinforces the following skills:

- ❑ Add lights to a model
- ❑ Edit light properties

1.   Open *lights.dwg*.

2.   Switch to the Render ribbon.

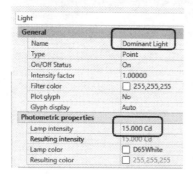

Select the **Point** light tool.

3.   Select the center of the circle on the left as the location for the point light.

Click ENTER to accept all the default settings for the light.

If you have difficulty placing the lights, freeze all the layers except for Light Locator.

4.   Select the point light.
Right click and select **Properties**.

Change the Name to **Dominant Light**.

Change the Lamp intensity to **15.00 Cd**.

Did you notice that the size of the light glyph updates?

Click ESC to release the selection.
5.   Enable **Full Shadows**.

Enable **Sky Background and Illumination**.

6.

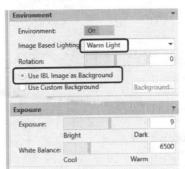

Click on **Sun Status**.
Click on **Adjust exposure settings**.

7.

Set Environment to **On**.

Set the Image Based Lighting to **Warm Light**.

Enable **Use IBL Image as Background**.

Close the palette.

8.

Click **Render to Size** on the ribbon.

The Render Window will appear.

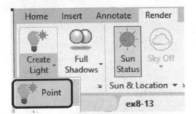

You will see a new rendering.

Close the Render Window.

9.

Select the **Point** light tool.

10.

Select the center of the circle in the middle of the room as the location for the point light.

Click ENTER to accept all the default settings for the light.

If you have difficulty placing the lights, freeze all the layers except for Light Locator.

11.
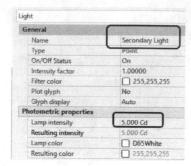

Select the point light.
Right click and select **Properties**.

Change the Name to **Secondary Light**.

Change the Lamp intensity to **5.00 Cd**.

Click ESC to release the selection.

12.
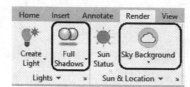

Switch to the Home ribbon.

Use **Thaw All** to thaw any frozen layers.

13.

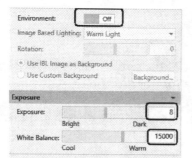

Switch to the Render ribbon.

Enable **Full Shadows**.
Enable **Sky Background**.

*If you can't enable the Sky Background, check that
the Viewcube is set to Perspective.*

14.

Click on **Sun Status**.

15.

→ Adjust exposure settings (recommended).

→ Keep exposure settings.

Click on **Adjust exposure settings**.

16.

Environment:	Off
Image Based Lighting:	Warm Light
Rotation:	0

Use IBL Image as Background
Use Custom Background   Background...

Exposure

Exposure:	8
Bright	Dark
White Balance:	15000
Cool	Warm

Set Environment to **Off**.

Set the Exposure to **8**.

Set the White Balance to **15000.**

Close the palette.

17.

Render to Size

Click **Render to Size** on the ribbon.

The Render Window will appear.

You will see a new rendering.

Close the Render Window.

Save as *ex9-12.dwg*.

Exercise 9-14:
Mapping Materials

Drawing Name: medallion.dwg
Estimated Time: 15 minutes

This exercise reinforces the following skills:

- ❑ Create a material
- ❑ Apply material to a face.

1.   Open *medallion.dwg*.

2.   Switch to the Render tab.

Click on Materials Browser.

3.   Select the **Create Material** tool at the bottom of the palette.

4.

Create the new material using the **Stone** template.

5.

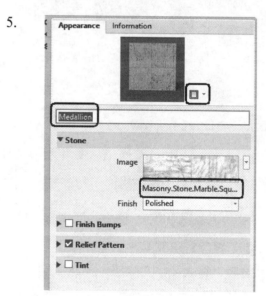

Change the name to Medallion.

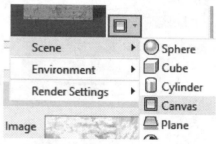

Change the Scene to **Canvas**.

Click on the image name.

6.

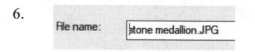

Locate the *stone medallion* file in the downloaded exercise files.

Click **Open**.

7.

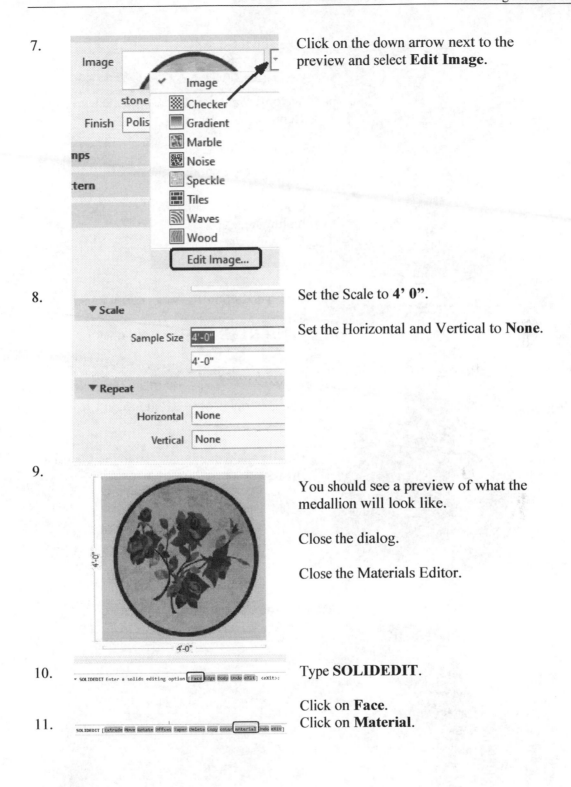

Click on the down arrow next to the preview and select **Edit Image**.

8.

Set the Scale to **4' 0"**.

Set the Horizontal and Vertical to **None**.

9.

You should see a preview of what the medallion will look like.

Close the dialog.

Close the Materials Editor.

10.

Type **SOLIDEDIT**.

Click on **Face**.

11.

Click on **Material**.

12.

Click on the disk.

`SOLIDEDIT Select faces or [Undo Remove ALL]:`

We only want the top face selected, so click **Remove**.

13.

Click on the bottom of the disk.

Click ENTER when only the top face is highlighted.

14.

`SOLIDEDIT Enter new material name <ByLayer>: Medallion`

Type **Medallion** for the material to be applied.
Click **ENTER** to exit the command.

15.

`[–][Top][Realistic]` Change the display to **Top Realistic**.

16.

The material is placed, but it didn't map properly.

Double click on the material in the Material Browser to open the editor.

17.

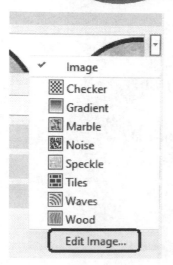

Select **Edit Image**.

18.  Adjust the Position and the Scale until the preview looks correct.

Close the dialogs.

19.  Save as *flower medallion.dwg*.

Extra: *Place the medallion in the beach scene you did in the previous exercise and see if you can create a nice rendering.*

Notes:

QUIZ 9

True or False

1. When rendering, it is useful to save named views using the View Manager.

2. Sky illumination can be used to add extra light to a scene.

3. You can use an image or a color to change the background of a scene.

4. To change the material used in a wall, you need to modify the wall style.

5. When mapping a material, you can only use grips to adjust the placement.

Multiple Choice

6. Materials can be displayed in the following views: (select all that apply)

    A. plans
    B. sections
    C. elevations
    D. renderings

7. Surface Hatches can be displayed in the following views: (select all that apply)

    A. model
    B. plan
    C. section
    D. elevation

8. Identify the tool shown.

    A. Expand Render Window
    B. Expand Tea Kettle
    C. Scale
    D. Render to Size

9. When there are no lights placed in a scene, the scene is rendered using:

    A. Default lighting.
    B. Sun & Sky.
    C. Nothing, it is really dark.
    D. Sun only, based on location.

10. Spotlights and point lights are represented by:

    A. Glyphs (symbols used to indicate the position and direction of the light)
    B. Nothing
    C. Blocks
    D. Linework

11. When mapping a material, you can adjust: (select all that apply)

    A. The X position
    B. The Y position
    C. Rotation
    D. Scale

12. Once you place a camera, you can adjust the camera using: (select all that apply)

    A. Properties
    B. The Adjust tool on the Render tab
    C. Using the MOVE and ROTATE tools
    D. You can't modify a camera once it has been placed

ANSWERS:

1) T; 2) T; 3) T; 4) T; 5) F; 6) A, B, C, D; 7) A, C, D; 8) D; 9) A; 10) A; 11) A, B, C, D; 12) A, B, C

Lesson 10
Collaboration

Drawing Management

Project Navigator allows drawing versioning for AutoCAD Architecture drawings. This ensures informed collaboration between multiple users and prevents unauthorized modifications to the drawings.

When a project involves multiple users working together on a set of drawings, it is essential to have a file locking and versioning mechanism to ensure that all changes are recorded and there are no access clashes. When you check out a drawing from the Project Navigator, other users only have read-only access to the drawing and can view the last saved version of the drawing, till you check the drawing back in. This ensures that there are no file access conflicts and each modification made to a drawing is recorded.

Changes being made to the checked out drawing are not dynamically visible to other users and as the checked out file is saved, no xref notifications are sent to others which can be an interruption to their work.

Hovering over a file in the Project Navigator shows you the status of the file, the name of the drawing, its location, the size of the file, the date and time the file was last checked in, the name of the user who last checked-in the file, and the check in version history with the check-in comments. Red checkmarks indicate that the drawing is checked out by another user, and a green checkmark indicates that the drawing is checked out by you.

When you check out a file, make modifications and check it back in, you are prompted to add comments on the changes you have made. This information is displayed in the Check-In Version History, along with your username, your domain or computer name, and the date and time of the check-in.

Every time you check-in a file, AutoCAD Architecture creates a backup file in the same folder where the drawing is stored, with the drawing name appended with check in time. If during the course of a project, you wish to revert to an earlier version of the drawing, the comments history makes it easier for you choose the backup file version you wish to roll back to.

If you open a drawing and make changes without checking it out, ACA will not create a backup file or update the version history. Even if you are a solo user on a project, it might be useful to check in and check out drawings, so that you can maintain a version history.

Exercise 10-1:

Set the Number of Backup Files in a Project

Drawing Name: Brown Medical Clinic
Estimated Time: 10 minutes

This exercise reinforces the following skills:

- ❑ Project Navigator
- ❑ Drawing Management
- ❑ Changing the appearance of the Project Browser
- ❑ Changing the number of backups

1.

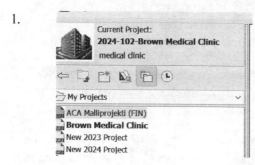

Launch **ACA (US Metric).**

Set the **Brown Medical Clinic** project current.

2.

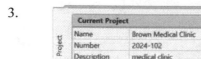

On the Quick Access toolbar, click **Project Navigator**.

3.

In the Project Navigator palette, Project tab, click the **Edit Project** icon, on the top right corner.

4.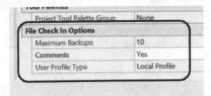

In the Project Properties dialog box, **Advanced→File Check In Options → Maximum Backups**.

Set the Maximum Backups to **10**.
Set Comments to **Yes**.

5.

Browse to your My Projects/Brown Medical Clinic folder.

Locate the *Commercial Template Project Bulleting Board.htm* file.

6.

Open using Notepad.

7.

Scroll down to the bottom of the document.

Modify the text to provide instruction on how to work in the Brown Medical Clinic folder.

8.

Type:
**This clinic must comply with VA standards and regulations.
Go to https://www.cfm.va.gov/til/projReq.asp for more information.**

Save the file as **Brown Medical Clinic.htm.**

9.

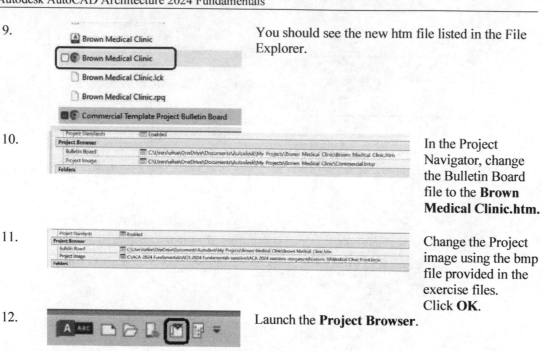

You should see the new htm file listed in the File Explorer.

10.

In the Project Navigator, change the Bulletin Board file to the **Brown Medical Clinic.htm.**

11.

Change the Project image using the bmp file provided in the exercise files. Click **OK**.

12.

Launch the **Project Browser**.

13.

Notice that the image and the text in the browser panel have changed.

Exercise 10-2:

Checking out and Checking in a Drawing

Drawing Name: Brown Medical Clinic
Estimated Time: 25 minutes

This exercise reinforces the following skills:

- ❑ Project Navigator
- ❑ Drawing Management
- ❑ Checking out a drawing
- ❑ Ceiling grids
- ❑ Checking in a Drawing

1. 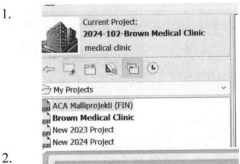 Set the **Brown Medical Clinic** project current.

2. 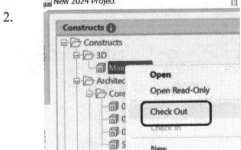 Open the Project Navigator.
Open the Constructs tab.
Highlight **Main Model**.
Right click and select **Check Out**.

3. 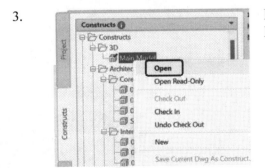 Highlight **Main Model**.
Right click and select **Open**.

4. Switch to the **Model** tab.

5.

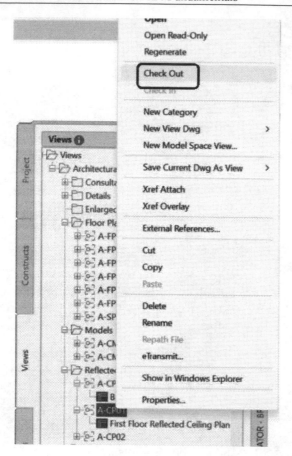

On the Project Navigator, open the **Views** tab.

Locate the **A-CP01**.
Right click and select **Check Out**.

Then Right click and select **Open**.

6.

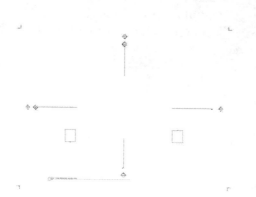

The drawing is empty, but is preloaded with elevations and a couple of external references.

Delete the two cyan rectangles.

Save and close the drawing.

7.

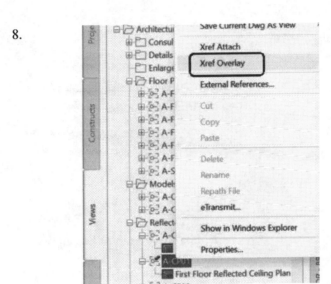

Switch back to the Main Model file.
Set the Layer State to **space planning items not visible**.

[−][Top][2D Wireframe]

Switch to a Top view.

8.

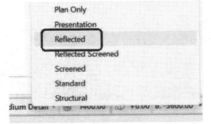

On the Project Navigator, open the **Views** tab.

Locate the **A-CP01**.
Right click and select **Xref Overlay**.

9.

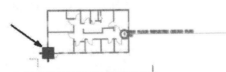

Select the **A-CP01.**
Locate the corner grid.
Move the external reference so the croner grid overlies the Main Model corner grid.

10.

Plan Only
Presentation
Reflected
Reflected Screened
Screened
Standard
Structural

Set the Display Configuration to **Reflected**.

11.  Select the **A-CP01.**

Select Edit Reference In-Place.

12. 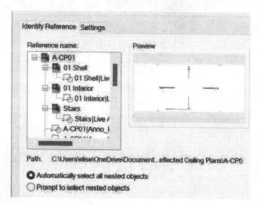 Click **OK**.

13. Move the view title block below the building model.

14. Adjust the length of the view title line.
Adjust the position of the view title attributes.

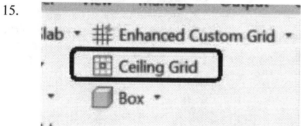

15. Select the **Ceiling Grid** tool from the Home ribbon.

16.

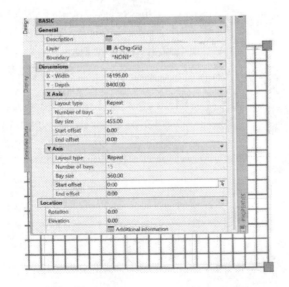

Click the lower corner of the building to place the ceiling grid.

17.

Click **Width**.

Set the Width to **16195**.

18.

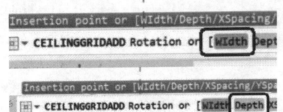

Click **Depth**.

Set the Width to **8400**.

Set the Rotation to **0**.

Click **ENTER** to exit the command.

19.

Select the grid.
Right click and select **Properties**.

Change the X Bay Size to **455**.
Change the Y Bay Size to **560**.

Click ESC to release the selection.

20.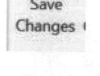

Save changes to the external reference.

21.  Click **OK**.

22.

Highlight **A-CP01** in the Project Navigator.

Right click and select **Check In**.

23. :−][SW Isometric][2D Wireframe]

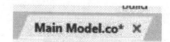

Open the **Main Model.co** tab.

Switch to a **SW Isometric** view.

24.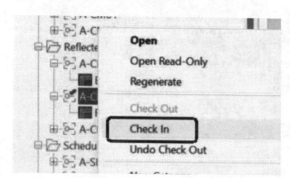

Select the ceiling grid.
Right click and select **Properties**.
Change the Insertion point Z to **3600**.
Click **ESC** to release the selection.

25.

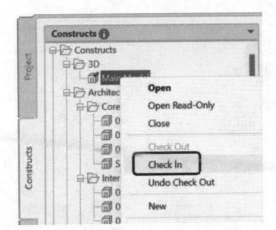

Save the file.

Highlight the **Main Model** in the Project Navigator.

Right click and select **Check In**.

26.

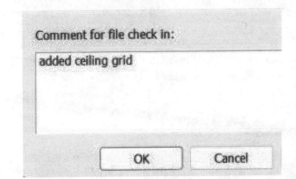

Type **added ceiling grid**.

Click **OK**.

27.

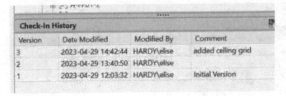

Expand the Check-in History section.
You can see the different versions that have been created and checked in.

Exercise 10-3:
Checking in a Drawing

Drawing Name: Brown Medical Clinic
Estimated Time: 35 minutes

This exercise reinforces the following skills:

- ❑ Project Navigator
- ❑ Drawing Management
- ❑ Elements

1.

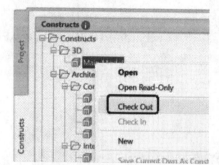

Highlight the **Main Model** construct.

Right click and **Check Out**.
Right click and **Open**.

2.

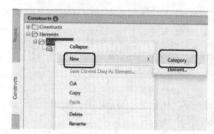

Highlight **Architectural** under Elements in the Constructs tab.
Right click and select **New→Category**.

3.

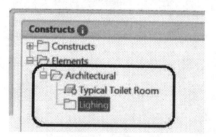

Name the new Category: **Lighting**.

4.

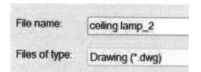

Open *ceiling lamp_2.dwg*.

This is a downloaded exercise file.

5.

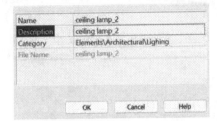

Highlight the **Lighting** category.
Right click and select **Save Current Dwg As Element**.

6.

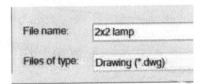

Click **OK**.

Close the drawing.

7.

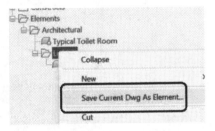

Open *2x2 lamp.dwg*.

This is a downloaded exercise file.

8.

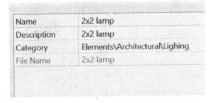

Highlight the **Lighting** category.
Right click and select **Save Current Dwg As Element**.

9.

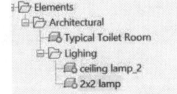

Click **OK**.

Close the drawing.

You should see two lighting elements listed.

10.

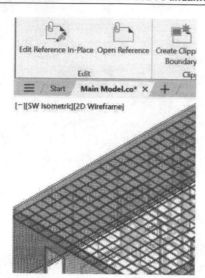

Select the ceiling grid.

Select **Open Reference** on the ribbon.

11.

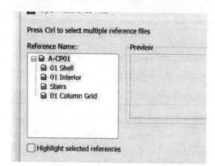

Click **Open**.

12.

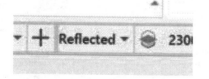

Set the Display Configuration to **Reflected.**

The ceiling grid is now visible.

13.

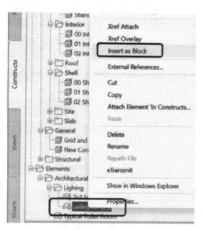

On the Project Navigator:

Locate the *ceiling lamp_2* element.
Right click and select **Insert as Block**.

14. 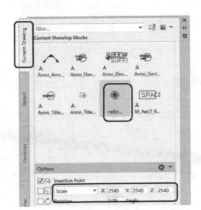 On the Blocks palette:

Open the **Current Drawing** tab.

Locate the *ceiling lamp_2*.

Change the Scale to **2540**.

15. 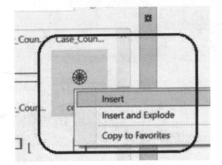 Highlight *ceiling lamp_2*.

Right click and select **Insert**.

16.  Place the block to the right of the ceiling grid.

17. 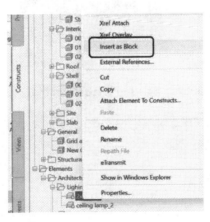 On the Project Navigator:

Locate the *2x2 lamp* element.
Right click and select **Insert as Block**.

18.

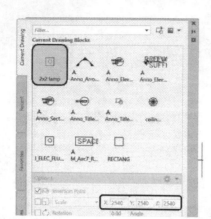

On the Blocks palette:

Open the **Current Drawing** tab.

Locate the *2x2 lamp*.

Change the Scale to **1**.

19.

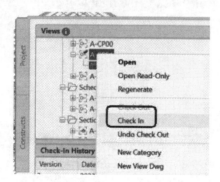

Highlight *2x2 lamp*.

Right click and select **Insert**.

Place the block outside the ceiling grid next to the other lighting fixture.

20.

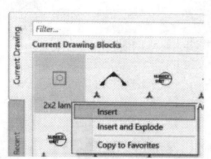

Save the file and close.

In the Project Navigator:

Highlight **A-CP01**.

Right click and select **Check In**.

21.

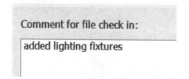

Type **added lighting fixtures** in the comment field.

Click **OK**.

22.

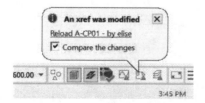

Inside the Main Model.co drawing, reload the external reference.

You should see the lighting fixtures that were added to the external reference.

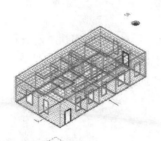

23.

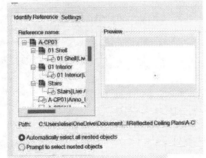

Select the ceiling grid.

Click **Edit Reference In-Place** from the ribbon.

Click **OK**.

24.

−][Top][2D Wireframe]

Switch to a **Top View**.

Set the Display Configuration to **Low Detail**.

25.

Move the *ceiling lamp_2* into the bathroom.

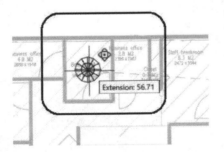

26.

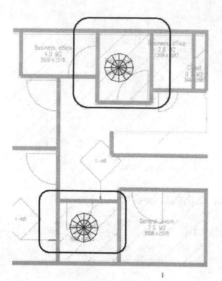

Copy the *ceiling lamp_2* block and place in the second bathroom.

27.

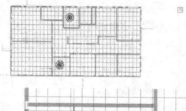

Set the Display Configuration to **Reflected**.

28.

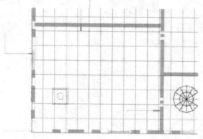

Move the 2x2 lamp onto the ceiling grid.

Create copies of the 2x2 lamp and place in the different rooms.

29.

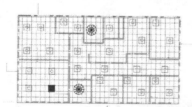

Create copies of the 2x2 lamp and place in the different rooms.

30.

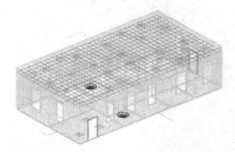

Switch to an Isometric view.

Some of the lighting fixtures may not be placed at the correct elevation.

31.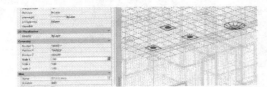

Select the lighting fixtures to be re-positioned.
Right click and select **Properties**.
Change the Position Z to **3600.**

32.

Click **Save Changes** on the ribbon.

33.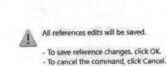

Click **OK**.

34.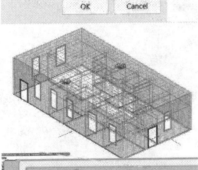

Save the *Main Model.co* file.

Close the file.

35.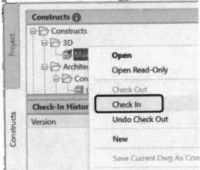

Highlight **Main Model** in the Project Navigator.

Right click and select **Check In**.

36.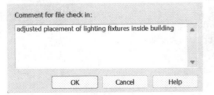

Type a comment: **adjusting the placement of lighting fixtures inside building**.

Click **OK**.

37. 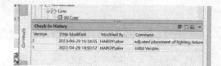 You can see the Check-In History.

Exercise 10-4:
Creating a Lighting Fixture Schedule

Drawing Name: Brown Medical Clinic
Estimated Time: 60 minutes

This exercise reinforces the following skills:

- ❏ Project Navigator
- ❏ Drawing Management
- ❏ Schedules
- ❏ Property Set Definitions
- ❏ External References

1. 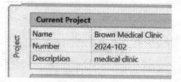 *(The Brown Medical Clinic should be the current project)*

2. 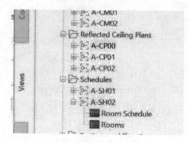 Open the Views tab on the Project Navigator.

   Delete the **A-CP00** and **A-CP02** views under Reflected Ceiling Plans.

3. 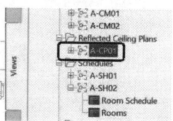 Check out and open **A-CP01**.

4.  Go to the Manage ribbon.
   Open the **Style Manager**.

5.

Under Documentation Objects:
Locate **GeneralOjects** under **Property Set Definitions**.

6.

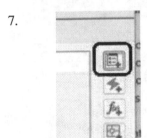

Select the **Applies to**: tab:

All Objects are enabled.

7.

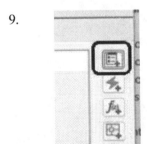

Click **Add Manual Property Set Definition**.

8.

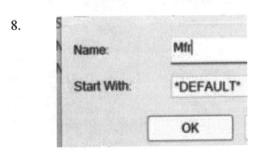

Select **Default** from the drop-down list.
Type **Mfr** in the Name field.
Click **OK**.
Change the **Format to Case-Sentence**.

9.

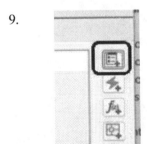

Click **Add Manual Property Set Definition**.

10.

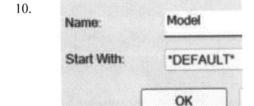

Select **Default** from the drop-down list.
Type **Model** in the Name field.
Click **OK**.
Change the **Format to Case-Upper**.

11.  Click **Add Manual Property Set Definition**.

12. 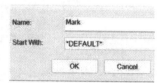 Select **Default** from the drop-down list.
Type **Mark** in the Name field.
Click **OK**.
Change the Format to **Integer**.

If you set the format to Integer, you can auto-number and renumber the mark values.

13.  Under Format: select **Number – Object**.

14.  Click **Add Manual Property Set Definition**.

15. 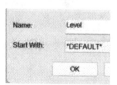 Select **Default** from the drop-down list.
Type **Level** in the Name field.
Click **OK**.
Change the Format to **Case - Upper**.

16.  Click **Add Manual Property Set Definition**.

17.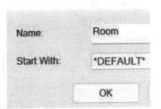

Select **Default** from the drop-down list.
Type **Room** in the Name field.
Click **OK**.
Change the Format to **Case - Upper**.

18. Click **Add Automatic Property Set Definition**.

19. This property set definition applies to all objects or styles and it will take some time to get a complete list o data sources.

Proceed?

Click **Yes**.

20.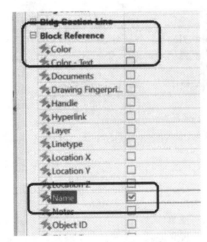

Locate Block Reference in the list.
Enable **Name**.
Click **OK**.
Change the **Format to Case-Sentence**.

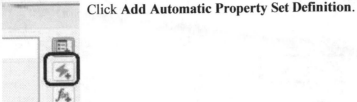

These are the property set definitions you should have.

If you click APPLY, the names of the property sets will automatically alphabetize.

21.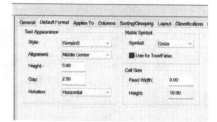

Highlight **Schedule Table Styles**.
Right click and select **New**.

22.

Rename it **Lighting Fixture Schedule**.

23.

Select the Default Format tab.

Set the Style to **RomanS**.

24.

Select the **Applies To** tab.

Enable **Block Reference**.
Enable **MInsert Block**.
Enable **Light**.
Enable **Multi-View Block Reference**.

25.

Select the Columns tab.

Add these columns:
- Quantity (enable Quantity column at the bottom of the dialog)
- Mark
- Name
- Mfr
- Model
- Room
- Level

26.

Highlight the **Name** column.
Click **Modify**.
Enable **Total**.
Click **OK**.

27.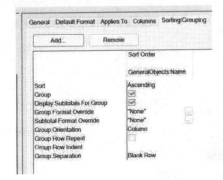

Select the Sorting/Grouping tab.
Click **Add**.
Sort by **GeneralObjects:Name**.
Enable **Group**.
Enable **Display Subtotals for Group**.
Add a **Blank Row** for Group Separation.

Click **OK**.

Save the file.

The file must be saved before you can create and use the schedule that was just defined.

28.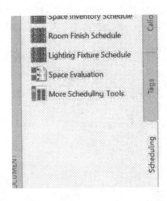

Switch to the Scheduling tab on the Document tools palette.
Reopen the Styles Manager.
Drag and drop the **Lighting Fixture Schedule** style onto the Scheduling tab of the tool palette.

29.

Click on the Lighting Fixture Schedule tool.
Click ENTER to use an external drawing file.
Click to place the schedule in the drawing.
Click ENTER.

30. 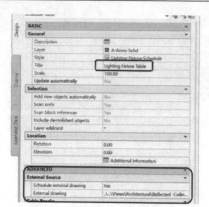 Select the Lighting Fixture schedule.
Right click and select Properties.
Change the Title to **Lighting Fixture Table**.
Set schedule external drawing to **Yes.**
Browse to the location and select the *A-CP01.dwg.*

31. Click **Update** on the ribbon.
Release the selection.

32.  Go to the Manage ribbon.
Open the **Style Manager**.

33.  Under Documentation Objects:
Locate **GeneralOjects** under **Property Set Definitions**.

34.  Open the Definition tab.
Click **Add Manual Property Set Definition**.

35. 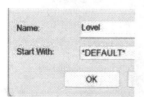 Select **Default** from the drop-down list.
Type **Level** in the Name field.
Click **OK**.
Change the Format to **Case - Upper**.

36.  Click **Add Manual Property Set Definition**.

37.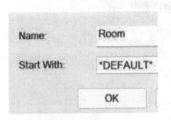

Select **Default** from the drop-down list.
Type **Room** in the Name field.
Click **OK**.
Change the Format to **Case - Upper**.

38.  Highlight the **Lighting Fixture Schedule**.

39.  Open the Columns tab.

Delete the columns referencing Space Objects.
Add columns for the Level and Room
properties from GeneralObjects.

Click **OK**.

40. The schedule updates.

41. [−][Top][2D Wireframe]

Switch to a **Top|2D Wireframe** view.

42. Select one of the 2x2 ceiling lights.

Right click and **Select Similar**.

Right click and select **Properties**.

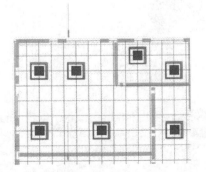

43.

All (28)

Block Reference (27)

External Reference (1)

DOCUMENTATION

Select **Block Reference** from the filter drop-down.

44.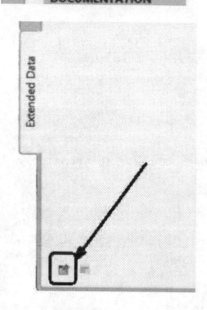

Select the **Extended Data** tab.

Click on **Property Sets**.

45.

☑ EquipmentObjects
☑ GeneralObjects

Enable the available property sets.

Click **OK**.

46.

Notes	
PROPERTY SETS	
EquipmentObjects	
Name	$0$2x2 lamp
Number	000
Description	--
GeneralObjects	
Color	256
Color-Text	BYLAYER
Description	/
Documents	
Hyperlink	
Layer	0
Linetype	BYLAYER
Notes	

Property sets have now been added to the selected blocks.

47.

Level	FIRST FLOOR
Linetype	BYLAYER
Mark	"VARIES"
Mfr	Wac lighting
Model	GFLED-2X2-CWT-010D-MV
Name	$0$2x2 lamp

Add **WAC Lighting** for the MFR.
Add **GFLED-2X2-CWT-010D-MV** for the Model.
Add **FIRST FLOOR** for the Level.

48.

Linetype	BYLAYER
Mark	001
Mfr	WAC LIGHTING
Model	GFLED-2X2-CWT-010D-MV
Name	$0$2x2 Lamp
Notes	
Room	BUSINESS OFFICE

Select each lighting fixture block and change the Mark Number value in the Properties palette.
Add the room location.

49.

Select the lighting fixture schedule.

Click **Update** on the ribbon.

50.

The schedule updates with the information.

51.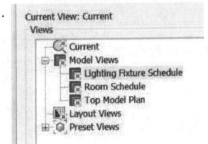

Go to the View Manager on the View ribbon.
Create a view for the lighting fixture schedule.

52.

Save and close all drawings.

53.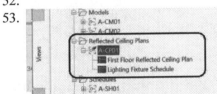

Highlight the **A-CP01** RCP view in the Project Navigator.

Right click and select **Check In**.

54.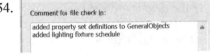

Type:
**added property set definitions to GeneralObjects,
added lighting fixture schedule**

Click **OK.**

55.

If you expand the Check-In History, you can see all the versions of the file.

Exercise 10-5:
Creating a Lighting Fixture Tag

Drawing Name: A-CP01.dwg, AECTag.dwg
Estimated Time: 15 minutes

This exercise reinforces the following skills:

- ❑ Project Navigator
- ❑ Drawing Management
- ❑ Tool Palettes
- ❑ Tags
- ❑ Property Set Definitions
- ❑ External References

1.

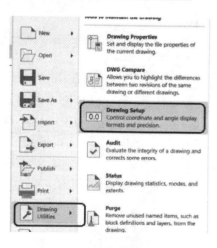

Open the Project Navigator.

Highlight the **A-CP01** view.

Right click and **Check Out**.
Right click and **Open**.

There is a lock icon displayed next to the file name to indicate that it is locked for editing.

2.

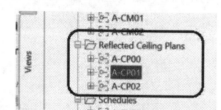

Go to the Application Menu.

Select **Drawing Utilities→Drawing Setup**.

3.

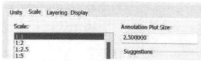

Select the Scale tab.

Set the Scale to **1:1**.
Set the Annotation Plot Size to **2.5**.
Click **OK**.

This ensures the schedule tag will scale correctly.

4.

Set the View scale to **1:50**.

5.

Launch the Document tool palettes.
Open the **Tags** tab.
Copy and paste a copy of the **Door Tag**.

6.

Rename the copied tag **Lighting Fixture**.
Drag and drop it below the Window Tag.
Right click and select **Properties**.

7.

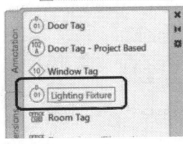

Change the Description to **Lighting Fixture tag**.
Set the Property data to **GeneralObjects.**
Set the Layer key to **EQUIPNO**.
Set the Tag location to *AECTag.dwg*
(browse to the exercise folder with the file).
Select the Tag Name: **LightingFixtureTag**.

Click **OK**.

If you want the tag available to other drawings/projects, save the tag to the schedule tags drawing in the ProgramData/Autodesk/ACA folder.
Select the lighting fixture tag tool.
Select a lighting fixture.
Click ENTER to place centered.

8.

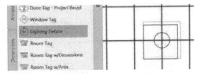

9.

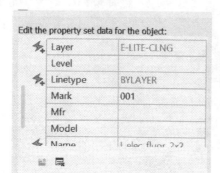

Type the Mark value if one is not displayed.

Click **OK**.
Click ENTER to exit the command.

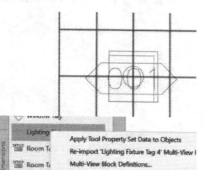

The tag is placed on the lighting fixture.

10.

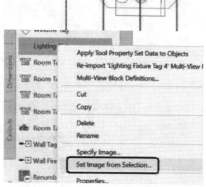

On the tool palette:
Highlight the **Lighting Fixture** tool.
Right click and select **Set Image from Selection**.
Then, select the tag that was placed.

The tool image will update.

11.

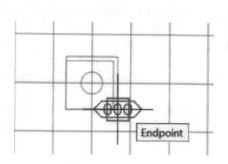

Click on the **Lighting Fixture** tool.
Select a light fixture.
Click **ENTER** to place the tag.

12.

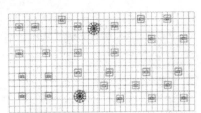

Check the property set data to ensure the Mark information is filled out.

Click **OK** to accept the property set data.

Select the next light fixture and repeat.

13.

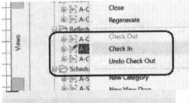

Save the drawing.

14.

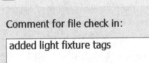

Open the Project Navigator.
Locate the A-CP01.dwg view.
Right click and select **Check In**.

15.

Comment for file check in:

added light fixture tags

Type **added light fixture tags**.

Click **OK**.

Exercise 10-6:
Version History

Drawing Name: A-CP01.dwg
Estimated Time: 10 minutes

This exercise reinforces the following skills:

- ❑ Project Navigator
- ❑ Drawing Management

1. 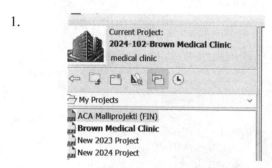 Set the **Brown Medical Clinic** project current.

2. Open the Project Navigator.
   Open the Views tab.
   Highlight **Version 4 – where the light fixtures were added to the ceiling grid.**
   Right click and select **Rollback to this version.**

3. Click **Yes.**

4.  A version has been added that shows it is back to Version 4.

5. 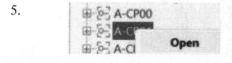 Open **A-CP01.dwg** view.

6. You see it is an earlier version of the reflected ceiling plan.

   Close the drawing.

7.

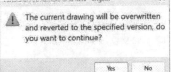

Highlight the version where you added the light fixture tags.
Right click and select **Rollback to this version.**

8.

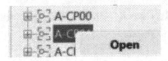

Click **Yes.**

9.

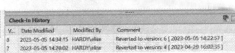

The new version is listed.

10.

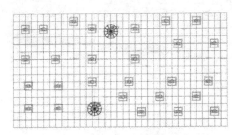

Open **A-CP01.dwg** view.

11.

The drawing now shows the lighting fixtures and tags.

Save and close.

Exercise 10-7:
Renumber Tags

Drawing Name: A-CP01.dwg
Estimated Time: 5 minutes

This exercise reinforces the following skills:

- ❑ Project Navigator
- ❑ Drawing Management
- ❑ Tags
- ❑ Re-number Data Tool

1.
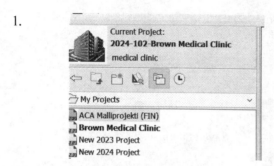
Set the **Brown Medical Clinic** project current.

2.
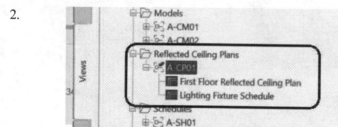
Open the Project Navigator.
Open the Views tab.
Highlight **A-CP01.**

Right click and select **Check Out**.
Right click and select **Open**.

3.

Launch the Document Tool Palette.

Open the Tags tab.

Locate the **Renumber Data Tool**.

This tool only works if the attribute uses an integer.

4.

Notice that ACA automatically knew which property set definition would be used to renumber the tags.

Click **OK**.

5.

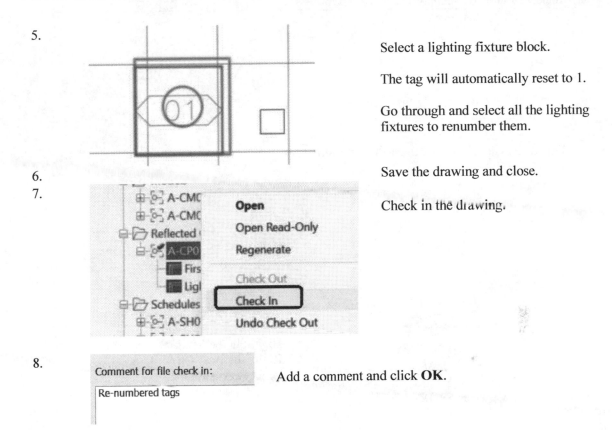

Select a lighting fixture block.

The tag will automatically reset to 1.

Go through and select all the lighting fixtures to renumber them.

6. Save the drawing and close.

7. Check in the drawing.

8. Add a comment and click **OK**.

Shared Views

A shared view allows users to share in the cloud a viewable, read only document. The recipient doesn't need an AutoCAD application, just a system with a current browser application and an internet connection. Viewers can mark up and add comments to the views. The shared document is outside of the project and any mark-ups and comments are not part of the document management system.

Exercise 10-8:
Shared Views

Drawing Name: Main Model.dwg
Estimated Time: 15 minutes

This exercise reinforces the following skills:

- Project Navigator
- Drawing Management

1.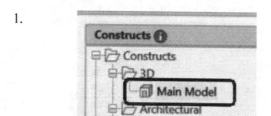

    Open the Project Navigator.

    Highlight the **Main Model 3D** construct.
    Right click and **Check Out**.
    Right click and **Open**.

2.

    Open the Collaborate ribbon.
    Click **Shared Views**.

3.

    Click **New Shared View**.

4. Enable **Share model view and all layout views**.
    Enable **2D views only**.

    Click **Share**.

5. Click **Proceed**.

6.

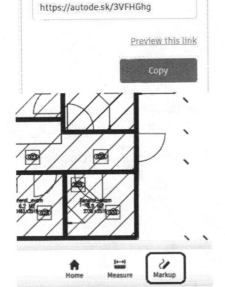

Click **View in Browser**.

7.

A browser window will open with the views.

Click the **Share** button located in the right.

8.

A link is displayed that you can copy and email to a team member for review.

Share   ✕

Share this design for 30 more days

▼ **Sharing Options**

☑ Section enabled

☑ Explode enabled

☑ Model Browser enabled

☑ Measure enabled

https://autode.sk/3VFHGhg

Preview this link

Copy

9.

Click **Markup** at the bottom of the browser window.

🏠 Home   ⟷ Measure   ✏️ Markup

10.

Click the **Comments** icon on the right side of the view.

11.

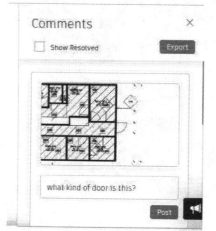

Place a cloud and arrow on the exterior door on the right of the building model.

12.

Click **Save**.

13.

In the Comments area:

Type **what kind of door is this?**

Click **Post**.

14.

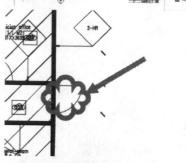

Back in the drawing, click the Refresh icon at the top of the Shared Views palette.

You will see the shared view that was created.

15.

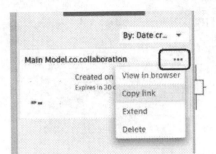

Click the … next to the view name.

Click **View in Browser**.

16.

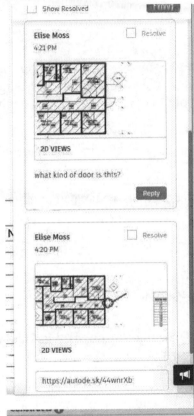

Expand the Comments section.

You will see the markup and the comment.

17.

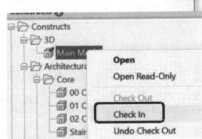

Close the browser window.
Save and close the drawing.

Highlight the **Main Model 3D** construct.
Right click and **Check In**.

18.

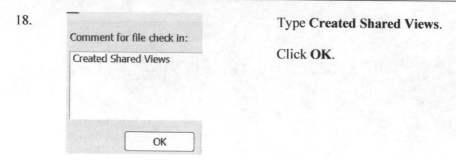

Type **Created Shared Views**.

Click **OK**.

Trace

Trace provides a safe space to add changes to a drawing in the web and mobile apps without altering the existing drawing. The analogy is of a virtual collaborative tracing paper that is laid over the drawing that allows collaborators to add feedback on the drawing.

Create traces in the web and mobile apps, then send or share the drawing to collaborators so they can view the trace and its contents.

Traces can be viewed in the desktop application but can only be created or edited in the mobile apps.

Markup Import

Markup Import attaches an image file or pdf to use as a trace for an existing drawing.

You can convert elements of the trace to AutoCAD elements using the Markup Import tool to make changes quickly and easily.

To create a markup, print out the drawing to be modified. Use a RED pen to indicate the desired changes. Then scan it to a pdf or take a picture with your phone and email it to the person who needs to make the changes.

Exercise 10-9:
Use Markup Import

Drawing Name: A-102 Floor Plans.dwg, *A-102 Floor Plan_trace1.pdf*
Estimated Time: 5 minutes

This exercise reinforces the following skills:

- ❑ markups
- ❑ Trace
- ❑ Project Navigator
- ❑ Drawing Management

1.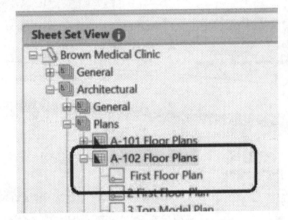

Launch the Project Navigator.
Open the Sheets tab.
Locate the **A-102 Floor Plans** drawing.
Right click and select **Check Out**.
Right click and select **Open**.

2.

Activate the Collaborate ribbon.

Click **Markup Import**.

3.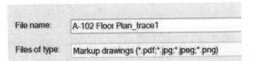

Locate the *A-102 Floor Plan_trace1.pdf* file downloaded from the publisher.

Click **Open.**

4.

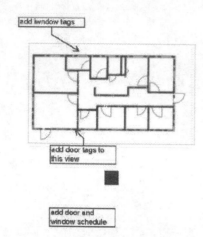

Zoom in to see the mark-up.

5.

MARKUPIMPORT Accept placement or [Move Align Rotate Scale Undo] <aCcept>:

Click **ENTER** to accept the placement.

6.

⚙ 🌑 Trace (Front)   🖳 ✕ ▬

Verify that Trace (Front) is enabled. This means the trace is active.

7.

add door and window schedule

do you want two separate schedules or one schedule with doors and windows?

Go to the Home ribbon and select the MTEXT tool.

Multiline Text

Add Mtext below the schedule note.

8.

⚙ 🌑 Trace (Front)   🖳 ✕

Click on the **Edit Trace** icon to exit the edit trace mode.

9.

Traces Palette | Markup Import

Traces

Switch back to the Collaborate ribbon.

Click on the **Traces Palette**.

10.

Traces

MODIFIED TODAY

Trace1
Elise Moss   5:18 PM

The trace is listed on the palette.

11.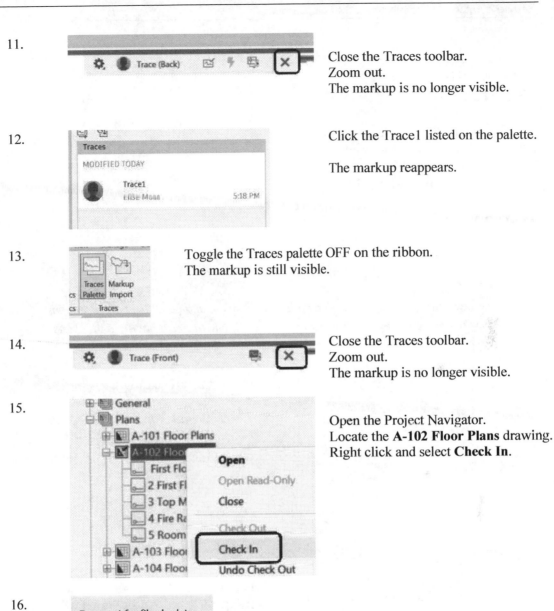

Close the Traces toolbar.
Zoom out.
The markup is no longer visible.

12.

Click the Trace1 listed on the palette.

The markup reappears.

13.

Toggle the Traces palette OFF on the ribbon.
The markup is still visible.

14.

Close the Traces toolbar.
Zoom out.
The markup is no longer visible.

15.

Open the Project Navigator.
Locate the **A-102 Floor Plans** drawing.
Right click and select **Check In**.

16.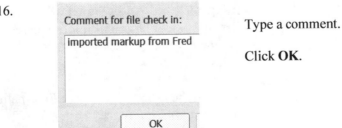

Type a comment.

Click **OK**.

Exercise 10-10:
Use Trace

Drawing Name: A-EL01_02.dwg, A-201 Building Elevations_02.dwg, mark-up.pdf
Estimated Time: 20 minutes

This exercise reinforces the following skills:

- ❑ markups
- ❑ Trace
- ❑ Project Navigator

You can share a drawing with a colleague or client and have them add notations for your review using the free mobile app available to smart phones and tablets.

1.

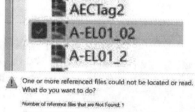

Open A-EL01_02.dwg.

This is included in the exercise files.

2.

One or more referenced files could not be located or read. What do you want to do?

Number of reference files that are Not Found: 1

→ Open the External References palette
→ Ignore unresolved reference files

Click **Open the External References palette**.

3.

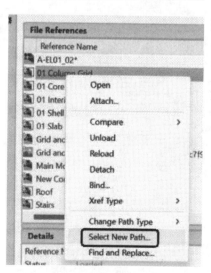

Bring up the XREF Manager.

Highlight the **01 Column Grid** file.
Right click and **Select New Path**.
Browse to the Structural folder where you are storing your project files.
Click **OK**.

You will be asked if you want to re-path all the other external references.
Click **Yes**.

All the path errors should be corrected now.

Close the XREF Manager.

4.

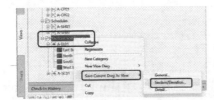

Open the Project Navigator.
Open the Views tab.
Highlight **Sections and Elevations**.
Right click and **Save Current Dwg As
View→Sections/Elevations.**

5.

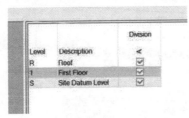

Click **Next**.

6.

Enable all the floors.
Click **Next**.

7.

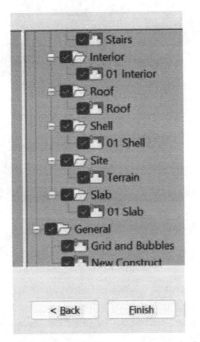

Click **Finish**.

8.

File name: A-201 Building Elevations_02

Files of type: Drawing (*.dwg)

Open *A-201 Building Elevations_02.dwg*.

This file is downloaded.
Click **Open the External References palette**.

9.

⚠ One or more referenced files could not be located or read.
What do you want to do?

Number of reference files that are Not Found: 1

→ Open the External References palette

→ Ignore unresolved reference files

10.

File References

Reference Name

🔲 A-201 Building Elevations_02

🔲 A-EL01

🔲 Grid and Bubbles

🔲 Main Model

🔲 New Construct

Highlight A-EL01.

11.

A-201 Building Elevations_02

A-EL01

Gri   Open
Ma   Attach...
Ne
        Compare   >
        Unload
        Reload
        Detach
        Bind...
        Xref Type   >
        Change Path Type   >
        Select New Path...
        Find and Replace...

Det
Refer
Status   Not Found

Right click and **Select New Path**.

12.

File name: A-EL01_02

Files of type: Drawing (*.dwg)

Select *A-EL01_02.dwg*.

Click **Open**.

13.

File References

Reference Name   ▲ Status
A-201 Building Elevations_02*   Opened
A-EL01   Loaded
Grid and Bubbles   Loaded
Grid and Bubbles|medical clinic-first floorf7c7f9d4 Loaded
Main Model   Loaded
New Construct   Loaded

All the errors are now corrected.

Close the XREF Manager.

Save the file.

14.

A-107 Reflected Ceiling Plans

        Collapse

        New   >

        Import Layout As Sheet...
Sect
Larg   Import Current Layout As Sheet...
Det
Sche   Delete
3D F

Open the Sheets tab of the Project Navigator.
Highlight Elevations.
Right click and select **Import Current Layout As Sheet**.

15.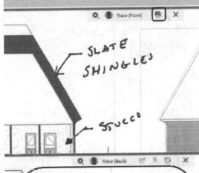

Open the Collaborate ribbon.

Click **Markup Import**.

16.

Select the *mark-up.pdf*.
Click **Open**.

This is a downloaded exercise file.

17.

Some callouts have been added.

Click the Markup Assist icon.

18.

If you hover over the text, you will see a tooltip.
Click at the edge of the dashed rectangular border.

19.

Markup Assist   ✕

SLATE
SHINGLES

Insert as Mleader

A Insert as Mtext

Update Existing Text

Fade Markup

The first word is identified.
Type SHINGLES below the word SLATE.

Click **Insert as MLeader**.

20.

You will be prompted to select a start point for the leader.

Place the leader.

Note that the word slate in the markup is now faded.

21.

If you hover over the second mark-up, you see that the Markup Assist does not recognize it.

22.

Click on the markup.
Click **Fade Markup**.

23.

Click on the SHINGLES markup.

Click **Fade Markup**.

24.

Click on the leader markup.

Click **Fade Markup**.

25.

Click the x on the toolbar to close the trace.

26.

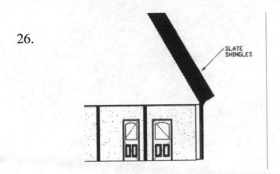

Save and close the drawing.

Notes:

QUIZ 10

True or False

1. A shared view is not linked to the drawing where the view originates.

2. You can install an AutoCAD app on your phone and use Traces to Markup a drawing.

3. A trace allows you to make edits directly in the drawing file.

4. You can have more than one trace in a drawing.

5. A drawing is checked out by one of your team members. If you click Refresh or Reload for the checked out drawing, you will see the changes.

Multiple Choice

6. You have received an AutoCAD drawing from the plumbing contractor for your building design. You want to change the color and linetype of the external reference using his drawing file so it is easy to identify. To do this:

   A. You create a new construct using his drawing.
   B. You create a new view using his drawing and change the colors and linetypes in the view.
   C. You create a new construct using his drawing, attach his drawing to the Building Model as an xref overlay, then apply a layer filter to control the appearance of his drawing.
   D. You open the plumbing drawing and modify the layer colors and linetypes.

7. A shared view can be accessed in a browser for:

   A. One week
   B. A user-specified amount of time
   C. A year
   D. 30 days

8. You have checked out a file and you are ready to check it in. ACA will provide a prompt. The prompt allows you to:

   A. Add comments regarding the changes that were made.
   B. Email your team members to let them know that you are checking in the drawing.
   C. Verify that the changes you made are correct.
   D. Save the drawing file as a new file.

9. You wish to revert to an earlier version of a drawing. You select the earlier version, right click and select Rollback to this version.

    A. ACA creates a new file based on the older version that you can open and adds it as the most recent version of the drawing.
    B. ACA deletes all saved versions of the file that were created after the selected version and makes the selected version the most recent version.
    C. ACA creates a new file based on the older version and deletes all saved versions that were created after the older version.
    D. ACA creates a new file based on the older version and purges all the changes that were made.

10. You open the Project Navigator and want to work on one of the construct files. However, you see a lock on the file. What does this mean?

    A. The file is currently set to Read-Only because it has been released.
    B. The file is checked out, open and is currently being edited – either by you or another team member.
    C. You do not have the security permissions to make changes to this file.
    D. The file is checked out and is unavailable for viewing.

11. You see a red check mark next to a file in the Project Navigator. This means:

    A. The external reference has lost its path and needs to have a new path specified,
    B. Changes have been made to the file and it needs to be reloaded.
    C. The file is checked out by another team member.
    D. The file needs to be audited.

ANSWERS:

1)   T; 2) T; 3) F; 4) T; 5) False – you will only see the changes once the drawing is checked in and you will see a notification when the drawing has been checked in 6) C; 7) D; 8) A; 9) A; 10) B; 11) C